A BEGINNER'S GUIDE TO MICROARRAYS

A BEGINNER'S GUIDE TO MICROARRAYS

edited by

Eric M. Blalock
University of Kentucky Medical Center, U.S.A.

KLUWER ACADEMIC PUBLISHERS
Boston / New York / Dordrecht / London

Distributors for North, Central and South America:
Kluwer Academic Publishers
101 Philip Drive
Assinippi Park
Norwell, Massachusetts 02061 USA
Telephone (781) 871-6600
Fax (781) 681-9045
E-Mail: kluwer@wkap.com

Distributors for all other countries:
Kluwer Academic Publishers Group
Post Office Box 322
3300 AH Dordrecht, THE NETHERLANDS
Telephone 31 786 576 000
Fax 31 786 576 254
E-Mail: services@wkap.nl

Electronic Services <http://www.wkap.nl>

Library of Congress Cataloging-in-Publication Data

A C.I.P. Catalogue record for this book is available
from the Library of Congress.

Title: A BEGINNER'S GUIDE TO MICROARRAYS
Editor: Eric M. Blalock
ISBN: 1-4020-7472-7

ABOUT THE AUTHORS

CHAPTER 1

Brian Ward is a Principle Investigator at the Life Science and High Technology Center in St. Louis. He earned his Ph.D. in Chemistry from Michigan State University (1984) studying structure-function relationships of metalloporphyrins. Prior to joining Sigma-Aldrich (1987), he did postdoctoral work studying drug-DNA interactions at Syracuse University (1986).

Kathryn Aboytes is a senior research scientist at the Sigma-Aldrich Life Science and High Technology Center in St. Louis, Missouri. She received the B.A. degree (1989) in microbiology and the M.S. degree (1992) in veterinary microbiology from the University of Missouri, Columbia.

Jason Humphreys is a research scientist at the Sigma-Aldrich Life Science and High Technology Center in St. Louis, Missouri. He received the B.S. degree (1997) in biology from the University of Evansville and the M.S. degree (2000) in biochemistry from the University of Illinois, Champaign-Urbana.

Sonya Reis is an associate research scientist for Sigma-Aldrich at the Life Science and High Technology Center in St. Louis, Missouri. Sonya received her B.S. in biology - medical sciences in 1999 from Southern Illinois University - Edwardsville.

CHAPTER 2

Levente Bodrossy is a research scientist at the Department of Biotechnology, Seibersdorf research, Austria. He is a microbiologist with a primary interest in applying microarrays and related technologies to the fast, parallel detection and identification of microbes. He recieved the M.S. degree (1994) in biology and Ph.D. degree (1997) in biophysics at the University of Szeged, Hungary. Prior to moving to Seibersdorf research he was Senior Lecturer at the Department of Biotechnology, University of Szeged, Hungary. He spent 2 years as visiting scientist at the University of Warwick, UK.

CHAPTER 3

Prior to TeleChem / arrayit.com **Todd Martinsky** served as director of education and consulting at the Codd and Date Consulting Group. Todd worked directly with Dr. E.F. Codd, the father of Relational Database Management Systems and inventor of the Relational Model. Since founding TeleChem, Todd has been instrumental in developing the ArrayIt Brand product line, which includes sample preparation, surface chemistry, hybridization and environmental control products. Todd has led the company to play a significant role in the microarray industry. Along with his daily technical and business direction of the ArrayIt

Brand Product line, Todd has established successful alliances with corporate partners in manufacturing, reagent, equipment and distribution. Todd is responsible for an educational outreach program that ensures that the broadly patented ArrayIt Brand Stealth Micro Spotting Device is applied in the field with optimal scientific and technological results. The Stealth Micro Spotting Device dominates as the most widely used microarray technology in the world.

CHAPTER 4

Robert Searles is the Manager of the Oregon Health and Science University Spotted Microarray Core in Portland, Oregon. He has an undergraduate degree in Biochemistry and Biophysics from Oregon State University and a doctorate from UCLA's Department of Biological Chemistry. He did postdoctoral work in molecular neuroscience. Following his post-doctoral work, he was recruited by the Division of Pathobiology at the Oregon National Primate Research Center to work in viral genomics. He subsequently worked as bioinformaticist for the OHSU West Campus and then established the OHSU cDNA microarray core. He has academic appointments as an Affiliate Assistant Scientist at the OHSU Vaccine and Gene Therapy Institute and as a Staff Scientist at the Oregon National Primate Research Center.

CHAPTER 5

Maureen Sartor is a bioinformatics research associate in the Center for Environmental Genetics at the University of Cincinnati. She received her B.S. degree (1998) in mathematics and her master's degree (2000) in Biomathematics (division of statistics department) from NC State University. After doing consulting work on the analysis of microarrays for Glaxo Wellcome, she now is responsible for statistical analysis of microarrays in her center and advises investigators on experimental design.

Mario Medvedovic, Ph.D. is a Research Assistant Professor in Center for Genome information at University of Cincinnati Medical Center. His research is focused on the use of statistical models in analysis of functional genomics and proteomics data.

Bruce J. Aronow, PhD is an Associate Professor and Director of the HHMI Genome Informatics Core for the University of Cincinnati College of Medicine. He is a member of the Divisions of Pediatric Informatics and Molecular Developmental Biology, the Department of Pediatrics in the College of Medicine, and the Department of Biomedical Engineering at the University of Cincinnati. His research efforts are devoted to the application of comparative and functional genomics methods to an understanding of biological processes and systems.

CHAPTER 6

Eric Blalock is an assistant professor in the Department of Molecular and Biomedical Pharmacology at the University of Kentucky Medical Center. He is a published author in the field of neuroscience, studying the effects of calcium regulation, aging, and epilepsy. For the past several years he has been working closely with the University of Kentucky Department of Statistics and Medical Center Microarray Core facility on the experimental design and statistical analysis of microarray data.

CHAPTER 7

Xuejun Peng is a research assistant in the Medical Center Microarray Core Facility and a PhD candidate in the Department of Statistics at the University of Kentucky. He earned his Medical Diploma (US MD equivalent) in 1994 and obtained his residence training in internal medicine and clinical genetics from 1994 to 1997 from the Hunan Medical University in Changsha, P.R. China. He received an M.S. degree in Statistics in 2000 and has been a Biostatistics consultant and analyst for the past several years.

Arnold J. Stromberg is an associate professor in the Department of Statistics at the University of Kentucky. He is Director of Data Analysis for UK Microarray Core Facility. In addition to his research interests in bioinformatics, he publishes on robust statistics and control chart methods.

CHAPTER 8

Willy Valdivia-Granda's research involves the development of computational techniques to analyze and model complex biological systems. At North Dakota State University, he is involved in the genomic characterization of plant responses to biotic and abiotic stresses. Willy is also the founder and CEO of Orion Integrated Biosciences, a corporation using micro and nanoarrays for the genomic characterization of organisms of medical, industrial and military relevance. Using DNA microarrays and computational tools, Willy is characterizing the infection mechanisms of malaria (*Plasmodium falciparum*). He is also actively involved in the development of advanced collaborative environments for education. He is the founder of the Virtual Conference on Genomics and Bioinformatics, and Chair and scientific advisor for the 2002 CHI Microarray Data Analysis International Conferences. In 2002, the Virtual Conference was broadcast without registration fees to more than 2000 researchers in 41 countries, and 33 states within the US, simultaneously using the Access Grid technology and live video streaming.

Contents

Preface

When our laboratory first began using microarrays (~1999) there were precious few books on the subject. In fact, most of our information came from vendors and word of mouth among colleagues. Microarrays have become an ever more integrated component of basic research, and from 1998 to 2001 researchers quadrupled each previous year's publication total. With 2002, a mere doubling of the previous year's scholarly publications indicates that the initial, exponential growth phase of microarray technology may finally be over. Furthermore, basic and clinical research journals now set the bar higher for publication of microarray studies- the Nature family of journals requires MIAME compliance (covered in Chapter 8) prior to publication, and very few top tier journals still accept microarray studies with no replication (Chapters 5, 6 and 7). Thus, microarrays may be moving from a 'hot' buzzword technology into the useful tool that its originators always intended.

In most facilities around the U.S. and in Europe, centralized cores provide a bridge between microarrays and the researchers interested in using them. Within these core systems, certain things have become apparent. First, as a core evolves, its members must explain the technology, its potentials and pitfalls, to the scientists from a broad variety of backgrounds. Second, several very good books covering in-depth aspects of microarray technology and data analysis exist. However, getting user-level information from such thorough treatises sometimes requires more time and effort than the investigator is prepared to invest. Third, there is no primer of microarray technology that covers all of the steps in microarray design and manufacture, as well as experimental design and data analysis.

Our book offers a broad, 'friendly' coverage of many of the most important aspects of microarray technology. We based our coverage on the questions asked of us by new microarray users in universities, laboratories, and microarray list servers. For instance, slide coating is a very important initial step in the preparation of spotted arrays, and has not been addressed thoroughly in any other microarray book. In Chapter 1, authors Kathryn Aboytes, Jason Humphries, Sonya Reis, and Brian Ward remove the black box from this process, and explain how different glass treatments interact with genetic material to form spots. Further, in Chapter 2, Bodrossy Levente provides a detailed 'how-to' guide for oligonucleotide probe design, as well as a powerful yet underused application of microarray technology- the typing and quantification of bacterial contaminants (this often overlooked application may have the most immediate impact on human health and safety of any of the proposed uses of microarrays). Robert Searles dedicates Chapter 3 to setting up and running a microarray core facility, complete with horror stories about here-today, gone-tomorrow vendors, and tried-and-true practices that can get a new facility up and running quickly. In Chapter 4, Todd Martinsky covers the care and use of robotic arrayers and print heads, and discusses the solutions that work best in spotting. He also includes examples of real world problems and fixes that work. In Chapter 5, Maureen Sartor, Mario Medvedovic, and Bruce Aronow cover data normalization, and clearly describe the ways in which technical error can result in misleading data, as well as how to check and control for its presence. In Chapter 6, I cover approaches for establishing differential gene expression and make recommendations as to the kind of experimental designs that will lead to useful data sets. In addition, I provide detailed, step-by-step procedures for analysis using Excel, and a section, unique to this book, on an automated procedure for identifying not only patterns of expression, but also the likelihood that those patterns would have arisen by chance (based on *post-hoc* statistical analysis). In Chapter 7, Xuejun Peng and Arnold Stromberg address the statistical impact of experimental design, as well as issues of power estimation ('how many chips do I need'), and multiple testing error/ correction. In addition, and unique to this book, proposals for experimental designs that maintain statistical power and reduce experiment cost are discussed. Finally, in Chapter 8, Willy Valdivia Granda gives a thorough overview of the available clustering methodologies along with detailed descriptions of their uses and the software available for these procedures. Further, he provides detailed information about different microarray database structures.

ACKNOWLEDGEMENTS

I thank the chapter authors for their patience, hard work, and dedication to this project, as well as Gretchen Stromberg for her excellent help in document preparation.

Chapter 1

SLIDE COATING AND DNA IMMOBILIZATION CHEMISTRIES

Kathryn Aboytes, Jason Humphreys, Sonya Reis and Brian Ward*
Sigma-Aldrich Corp., P.O. Box 14508, St. Louis, MO 63178

INTRODUCTION

Printing DNA microarrays on glass microscope slides originated with the Brown group in the mid 1990s (Schena et al., 1995). Since then, the use of microarrays in hybridization based assays such as measuring comparative gene expression levels, has escalated precipitously. The choice of using glass as a microarray substrate stems from its low fluorescence, low cost, high heat resistance, and rigidity. With few exceptions (Kumar et al., 2000), it has been necessary to coat the glass surface to facilitate DNA immobilization/spotting. Furthermore, before glass can be coated, it must be prepared and cleaned to accept that coating. In brief, a rigorous glass-cleaning step is followed by slide coating, which is in turn followed by DNA immobilization using an attachment chemistry that is matched with the slide coating. The purpose of this chapter is to review current approaches for preparing glass slides for coating, coating the slides and immobilizing appropriately modified nucleic acids onto coated glass surfaces. In the end, this chapter should enable the scientist to choose an appropriately matched set of slide and immobilization chemistries, thus clarifying the seemingly myriad array of immobilization strategies.

GLASS PROPERTIES

Glass microscope slides, available from a variety of vendors, include sodalime "soft" glass, borosilicate "hard" glass, and to a lesser extent pure silica (Birch, 2000). Borosilicate glass is a heat resistant glass made by adding boric oxide and alumina to the glass mix. Decreasing the thermal sensitivity helps borosilicate glass retain its shape when heated, preventing it from flowing or softening like sodalime glass. Sodalime glass is composed of quartz sand (silica), soda (sodium oxide), lime, and other oxides. It has a relatively low work temperature (or working point- the temperature at which glass is soft enough to be shaped), and becomes fluid when heated. The soda added to the silica increases the heat sensitivity (and lowers the working point). Lime is added for strength. Sodalime glass is the most common glass substrate and contains approximately 13% sodium oxide (Birch, 2000). This, when exposed to ambient air, will react to form sodium hydroxide which may interfere with coating (Birch, 2000).

Glass surfaces are often represented as in Figure 1. This representation to the non-glass expert likely portrays glass as an inert flat surface, void of structure and relatively poor in chemistry, to which a coating will be applied that will enable DNA attachment. While the simple silanol picture is useful for conceptualizing coating chemistries, the surface of glass is not as simple as the cartoon might suggest. Glass has layers of adsorbed water and for smooth glass has nanometer-size hill

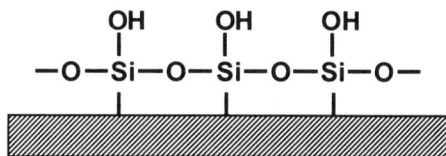

Figure 1. Typical representation of a glass surface.

and valley topology. The combination of structural roughness and highly polar surface groups makes glass an ideal substrate, binding a variety of materials by adsorption. In fact, it is only with great care that glass can be kept free of environmental contaminants that adulterate the surface, thereby causing it to be unsuitable for surface modification. The force one seems to be combating is the glass-air interface. That is, since glass is extremely hydrophilic and air is hydrophobic it seems reasonable that it is thermodynamically favorable to adsorb molecules to the surface that will "neutralize" the opposing polarities at the interface. Further, one might expect that amphiphilic molecules (those containing both hydrophilic and hydrophobic regions; i.e. everything "organic" except water and hydrocarbons) would be best at performing this task. Thus, one should expect that environmental contaminants (sometimes referred to as environmental organics) present in low concentrations in the environment would eventually adsorb to the glass and cause it to become more hydrophobic. In fact, it is common to observe surface energy changes in as little as 24 hours in some environments.

It is a fairly common observation that scrupulously clean glass slides become more hydrophobic with time, i.e. drift to higher and higher contact angles. The contact angle is an important index of glass slide properties and is discussed in detail in Surface Analysis Methods. Additionally, the rate of this drift is dependent upon storage conditions. While one may be tempted to suggest that increasing hydrophobicity could be the result of surface dehydration (Englander et al., 1996), it is only under carefully controlled and extreme conditions that glass surfaces actually dehydrate. We recently illustrated this by sealing one group of clean-hydrophilic slides in a new glass jar and another group in an identical jar containing silica desiccant. After several weeks the slides without desiccant were more hydrophobic, while those with desiccant remained unchanged (B. Ward, unpublished observations) demonstrating that the silica desiccant was ineffective at dehydrating the glass surface but was able to prevent the hydrophobic drift seen under normal storage conditions.

While interesting in its own right, these results clearly demonstrate the need for storing slides in a controlled environment, including inert storage containers, to avoid surface contamination by environmental organics. Common storage methods include placing the slides in an inert gas atmosphere such as argon, or within a vacuum chamber. Either method will prolong the half-life of the slide; however, no matter how slides are stored, it is a good practice to check surface hydrophobicity, typically by contact angle measurement, prior to using or coating any stored slide. Further, it is common practice to coat only freshly cleaned slides.

GLASS CLEANING

Preparing a clean surface can be the greatest challenge in the microarray slide generation process. The level of cleaning is very dependent on the glass source; a poor source will require more cleaning and may still generate a less than ideal surface for coating. There are numerous derivations and alternatives beyond the cleaning methods discussed below, however these are the most abundant from published literature and unpublished academic sources. On average, a glass microscope slide prior to cleaning will have 30% carbon contamination. Such contamination will adversely affect any coating procedure, rendering a non-uniformly coated slide unsuitable for consistent microarray printing. Many methods exist for removing glass contaminants but can be generally classified as acid, alkali or physical cleaning methods. Table 1 contains a summary of the advantages and disadvantages of the various methods. Contact angle, X-ray photon microscopy (XPS), and atomic force microscopy (AFM) can all be used to determine the effectiveness of the cleaning process. Success with any method should result in a glass surface with a very low contact angle (4-10 degrees) and overall uniform wettablility. The other more detailed, and expensive, i.e. XPS and AFM, are more likely to be used for process validation than routine cleanliness measurement. The combination of cost, effectiveness, and resources should all be weighed to determine the method of choice.

Table 1. Slide cleaning method summaries

Method	Advantage	Disadvantage
Chromerge	Consistently effective cleaning.	Banned in many locations because of toxicity. High disposal expense.
Piranha Solution	Consistently effective cleaning.	Very reactive, and high disposal expense.
Hydrofluoric Acid	Very effective at removing contaminants. Fast reaction time.	Highly toxic, and slide damage can occur if process goes for an extended period.
Alkaline	Consistently effective cleaning.	Slower reaction, requiring longer incubation periods.
Ultrasonication	Clean surface with little strong solvent disposal.	Initial equipment requirement, and not as effective as the Piranha solution (Shirai 2000).
Plasma	Very effective cleaning method with a fast reaction time.	Requires a reactor
Pyrolysis	No solvent requirement.	Reactor/High heat environment. Large contaminants may produce soot residue.
Solution Sprayer	Effective for macromolecule removal only.	Not effective against smaller particle/organic contamination.
UV/ozone	Effective at removing thin film contaminants, very fast reaction time.	Thicker layers of contaminants will not be effectively cleaned. Initial equipment cost is high.
Laser	Effective cleaning method with a fast reaction time.	Initial equipment requirement, and possible recontamination if vacuum not applied.

Acid

Most acid cleaning methods consist of the combination of an acid and an oxidizing agent. Chromerge™ (potassium permanganate/sulfuric acid) was used early on but, though effective, it is rarely used today because of its toxicity to the environment and the associated high cost incurred from disposal (Birch, 2000; Zammatteo et al., 2000). The most common acid cleaning solution is hot piranha solution, composed of concentrated sulfuric acid and hydrogen peroxide. The mechanism of cleaning with either of these methods is oxidation and solubilization of glass-adsorbed organics. Hydrofluoric acid (HF) cleans glass by etching, removing silyl moieties and along with them any adsorbed contaminants. Though effective, HF etching attacks the glass and so is extremely time dependent; the longer the acid is left to react with the slide the rougher the surface will become. Hydrofluoric acid is extremely toxic and its use requires extreme caution. The acid can neutralize nerve endings and decalcify bone (Birch, 2000). Exposing skin areas larger than the palm of a hand may cause heart failure (Birch, 2000). A variety of other strong and weak acids have been tested but do not perform at the level of the above three. While acids are an effective method for removing most contaminants, chromerge and piranha are ineffective at removing silica particles. Though HF is able to remove the silica particles it is less commonly used because of its toxicity. Acid cleaning is a common small lab cleaning method, but higher capacity cleaning with acid is usually avoided because of the amount of waste product.

Acid Cleaning Protocols

Chromerge: Dissolve 20 grams potassium dichromate ($K_2Cr_2O_7$) in 90 ml water then slowly add approximately 500 ml concentrated sulfuric acid. Acid addition is extremely exothermic and so caution must be taken. The chromerge solution will remain good until the color changes from brown to green indicating the chromic acid has been spent. Slides are washed by a 20 minute cleaning followed by 20 minutes wash in 1:1 HCl:water to remove chromium ions (Birch, 2000). The slides are dried using airflow across the surface or baking.

Piranha: 2:1 and 4:1 concentrated sulfuric acid: 30% peroxide are most frequently used (Seeboth and Hettrich, 1997; Birch, 2000; Shirai et al., 2000; Zammatteo et al., 2000; Benters et al., 2002;). Temperature varies from room temperature to 90 °C along with the cleaning time from 10 minutes to 2 hours (Gray, et al., 1997; Seeboth and Hettrich, 1997; Birch, 2000; Shirai et al., 2000; Zammatteo et al., 2000; Benters et al., 2002). Reversing the acid:oxidizer ratio has been shown to decrease the effectiveness of the cleaner (Shirai et al., 2000). Multiple water rinses followed by air or heat drying finalizes the procedure. Piranha solution is highly effective at removing carbon contaminants (Shirai et al., 2000). Care should be taken when using hot piranha solution. The combination of concentrated acid, strong oxidizing agent, and heat necessitate the use of a fume hood as well as proper protective equipment for the operator.

Hydrofluoric acid: 5% or less HF is generally used at room temperature for 10 minutes (Seeboth and Hettrich, 1997; Birch, 2000; Lee et al., 2002) followed by extensive water rinses.

6

Alkaline

Like hydrofluoric acid, alkaline glass cleaning works by etching, though not as rapidly. Sodium hydroxide and potassium hydroxide solutions greater than pH 9 are commonly used (Kumar et al., 2000; Erdogan et al., 2001). Incubation time is dependent on the level of glass contamination. One hour cleaning is generally sufficient to remove most contaminants. The addition of an alcohol e.g. ethanol, will help remove adsorbed organics from the surface, thus increasing the efficiency of the alkaline cleaning process. Post cleaning the slides should either be neutralized with a quick rinse in hydrochloric acid, and/or successive water rinsing to halt the glass etching (Erdogan et al., 2001). Air drying or baking is used to finalize the cleaning. Alkaline solutions are readily available to most labs and disposal is less of an issue than with acid solutions.

Sodium Hydroxide
1-3.5 N Na/KOH/ 0-70% EtOH 2 hrs followed by extensive water rinsing.

Physical

Physical cleaning methods either remove surface adsorbates by force, e.g. ultrasonic and spray cleaning, or by a process akin to burning the surface contaminants, e.g. plasma, pyrolysis, UV/ozone and laser. Their use is limited by the availability of specialized equipment.

Ultrasonic

Ultrasonication uses cavitation as an efficient and powerful cleaning tool. High frequency sound waves in a liquid cleaning solution generate high/low pressures zones. In the negative pressure zones, the solution boiling point decreases and microscopic vacuum cavities are formed. As the sound waves move, the same zone becomes positive, causing the bubbles to implode. Cavitation exerts enormous pressures (10,000 lbs/sq in) and temps of 20,000° F on a microscopic scale (Williams and Randall, 1994). Temperature is a critical part of ultrasonic cleaning. Increased temperatures result in increased cavitation intensity, leading to more efficient cleaning as long as the cleaning solution boiling point is not approached. Boiling "short circuits" cavitation (Williams and Randall, 1994). Ultrasonic cleaning is capable of removing one and 0.5 micron size particles with 95% and 84% efficiency respectively (Awad, 1996). A surfactant is generally included in the cleaning solution to help transmit the cavitation energy (Birch, 2000; Belosludtsev et al., 2001). If the cleaning solution is not properly degassed the efficiency of the cleaning will decrease because the bubbles will absorb energy (Birch, 2000). Multiple frequencies are recommended to increase the uniformity of the cleaning (Birch, 2000). A single frequency may lead to "blind" spots, which were not effected/encountered by the cavitation waves (Birch, 2000). Ultrasonication is frequently combined with the other cleaning methods as part of the wash protocol.

Solution Sprayer

High-pressure spray directed at the glass surface can effectively remove some contaminants and can be the precursor to many cleaning processes. Its fast process and minor equipment requirements make it an attractive method for cleaning. However, Stowers demonstrated that spray cleaning is unable to remove particles smaller than 5 microns (Stowers, 1978). The purity of the water used during the spray process is also critical. Though spray processes are able to quickly remove dust particles and other loosely adsorbed species, more stringent methods are required to achieve a uniformly cleaned surface.

Plasma

Plasma is defined as a partially or completely ionized gas with an approximately equal number of positively and negatively charged particles. It has a state of matter similar to gas and liquid. Plasma can be thought of in two forms, high temperature plasma and low temperature plasma. Most cleaning procedures are done at low temperature, where ionized gases are generated at low pressures. Gases are ionized by applying low frequency, radio frequency (RF), or microwave frequency energy to a gas (Williams and Randall, 1994) with the amount of energy controlling the efficiency of the cleaning. Radio frequency is most commonly used. During exposure to the RF, diatomic oxygen molecules are split into extremely reactive monatomic oxygen (Guiseppi-Elie, 2000). Organic contaminants that come in contact with monatomic oxygen are immediately oxidized to pyrolysis products (water vapor, carbon monoxide and carbon dioxide) that are carried away in the vacuum stream (Guiseppi-Elie, 2000). Plasma treatment can also be used to etch or ablate the surface. Gas selection and plasma treatment reaction time will determine the level of ablation. Most standard cleaning procedures are finished in less than 1 minute if at least 60 watts of RF power is supplied. As exposure time increases the surface roughness will also increase (Choi et al., 2001).

Pyrolysis

Exposing glass to 300+ °C will desorb and degrade surface contaminants while producing few waste products. Heavily contaminated slides may acquire a burn residue and so it is advised that macromolecule contamination not be removed via this method (Birch, 2000). Caution should be taken when performing pyrolysis on soda-lime glass, where moisture may cause the leeching of alkalis resulting in a sodium hydroxide residue (Birch, 2000).

UV/Ozone

UV/ozone treatment is an excellent method for removing hydrocarbon contaminants. This method desorbs surface contaminants by photosensitized ozone oxidation of the adsorbed molecules (Vigs, 1993). An exposure time of 10 to 25 minutes removes most organics (Vigs, 1993; Guiseppi-Elie, 2000). This is unfortunately a supplementary cleaning step for heavily contaminated substrates; dust particles and thick layers of contaminants are not effectively removed. Only already fairly clean slides benefit from this treatment. Hazards include possible exposure to UV radiation, a known carcinogen, and overexposure to toxic levels of ozone. Secondary UV lamps may be setup to reduce the ozone. If organic contamination is a major problem, then UV/ozone treatment should provide a definite cleaning benefit. While it is possible to build an instrument (Vigs, 1993), instruments are commercially available.

Laser

UV lasers have proven to be an effective tool for cleaning particulate and organic contaminants. The ideal laser for cleaning will generate short pulse (20 - 100 ns) of light in the UV range (Kimura et al, 1994). UV is desired over other wavelengths because a majority of organic molecules absorb UV light and, most importantly, UV wavelengths have been determined experimentally to be the most efficient. Excimer lasers, which consist of a rare gas/halogen gas mixture laser medium, are commonly used. Excimer lasers also have the capability of producing a high pulse repetition (100-1000 hz), which produces a shorter, more efficient cleaning process (Kimura et al, 1994). The laser works via two major mechanisms. Mechanism one is ablation of electrostatically bound particles such as dust, water droplets, and other macromolecules that are not transparent and absorb large amounts of UV energy (Williams and Randall, 1994). The second method, sub-surface photochemical dissociation, allows the UV energy to pass through the contaminant. The surface directly beneath it absorbs the energy and creates a microexplosion, dissociating the particle from the surface (Williams and Randall, 1994).

SLIDE COATING CHEMISTRIES

DNA immobilization/spotting, unless one chooses to make glass reactive DNAs (Kumar et al., 2000), requires a modified glass surface. Glass can be coated either adsorptively, as with poly-l-lysine, or covalently, as with functionalized silanes. Adsorption relies upon the glass surface interacting with the adsorbing molecules (typically polymers) via a combination of H-bonds, coulombic interactions and van der Waals' forces. Functionalized silanes are used to introduce a variety of functional groups to glass slides with epoxy-silane and amino-silane being the most common (Guo et al., 1994; Lamture et al., 1994; Schena et al., 1995). These and other alkoxysilanes such as (3-mercaptopropyl)trimethoxysilane (Möller et al., 2000), bind covalently to glass via silylethers, leaving the alkylfunctional group available for further manipulation. The resulting silanized slides can be used directly to bind nucleic acids or be further reacted with a variety of homo- or heterobifunctional cross-linkers to introduce a variety of surface chemistries.

In practice, slides can be coated by dipping them into a solution of the coating reagent or by vapor deposition. While most coatings can be deposited by dipping, only reagents that are sufficiently volatile are able to modify the slide surface via vapor deposition. For this reason, poly-l-lysine slides can only be prepared using a dipping process.

Coating slides with silanes is most often accomplished using substituted alkoxysilanes of the general formula

$$R_nSi(OR')_{4-n}$$

where R is an organic moiety (likely bearing a functional group that will be utilized in the immobilization of the nucleic acid) and OR' is a simple alkoxy substituent such as methoxy ($-OCH_3$) or ethoxy ($-OCH_2CH_3$). Further, the silanes most encountered in microarray slide coatings are of the general formula $RSi(OR')_3$. Coating slides with silane can be performed by dipping the slides into a solution of the silane or by vapor deposition. Figure 2 shows the series of reactions occurring when glass is coated with a trialkoxysilane in a water-containing medium (Arkles, 1977). The alkoxy substituents (-OR') hydrolyze in water containing media to form silanols and alcohols. Silanols then condense to form oligosilyl ethers. The size of the silyl ethers is dependent upon the solubility and size of the silane, governed by R, with less soluble and/or small R silanes favoring larger polymers and highly soluble and/or bulky R silanes favoring smaller oligomers. Stability of aqueous silane solutions range from hours to weeks depending on the subtituent R. Covalent attachment of the silane to a glass substrate is initiated via a hydrogen bonded intermediate complex between glass adsorbed water molecules and the oligosilyl ether. Bond formation finally occurs with loss of water during a "curing" step to form single rather than the possible three-silylether bonds with the glass.

Slides can be coated from aqueous solutions, organic solutions or by vapor deposition. In organic solvents, water required for alkoxy group hydrolysis can

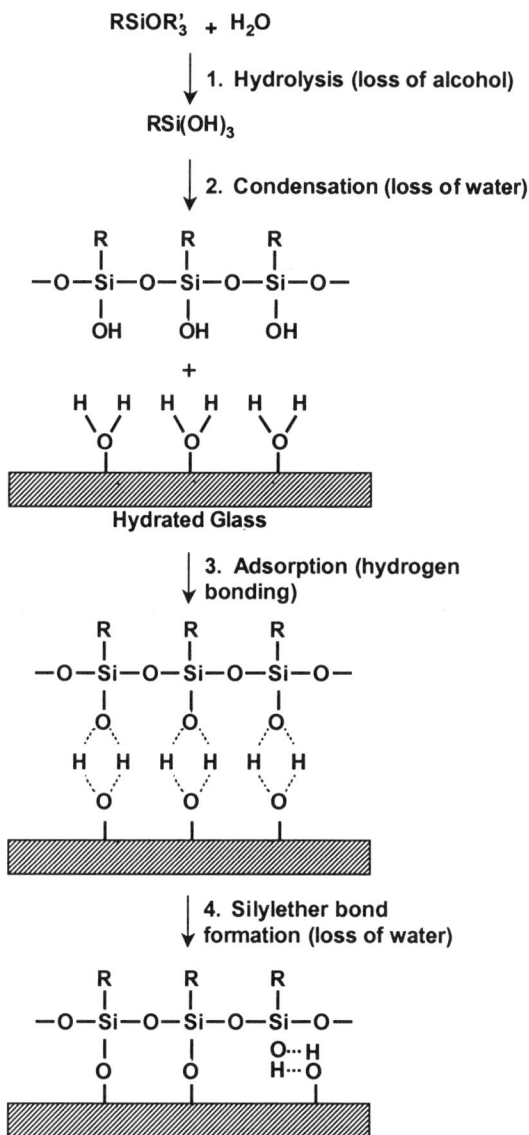

Figure 2. Sequence of reactions involved in coating glass with trialkoxysilanes in water containing solvents (Arkles 1977).

come from small quantities of water already present in the solvent (e.g 95% ethanol/ 5% water), water added to the solvent, from air or from water molecules adsorbed onto the slide surface (Arkles, 1977). Vapor deposition, exposing the glass substrate to silane vapor, offers another convenient route to silane coating. Water necessary for alkoxyl hydrolysis originates from substrate adsorbed water molecules and the atmosphere.

Below are listed brief protocols for coating slides with poly-l-lysine and aminopropyltriethoxysilane (APS). Most coatings can be applied to glass slides with adjustment/optimization of the below protocols.

Slide coating protocols

Poly-l-lysine: Freshly cleaned slides are dipped into ca. 0.01-0.02% poly-l-lysine/0.1X PBS for 30-60 minutes followed by extensive rinsing with 18 Mohm water. Many labs report that it is necessary to allow the slides to age at least two weeks before printing.

Aqueous APS coating: Place slides in a 0.5-2% aqueous aminopropyltriethoxysilane solution for 5-10 minutes followed by extensive water rinsing and drying/curing. (United Chemical Technologies, Inc., Bristol, Pennsylvania. http://www.unitedchem.com/PDF/Petrarch%202.pdf)

Organic solvent APS coating: Dip slides into 2% Aminopropyltriethoxysilane in acetone solution for 10 seconds followed by water rinsing and drying/curing.

(http://www.nottingham.ac.uk/~mbzspd/methods/Silane_and_Poly-L-Ly.html).

Vapor deposition APS coating: Incubate slides at least two hours in a 2 liter desiccator containing 50 μl of Aminopropyltriethoxysilane. (Revenko and Hansma, 1996). Further rinsing etc. is unnecessary.

DNA IMMOBILIZATION CHEMISTRIES

DNA has been immobilized onto glass surfaces by non-covalent and covalent means. The most common non-covalent methods bind nucleic acids to aminopropylsilane or poly-lysine slides (Schena et al., 1995; Duggan et al., 1999). Great attention has however been focused on covalent attachment of PCR products or pre-synthesized oligonucleotides to glass microarray slides. With few exceptions (e.g. copolymerization and diazotization), covalent attachment borrows chemistry from the bioconjugationist's toolbox. Several factors should be considered when selecting the attachment strategy for preparing microarrays. In general, the attachment chemistry should be robust, efficient and stable. Below and in Table 2, we briefly review attachment chemistries used to attach DNA to glass slides for microarray experiments. For simplicity, the bond forming/immobilization reaction is written in a generic fashion where the substituents R, R' etc. refer to the presumed non-participatory DNA and glass/silane moieties.

Table 2. Slide and DNA Linkage Chemistry

Chemistry	Surface Treatment[(ref)]	5' Modifi-cation	Cross-Linker/ Activation Process
Non-covalent			
Electrostatic, hydrophobic, UV cross-link	Aminopropyl-silane[18, 51]	None	None
	Poly-L-Lysine[18,21,55]	None	None
Hydrophobic	Vinyl silane[3]	None	None
Biotin-Streptavidin	Streptavidin[44]	Biotin	None
Amine			
Aldehyde	Aldehyde silane[40, 56]	Amine	None
	Aminopropyl-silane[71]	Amine	Glutaraldehyde
	Polyacrylamide Gel Pad[33, 35, 48, 64, 70]	Amine	Various
	Aminopropyl-silane[72]	Amine	TETU/PCC
	Amino-silane[2]	Amine	Agarose/ $NaIO_4$
	Semicarbazide silane[47]	Benzaldehyde	None
Epoxide	Epoxy-silane[13]	None	None
	Epoxy-silane[1, 36, 42, 60]	Amine	None
Isothiocyanate	Aminopropyl-silane[7, 9, 15,16, 39,45]	Amine	PDITC
EDC (carbodiimide)	Amino derivatized[31]	Succinylated	EDC
	Amino-silane[1]	Phosphate	EDC
Imidoester	Aminopropyl-silane[7]	Amine	DMS
NHS-esters	Aminopropyl-silane[7, 24]	Amine	DSO; DSC
Dendrimers, dendritic linkers	Aminopropyl-silane or epoxy-silane[7, 8, 9]	Amine	Activation and construction of dendrimeric linkers or attachment of starburst dendrimers
Thiol			
Silanized oligonucleotides	None[34, 39]	Thiol	Coupling with mercapto-silane via disulfide bond
	None[34]	Thiol	Coupling with amino-silane SPDP or SIAX cross-linkers

Table 2. Continued

Thiol	Amino-silane[1, 15, 16, 46, 52]	Thiol	EMCS; SMPB; MBS; SMCC; GMBS; MPS; SIAB
	Mercapto-silane[52]	Thiol	None; Disulfide formation
	Mercapto-silane[39, 50]	Acrydite	None; Michael addition
Others			
Diazotization	p-amino-phenyl trimethoxy-silane (ATMS) [19]	None	Nitrous acid
Copolymerization	Acrylic-silane (bind-silane) [43, 50]	Acrydite	Co-polymerization with acrylamide
	None[34]	Acrydite	Co-polymerization with acrylic-silane

Abbreviations: TETU= 2',2',2'-trifluoroethyl-11-(trichlorosilyl)undecanoate; PCC= pyridinium chlorochromate; PDITC= 1,4-Phenylene diisothiocyanate; EDC= 1-ethyl-3-(3-dimethylaminopropyl)-carbodiimide hydrochloride; DMS= Dimethylsuberimidate; DSO= Disuccinimidyloxalate; DSC= Disuccinimidyl-carbonate; SPDP= N-Succinimidyl-3-(2-pyridyldithiol)-propionate; SIAX= Succinimidyl-6-(iodoacetylamino)-hexanoate; EMCS= N-(6-maleimido-caproxy) succinimide; SMBP= Succinimidyl 4-[malemidophenyl]butyrate; MBS= m-maleimido-benzoyl-N-hydroxy-succinimide ester; SMCC= succinimidyl 4-(N-maleimidomethyl) cyclohexane-1-carboxylate; GMBS= N-(γ-maleimido-butryloxy) succinimide ester; MPS= m-maleimidopropionic acid N-hydroxysuccinimide ester; SIAB= N-succinimidyl (4-iodoacetyl) aminobenzoate

Non-covalent

Adsorption of nucleic acids to a glass substrate by non-covalent means is defined here as a process by which the DNA is immobilized without designing a linkage that produces a covalent linkage from the glass to the DNA. The most prevalent method for non-covalent DNA immobilization is printing on polycationic surfaces (e.g. aminopropylsilane and poly-l-lysine). Additional non-covalent strategies include pH dependent adsorption to hydrophobic surfaces and biotin-streptavidin mediated immobilization. Each of these methods is summarized below.

14

Polycationic Slide Surfaces

Spotting DNAs onto amine modified glass surfaces (e.g. APS or poly-l-lysine) is considered to be a non-covalent process (Belosludtsev et al., 2001) given that the immobilized polyanionic DNA first encounters and interacts with the polycationic surface via coulombic attraction (Figure 3). After printing, the DNA is effactually locked to the surface by ultraviolet irradiation or baking. The molecular mechanism by which the DNA is locked to the surface is not well characterized though it has recently been demonstrated by Surface Enhanced Raman Spectroscopy (SERS) that adsorption perturbs DNA and polyamine structure and that the interaction between the two polymers is sequence dependent with dAMP sites demonstrating the strongest interaction with a polycationic surface coating (Sanchez-Cortes et al., 2002). Though one may speculate that photocrosslinking causes TT photodimers (Leonard et al., 1973) to form an interlocking weave with the slide surface, immobilization on an amine modified slide surface is likely similar to photoinduced crosslinking of nucleic acids to amine bearing membranes (Church and Gilbert, 1984; Reed and Mann, 1985). That is, irradiation presumably covalently links the nucleic acid to the surface via nucleophilic attack by surface amine groups on the excited state of thymine (Saito et al., 1981; Church and Gilbert, 1984). For free and N1 substituted thymines, irradiation in the presence of primary amines results in a ring opened adduct. This adduct cyclizes upon heating, displacing the original N1 substitution (Saito et al., 1981). In the case of polynucleic acids, displacement necessarily results in producing an abasic site and thus loss of the DNA-surface covalent linkage. Similarly, the mechanism by which baking locks DNA to the surface is presumably through nucleophilic attack- eg. via Michael addition (i.e. reaction of nucleophiles with α-β unsaturated electrophiles, Figure 4) to pyrimidine C6 (Bradshaw and Hutchinson, 1977) of the ground state nucleobases by surface amines. Though the chemistry of attachment is speculative and seemingly not thoroughly understood, nucleic acids bound to these surfaces are expected to be conformationally restricted through binding at multiple attachment points to the surface and therefore may be less available for hybridization than covalent end attachments. Additionally, the printed DNA will be susceptible to removal under hybridization and wash conditions that involve high salt and (often) high temperatures.

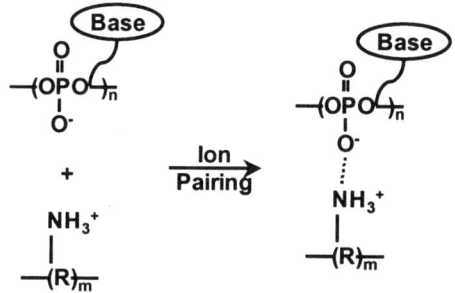

Figure 3. Immobilization of DNA onto polycationic $((RNH_3^+)_m)$ surfaces. R = lysine, aminopropyl silane glass etc.

RHC▪CHCX   +   **R'''N H$_2$** $\rightleftharpoons$ **RHC CH$_2$CX**
                                                        R'''N H

α, β unsaturated   **1°**
electrophile   **Amine**

CX	X
Aldehyde	=O, H
Ketone	=O, R'
Amide	=O, NR'R"
Nitrile	≡N

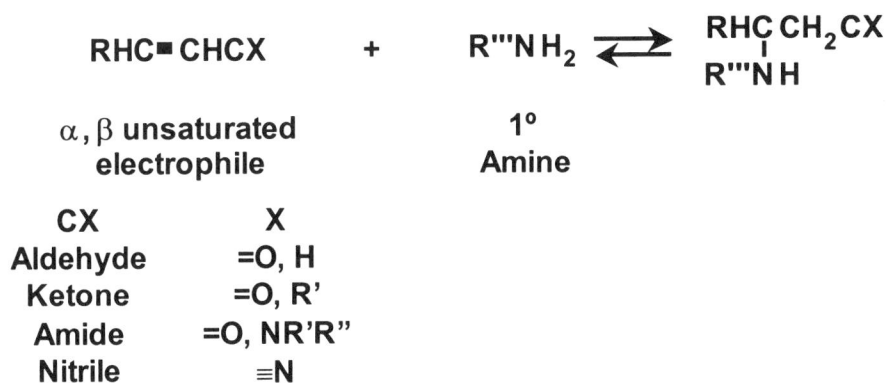

Figure 4. Michael addition. Reaction of a nucleophile, amine shown, with the β carbon of an α,β unsaturated electrophile (CX).

pH Specific Binding to Hydrophobic Surfaces

The pH-dependent specific binding of DNA by its extremities to a hydrophobic glass surface has been reported by Allemand et al. (1997). Specific binding of DNA is shown to occur at pH 5.5 in MES buffer on glass surfaces coated with a silane containing a vinyl (-CH=CH$_2$) end group. At low pH the DNA adsorbed strongly and nonspecifically whereas at high pH adsorbed weakly or not at all. Zammatteo et al. (2000) discusses the effect of pH on the binding of DNA to a neutral aldehyde-derivatized hydrophobic glass surface. At low pH the DNA bases become protonated, inducing DNA melting and exposing the hydrophobic core of the helix (Zammatteo et al., 2000). Increasing the hydrophobicity of the DNA at low pH, thus increases binding to a hydrophobic surface.

Biotin-Streptavidin

The strong and specific binding of biotin to streptavidin has recently been exploited for the non-covalent immobilization of DNA for a microarray experiment (Miyachi et al., 2000). Biotinylated oligonucleotides have been bound to streptavidin coated slides with a reported decrease in nonspecific DNA binding as compared with poly-L-lysine coated substrates (Miyachi et al., 2000).

Covalent

While non-covalent attachment is at present predominant, it suffers from a lack of complete understanding with respect to the mechanism of binding. It is however almost without question impossible for the poly anionic nucleic acid to bind to a poly cationic surface and be completely unencumbered for hybridization. That is, it must cost energy for the polycation-single stranded DNA to rearrange to a polycation-double stranded DNA during hybridization. For that reason it has been important that methods aimed at specific-covalent attachment of DNA to glass slides be developed.

Covalent end coupling chemistries offer several advantages over non-covalent attachment. Covalent strategies presumably allow the DNA to orient itself in reference to an engineered oligonucleotide end modification (5' typical). This should increase the availability of slide-immobilized (probe) sequences for hybridization with target, because binding is not occurring through the probe's nucleotide bases or backbone (Ghosh and Musso, 1987; Beier and Hoheisel, 1999). Covalent attachment permits more stringent washing, less non-specific binding and offers a potential for stripping and re-hybridizing (reusing) arrays.

Further, introduction of spacer atoms in the immobilization tether places the oligonucleotide physically further from the glass surface, reducing the substrate's contribution to steric hindrance. The choice of spacers can affect the relative hydrophilicity of the immediate environment around the attached oligonucleotide. A variety of authors have looked at the effect of the length and composition of spacer molecules between the glass slide and the presenting reactive group. The use of spacers can increase the surface loading capacity; with longer spacers the strength of the hybridization signal is expected to improve (Maskos and Southern, 1992; Guo et al., 1994; Shchepinov et al., 1997; Beier and Hoheisel, 1999; Southern et al, 1999; Afanassiev et al, 2000). Afanassiev et al. (2000) demonstrated that long 5' hydrocarbon spacers (e.g. C6 and C18) improved the hybridization signal strength. Charged groups (positive or negative) within the spacers reportedly diminished the hybridization yield (Shchepinov et al., 1997).

A variety of 5' oligonucleotide modifications are commercially available for covalently attaching DNAs to microarray substrates. These modifications include amine, thiol, and acrylic (Acrydite™) groups, which for the most part are stable. Plates of oligonucleotides or PCR products for printing microarrays can be stored frozen and re-used as needed.

To be most effective, the slide surface and DNA modifications must be appropriately matched. In the next sections we individually summarize the various chemistries involved with immobilizing DNA on slide surfaces. In general, an ideal slide surface will be stable under ambient laboratory conditions, i.e. not be particularly susceptible to degradation or inactivation by air or moisture. Further, methods that use a single-step attachment are generally preferred to those requiring cumbersome activation steps with the printing solution not requiring labile additives such as catalysts, co-reactants or cross-linkers.

Amine Chemistry

Amines are nucleophilic moieties that react with electrophilic carbon atoms to form, in most cases, stable carbon-nitrogen bonds. This elementary principle has been capitalized upon for decades to conjugate, cross-link, label and immobilize biomolecules (Hermanson, 1996). The reactions so far used for nucleic acid immobilization on glass slides include reductive amination and nucleophilic displacement to produce secondary amines and condensation reactions resulting in amide bond linkages.

Reductive Amination

Amino modified DNAs can be immobilized to aldehyde modified slides via reductive amination (Figure 5). An aldehyde is an oxygen atom double bonded to the carbon at the end of a hydrocarbon chain, generally represented RCHO. Aldehydes react with amines to form Schiff's bases (RCHNR'). Though acid labile, Schiff's bases can be converted to stable secondary amines (C-N) by reduction, e.g. with sodium borohydride (NaBH$_4$) (Hermanson, 1996). Schiff's base reduction is however likely not necessary (Todd Martinsky personal communication),

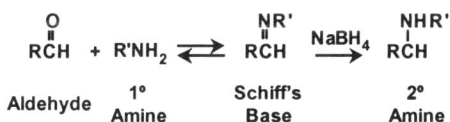

Figure 5. Reductive amination reaction scheme: reaction of primary amines with aldehydes to form Schiff's bases followed by Sodium Borohydride reduction. R is typically an aldehyde silane modified surface and R' is a 5' amino oligonucleotide.

assuming the printed surface will remain non-acidic. Schiff's bases are readily hydrolyzed at acid pHs. Further, the requirement for amine tethers is also in question. That is, we have found it difficult to discriminate between amine modified and unmodified long oligonucleotides on aldehyde surfaces (K. Aboytes, unpublished results). If reduction is carried out with sodium borohydride, along with the Schiff's bases, remaining aldehyde groups will be reduced to primary alcohols (RCHO→RCH$_2$OH).

Aldehyde functionalized silylated glass slides are available commercially (Schena et a l., 1996) o r may b e prepared b y one of several m ethods. The most straightforward method would be to treat clean glass slides with a triethoxysilane aldehyde such as 4-(triethoxysilyl)butanal using an adaptation of one of the methods described above for aminopropyltriethoxysilane (APS). Alternatively, aminosilanized surfaces can be modified with the homobifunctional cross-linker glutaraldehyde (Yoshioka et al., 1991). Zammatteo et al., (2000) prepared aldehyde s urfaces by f irst b inding t he silane, 2',2',2'-trifluoroethyl-11-(trichlorosilyl)undecanoate (TETU) to clean glass slides. The ester groups were reduced t o a lcohols (RCH_2OH) f ollowed b y o xidation w ith p yridinium chlorochromate (PCC) ($RCO_2R' \rightarrow RCH_2OH \rightarrow RCHO$). Their recommended buffer for binding the aminated DNA to the aldehyde-derivatized surface was 0.1M MES, pH 6.5. Since aldehydes are easily oxidized to carboxylic acids ($RCHO \rightarrow RCO_2H$), aldehyde modified surfaces should be protected from air (Zammatteo et al., 2000). An advantage of binding amino-modified nucleic acids via reductive amination is that the first step proceeds without the need of preparing short-lived reagents such as EDC (see condensation section below), a coupling reagent which needs to be freshly prepared and so could be a source of variability from sample to sample. Afanassiev recently reported an increased immobilization capacity by preparing glass slides with periodate oxidized agarose film (Afanassiev et al., 2000). Agarose, being a polysaccharide, contains 1,2 diols and so is oxidized by $NaIO_4$ to y ield aldehydes ($RCH(OH)CH(OH)R' \rightarrow RCHO + O HCR'$). Similarly, aldehyde bearing polyacrylamide gel pads on glass slides have been used to immobilize amine-modified oligonucleotides (Proudnikov et al., 1998). LaForge et al. (2000) reported the re-use of chips prepared by this method six times without visible changes in hybridization properties. The method was used similarly for gene polymorphism analysis (Yershov et al., 1996). Key features of this method are the low fluorescent backgrounds of polyacrylamide gels, increased capacity for immobilization and the reliability for discrimination of perfect duplexes from mismatches (Yershov et al., 1996).

Semicarbazide

Semicarbazides ($RNHCONHNH_2$), like primary amines, re act with aldehydes to form Schiff's base like semicarbazones; (Figure 6). Podyminogin et al. (2001) used this chemistry to couple a benzaldehyde-modified oligonucleotide with a

Figure 6. Reaction of semicarbazides with aldehydes to form semicarbazones. R is the modified glass surface and R' is 5' modified oligonucleotides.

semicarbazide (SC) silanized glass surface. Unlike primary amine derived Schiff's bases, the electrophilic benzaldehyde moiety reacts with the semicarbazide glass surface to form stable semicarbazone bonds, a property which obviates the need for post coupling reduction. Another advantage to this chemistry is that benzaldehyde-modified oligonucleotides and semicarbazide-silanized glass slides are stable and require no special storage precautions (Podyminogin et al., 2001).

Nucleophilic Displacement

Primary amines react with epoxides, 3-membered cyclic ethers, to form secondary amino alcohols via nucleophilic displacement (Figure 7). Epoxide activated glass slide surfaces can

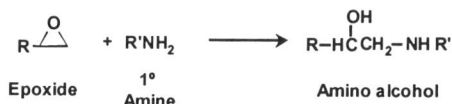

$$R \overset{O}{\triangle} + R'NH_2 \longrightarrow R-HC\underset{OH}{|}CH_2-NHR'$$

Epoxide   1° Amine   Amino alcohol

Figure 7. Nucleophilic attack of epoxides by amines to form amino alcohols. R is epoxide modified glass and R' is 5' modified DNA.

be readily prepared by applying 3-glycidoxypropyltrimethoxysilane (epoxy-silane) to clean glass slides. Although epoxy-silane may be applied by vapor deposition (Mandenius et al., 1986), most microarray researchers report use of a dip method. The slides are soaked for several hours at 80°C in a solution of anhydrous xylene, 3-glycidoxypropyltrimethoxysilane, and *N,N*-diisopropylethylamine followed by washing, drying and reduced pressure desiccated storage (Maskos and Southern, 1992; Lamture et al., 1994; Beattie, 1995; Möller et al., 2000). Covalent attachment of amine-modified oligonucleotides to epoxy-silane surfaces has been accomplished using high pH solutions such as 0.1M KOH or 0.1M NaOH (Shumaker et al., 1996; Möller et al., 2000). Another use of epoxy-silane treated slides involves the binding of unmodified oligonucleotides through what is likely a non-covalent interaction (Call et al., 2001). Unmodified oligonucleotides, as opposed to amino-modified oligos, were not stably bound to the epoxy-silane derivatized surface as demonstrated by washes at pH > 10 (Call et al., 2001). Adessi et al., (2000) examined methods for attaching oligonucleotides as PCR primers by their 5' ends to glass supports. Although they found the attachment of 5'-amino modified oligonucleotides on epoxy-derivatized slides to be relatively thermostable they reported problems with reproducibility due to the sensitivity of the epoxy ring to moisture.

Condensations

Amines (RNH$_2$) react with acids (R'X(Y)OH) or their equivalents (R'X(Y)Z) to form amides (R'X(Y)NHR). Since the initial starting materials for these reactions are acids, formation of an amide requires a net loss of water from the parental acid. That is, the net reaction between an amine and an acid to produce an amide is:$RNH_2 + R'X(Y)OH \rightarrow R'X(Y)NHR + H_2O$ and so is generically termed a condensation. Though reaction of an amine with an acid equivalent does not on the surface result in loss of water, at some point the preparation of the acid equivalent involved a dehydration or removal of water (or OH⁻) from the starting acid (i.e. $R'X(Y)OH + HZ \rightarrow R'X(Y)Z + H_2O$). Simple acid species such as carboxylates (RC(O)OH) and phosphates (RO(R'O)P(O)OH) are not in themselves reactive with amines. For this reason, acids must be either preactivated, e.g. isothiocyanates, NHS esters and imidoesters, or activated *in situ* with coupling agents, e.g. carbodiimides. Such coupling is often accomplished using crosslinkers, i.e. bifunctional molecules that have functional group-specific reactive moieties (described in greater detail below). Beier and Hoheisel (1999) recently compared cross-linkers utilizing different functional groups to condense with amines. Their work demonstrated that 1,4-phenylene diisothiocyanate (PDITC), dimethyl suberimidate (DMS), disuccimidyl carbonate (DSC) and disuccimidyl oxylate (DSO) are suitable for oligonucleotide and PNA-oligomer immobilization whereas PDITC and DMS were superior for amino-linked PCR fragments.

Isothiocyanate

Isothiocyanates (RN=C=S) react with amines to form substituted thioureas (Hermanson 1996) (Figure 8A). The use of isothiocyanate chemistry has been described for the covalent immobilization of 5'-amino modified oligonucleotides (Guo et al., 1994 ; Beier and Hoheisel, 1999; Möller et al., 2000; Lindroos et al., 2001). Isothiocyanate glass slide surfaces are prepared by reacting the homobifunctional cross-linker 1,4-phenylene diisothiocyanate (PDITC) (Figure 8B) in 10% anhydrous pyridine/DMF with an amine modified glass surface. Immobilizing the amine modified DNA is typically performed under moderately basic conditions (e.g., 100 mM sodium carbonate/bicarbonate, pH 9.0) (Guo et al., 1994; Möller et al., 2000). Beier and Hoheisel (1999) demonstrated that PDITC slides are suitable for oligonucleotide, PCR product and PNA-oligomer immobilization.

NHS Ester

NHS esters react with amines to form amides (Figure 9). These active

A

RN=C=S + R'NH₂ ——→ RN–C–NR'

$$RN\!=\!C\!=\!S \quad + \quad R'NH_2 \quad \longrightarrow \quad \underset{H}{R}N\!-\!\overset{\overset{\displaystyle S}{\|}}{C}\!-\!\underset{H}{N}R'$$

Isothiocyanate   1°   Substituted
                  Amine   Thiourea

B

$$S\!=\!C\!=\!N\!-\!\langle\ \rangle\!-\!N\!=\!C\!=\!S$$

**1,4-phenylene diisothiocyanate
(PDITC)**

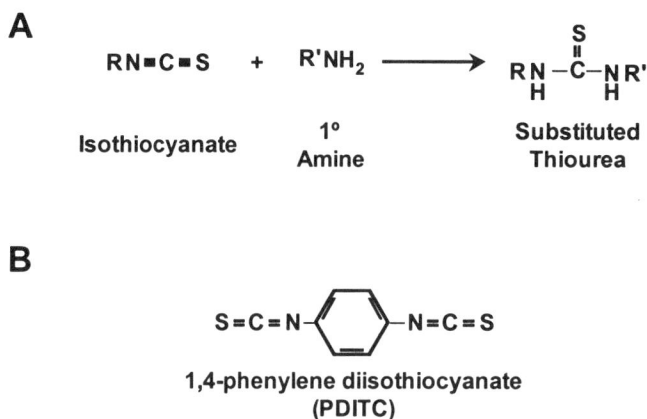

Figure 8. **A**. Reaction of an isothiocyanate moiety with an amine to form a substituted thiourea. R is phenyl thiourea glass (prepared from PDITC and amine modified glass surface), R' is 5' modified DNA. **B**. Structure of 1,4-phenylene diisothiocyanate (PDITC).

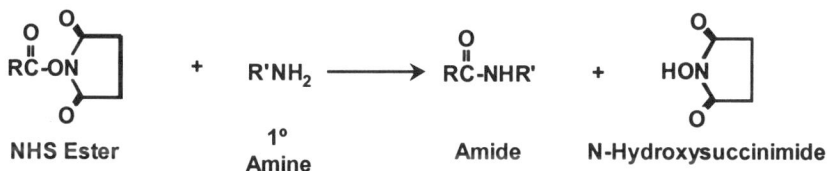

$$RC\!-\!ON \quad + \quad R'NH_2 \quad \longrightarrow \quad RC\!-\!NHR' \quad + \quad HON$$

NHS Ester   1°   **Amide**   **N-Hydroxysuccinimide**
                 Amine

Figure 9. NHS ester/amine condensation. R = derivatized glass, R' modified DNA.

esters can react with sulfhydryl and hydroxyl groups to form thioesters and ester linkages, respectively. These bonds, however, are fairly labile and hydrolyze in aqueous environments (Hermanson, 1996).

NHS esters have been prepared via bifunctional crosslinkers or by carbodiimide mediated condensation of N-hydroxysuccinimide with a carboxyl-modified substrate. Two homobifunctional cross-linkers, DSC and DSO (Figure 10) have been evaluated by Beier and Hoheisel (1999) on prepared NHS ester functionalized glass slide surfaces. DSC is the di-N-hydroxsuccinimide of carbonic acid (HOCOOH), DSO is the di-NHS ester of oxalic acid (HOCOCOOH). DSC and DSO activated surfaces were each found to be suitable for covalent attachment of DNA and PNA oligomers. Ghosh and Musso (1987) compared direct EDC amine coupling (see EDC below) with NHS ester coupling on CPG. They treated carboxyl derivatized controlled pore glass (CPG) with dicyclohexylcarbodiimide and N-hydroxysuccinimide to prepare the NHS-

22

activated carboxyl surface. They reported lower attachment efficiencies with the NHS ester surfaces than with diimide coupling reactions due to competing hydrolysis of the activated esters.

Imidioesters

Imidoesters (RC(NH)OR') are amine-reactive masked carboxylic acids that form amidines upon reaction with amines (Figure 11). This chemistry has been used to attach 5'-amine modified oligonucleotides to amine modified microarray surfaces via the cross-linker, dimethylsuberimidate,

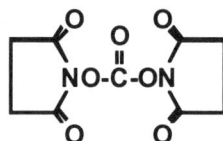

Disuccimidyl carbonate (DSC)

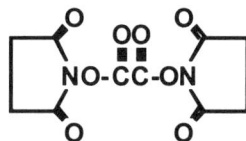

Disuccimidyl oxylate (DSO)

Figure 10. Structures of disuccimidyl carbonate (DSC) and disuccimidyl oxylate (DSO) homobifunctional crosslinkers. Crosslinkers are used to prepare NHS ester derivatized glass slides from amine-modified substrates.

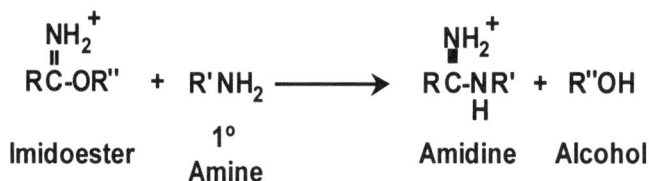

$$\underset{\text{Imidoester}}{\overset{\overset{+}{\overset{NH_2}{\|}}}{RC\text{-}OR''}} + \underset{\underset{\text{Amine}}{1^o}}{R'NH_2} \longrightarrow \underset{\text{Amidine}}{\overset{\overset{+}{\overset{NH_2}{\|}}}{\underset{H}{RC\text{-}NR'}}} + \underset{\text{Alcohol}}{R''OH}$$

Figure 11. Reaction of imidoester with amine to form an amidine linkage. R = modified glass surface, R'= modified nucleic acid, R''= alkyl (CH 3 , C 2 H 5 typical).

$$\underset{\text{Dimethyl suberimidate (DMS)}}{\overset{\overset{NH}{\|}\qquad\overset{NH}{\|}}{CH_3O\text{-}C(CH_2)_7 C\text{-}OCH_3}} \xrightarrow[\text{2. R'NH}_2]{\text{1. R NH}_2} \overset{\overset{NH}{\|}\qquad\overset{NH}{\|}}{RHN\text{-}C(CH_2)_7 C\text{-}NHR'}$$

Figure 12. Structure of homobifunctional crosslinker dimethylsuberimidate used for preparing imidoester derivatized surfaces beginning with amine modified glass. R=amine modified glass, R' amine modified DNA.

DMS (Figure 12). DMS is a homobifunctional crosslinker containing imidoester groups at each end of the molecule resulting in an eight-atom bridged tether. Beier and Hoheisel (1999), demonstrated that DMS was suitable for oligonucleotide and PNA-oligomer immobilization.

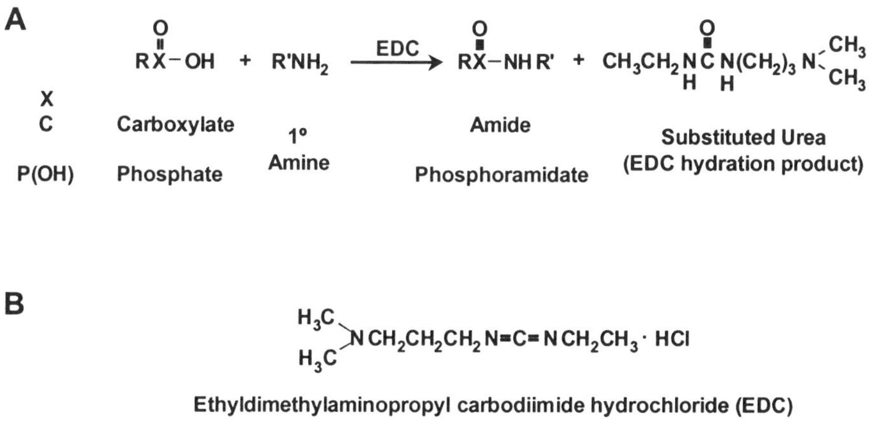

A

X				
C	Carboxylate	1°	Amide	Substituted Urea
P(OH)	Phosphate	Amine	Phosphoramidate	(EDC hydration product)

B

Ethyldimethylaminopropyl carbodiimide hydrochloride (EDC)

Figure 13. **A.** 1-ethyl-3-(3-dimethylaminopropyl)-carbodiimide hydrochloride (EDC) mediated condensation between carboxylates (RCO$_2$H) and phosphates (ROPO$_2$OH) and an amine. Carboxylic acid: R modified glass or modified DNA, R' vice versa. Phosphate: R = DNA, R' = amine modified surface. **B.** EDC structure.

EDC

The use of EDC is an older method, initially described by Gilham (1968) to attach DNA to cellulose membranes via the 5'-phosphate group. It is used to covalently attach nucleic acids to solid supports, and involves the activation of a an acid moiety (carboxylic or phosphoric X= C, P(OH) Figure 13A respectively) with the water soluble diimide 1-ethyl-3-(3-dimethylaminopropyl)-carbodiimide hydrochloride, EDC (Figure 13B). Subsequent reaction with an amine forms stable amide bonds (Ghosh and Musso, 1987; Hermanson, 1996). Immobilization of nucleic acids can proceed with either the surface or the DNA bearing the amine.

Five prime -amine modified nucleic acids have been attached to carboxylated slide surfaces, prepared by reaction of APS slides with succinic anhydride, via amide bond formation with this chemistry. Ghosh and Musso (1987) demonstrated effective end-attachment using 5'-phosphorylated and 5'-amino modified oligos to controlled pore glass (CPG) beads. 5'-phosphorylated oligonucleotides were linked via phosphoramidate bonds to amine modified CPG. 5'-aminohexyl modified oligonucleotides were linked via amide bonds to carboxyl modified CPG. They reported that the EDC coupling methods using carboxyl-modified supports resulted in lower levels of non-specific binding than with amine-modified CPG supports.

Walsh et al. (2001) compared EDC versus an isothiocyanate (PDITC on aminopropylsilane) for the immobilization of single-stranded 5'-amine modified DNA to glass beads. They found that a one-step, EDC mediated coupling with succinylated or PEG-modified beads in 0.1M MES (2-[N-morpholino]ethanesulfonic acid) buffer, pH 4.5, resulted in the highest immobilization efficiency. Using the same chemistry but changing the locations of the reactive groups, Joos et al. (1997) attached 5'-succinylated oligonucleotides to aminophenyl and aminopropyl silane derivatized glass slides in a one-step carbodimide-mediated condensation reporting that up to 90% of the immobilized sequences were available for hybridization. One potential problem with using EDC in a microarray print solution would be that EDC is water labile (Hermanson, 1996).

Dendrimeric Linkers

The above condensation chemistries have been utilized to couple nucleic acids to dendrimer, highly branched tree-like polymers originally prepared by cyclic synthesis (see for example Figure 14 and Tomalia et al., 1990), modified slides. Dendrimer-like linkers were introduced by Beier and Hoheisel (1999) as a flexible support chemistry that could be used in a controlled fashion to increase loading capacity, present a platform for covalent attachment, and modify the hydrophobicity and charge of glass and polypropylene surfaces. Dendrimeric linker structures were synthesized on the slide surface by a four-step reaction increasing the number of reactive sites available for nucleic acid binding by a factor of 10. Starting with an aminated support (APS), the first reaction was an acylation of the surface-bound amino-groups with an acid chloride, either 4-nitrophenyl-chloroformate or acryloylchloride. The acylated support was then reacted with a polyamine such as tetraethylenepentamine to produce a branched, dendrimeric structure. The branched amines were then acylated with acryoylchloride followed by Michael addition with 1,4-bis(3-aminopropoxy)-butane. A variety of linker structures can be synthesized by repeating the process of acylation and alkylation with a variety of polyamines. The slides were finally activated using one of four homobifunctional cross-linkers, phenylenediisothiocyanate (PDITC), disuccinimidylcarbonate (DSC)

A

B

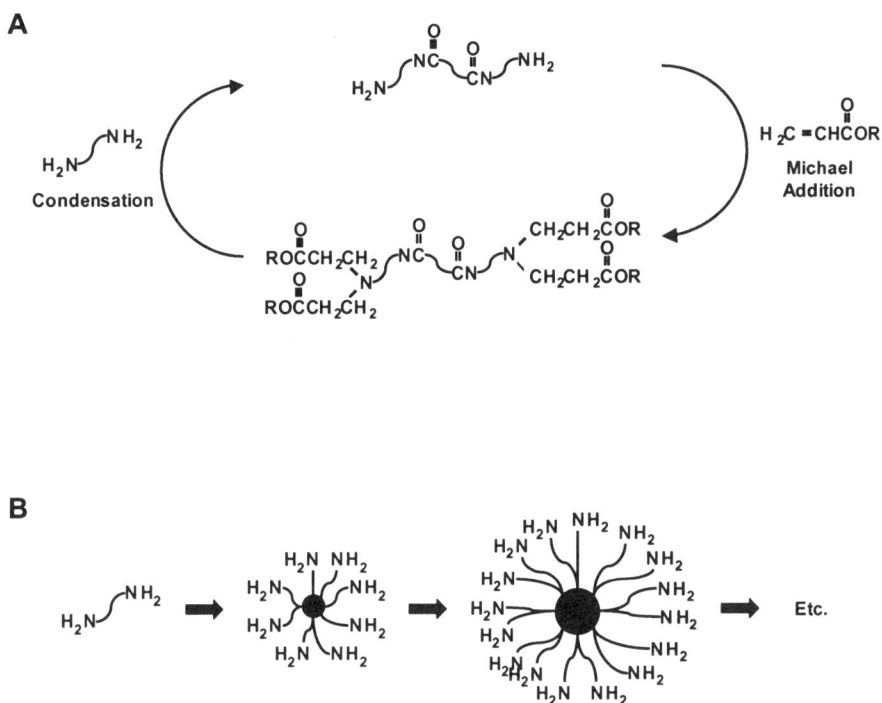

Figure 14. Starburst Dendrimers. *A.* cyclic synthesis involving Michael addition of diamine to acrylic ester followed by condensation between the resulting polyester and diamine. *B.* schematic representation of generational dendrimer synthesis.

disuccinimidyloxalate (DSO), or dimethylsuberimidate (DMS). PDITC, DMS, DSC and DSO were reported to work well for oligos whereas PDITC or DMS were superior for amine modified PCR products. One explanation for this difference is that these reactive groups may be less labile to the Tris buffer found in PCR amplification buffers.

In evaluating the coupling efficiencies of amino vs. unmodified oligonucleotides, amino vs. ordinary PCR-products, and PNA oligomers to these activated supports, Beier and Hoheisel (1999) found that only 5'amine modified nucleic acids were covalently attached. That is, oligonucleotides, PCR products and PNA oligomers devoid of amine modification were removed after a couple of rounds of stripping and reprobing. They concluded that little covalent linkage occurred with the nucleic acid exocyclic amino-groups. It is worth noting that the PNA oligomers showed increased signal intensities compared to oligonucleotide sequences due to their higher binding affinity in hybridization. In order to create a more hydrophilic surface, hydrophilic amines like 1,4-bis(3-

aminopropyl)butane could be incorporated into the dendrimeric linker synthesis. It was reasoned that the more hydrophilic surface would facilitate the approach of a hydrophilic DNA-probe molecule. Nucleic acids covalently bound to this type of array surface were found to be stable for more than 30 cycles of stripping (Beier and Hoheisel, 1999).

Commercially available synthetic PAMAM starburst dendrimers covalently attached to glass slides have recently been described as a method to increase the capacity for binding amine-modified biomolecules such as oligonucleotides (Benters et al., 2002). Starburst dendrimers containing 64 amine groups in the outer sphere were attached to an APS surface activated with homobifunctional cross-linkers such as disuccinimidylglutarate (DSG) or 1,4-phenylenediisothiocyanate (PDITC). The immobilized dendritic monomers were then activated with DSG or PDITC to produce a surface reactive to amine modified oligonucleotides. Hybridization signals were reported to be significantly higher than with aminopropylsilane or poly-L-lysine glass slide surfaces, perhaps because of decreased steric hindrance due to the long flexible chain between the surface and the oligonucleotide (Benters et al., 2002). PAMAM starburst dendrimers modified with glutaric anhydride and activated with N-hydroxysuccinimide were also described (Benters et al., 2002). The use of activated dendrimer surfaces provides a method for efficient immobilization of amine-modified nucleic acids and permits multiple hybridization experiments without significant loss of signal intensity (Benters et al., 2002).

Thiol Chemistry

Thiols, like amines, are nucleophilic agents that react with electrophiles. Unlike amines however, thiols undergo facile redox reactions that have been used extensively for bioconjugations. Thiol-modified oligonucleotides can be immobilized onto a glass support by nucleophilic attack, heterobifunctional cross-linkers or thiol/disulfide exchange reactions. An advantage to thiol chemistries is that the linkage chemistry is generally thiol group specific and so the nucleic acid backbone and bases, being devoid of sulfhydryls, cannot contribute to the attachment (Kumar et al., 2000).

Nucleophilic Attack

Thiols react with $\alpha\beta$ unsaturated esters (e.g. acrylic moieties) to form Michael addition product thioethers (Figure 15). Rehman et al. (1999) and Lindroos et al.

$$\text{RNHCC=CH}_2 \quad + \quad \text{R'S–H} \quad \longrightarrow \quad \text{RNHCC–CH}_2\text{–SR'}$$

N-Substituted Methacryamide	Thiol	Thioether

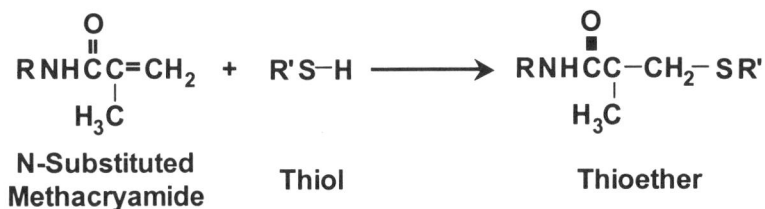

Figure 15. Thioether formation by Michael addition of a thiol (RSH) to acrylic amides.

(2001) immobilized 5' acrylamido (Acrydite™) oligonucleotides on thiosilane surfaces.

Heterobifunctional Cross-linkers

Heterobifunctional cross-linkers possess reactive centers capable of reacting with two chemically distinct functional groups, e.g. amines and thiols (Figure 16). The linkers serve two purposes: they covalently bind two distinct chemical entities which otherwise would remain un-reactive towards one another, and they act as physical spacers which provide greater accessibility and/or freedom to each of the linked biomolecules (Chrisey et al., 1996a, b). Several groups have prepared thiol reactive slides via the heterobifunctional cross-linker approach. Figure 17 shows the reactions and structures of the various heterobifunctional crosslinkers that have been used to immobilize nucleic acids on glass slides. It should be noted that these crosslinkers all take advantage of chemistry discussed

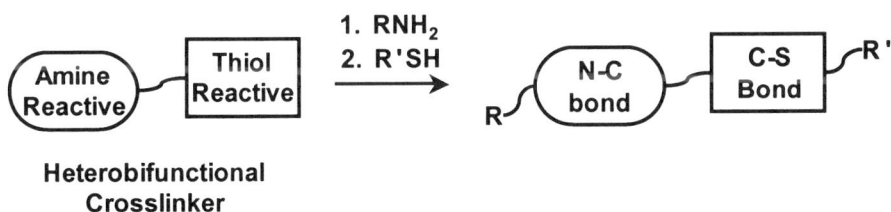

Figure 16. Schematic representation of heterobifunctional crosslinker reaction between amines and sulfhydryl bearing molecules.

28

individually elsewhere in this chapter, i.e. amine reactivity is an NHS ester condensation, while the thiol specific reactions are Michael additions (Figure 17A) and disuflide exchanges (Figure 17B). Segregation of heterobifunctional crosslinkers from the individual chemistries was done here for the sake of simplicity, because of their common use in bioconjugations, and because the use of such molecules does not *a priori* require a specific synthetic scheme. That is, one could conceivably prepare a modified slide surface with the crosslinker followed by modified oligo printing or react a DNA modification moiety with the crosslinker followed by immobilization on a slide surface. An example of this

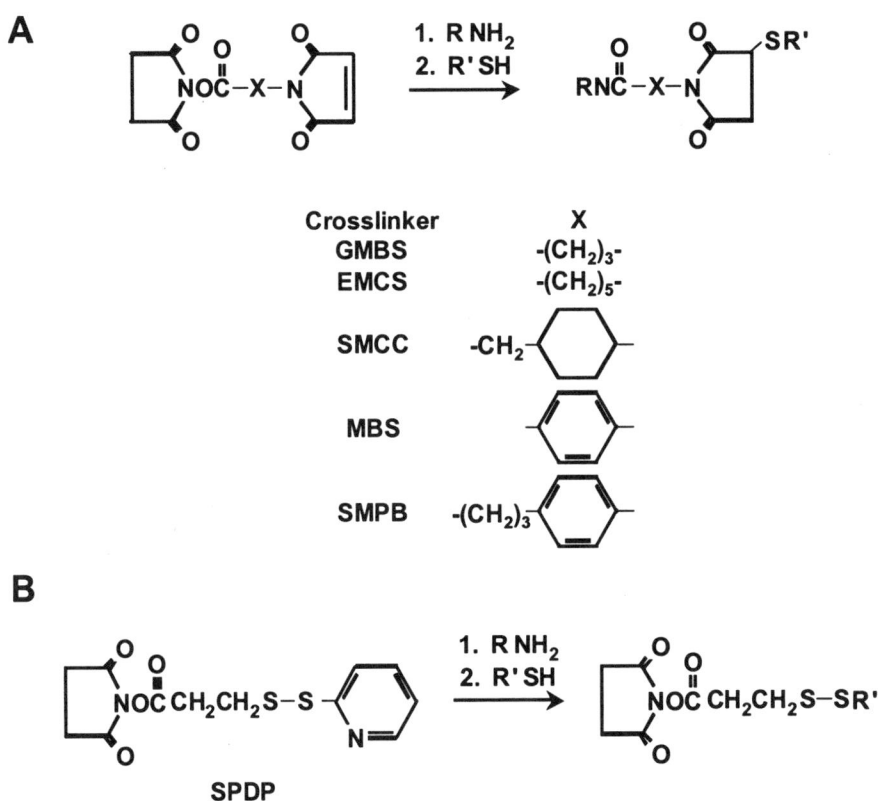

Figure 17. A. Reaction and structures of NHS/maleimide heterobifunctional crosslinkers with amines and thiols. Crosslinkers: N-(4-maleimido-butryloxy) succinimide ester (GMBS), N-(6-maleimido-caproxy) succinimide (EMCS), succinimidyl 4-(N-maleimidomethyl) cyclohexane-1-carboxylate (SMCC), m-maleimido-benzoyl-N-hydroxy-succinimide ester (MBS) and, Succinimidyl 4-[malemidophenyl]butyrate (SMPB). *B.* Reaction of NHS ester/disulfide heterobifunctional crosslinker N-Succinimidyl-3-(2-pyridyldithiol)-propionate (SPDP) with amines and thiols. R is derivatized glass or DNA, R' is vice versa.

second strategy is discussed below for printing silanized oligonucleotides on unmodified glass slides.

Crosslinking DNA Prior to Slide Surface Immobilization

Okamoto et al. (2000) treated APS glass substrates with the heterobifunctional cross-linker N-(6-maleimidocaproxy)succinimide (EMCS). EMCS has an NHS at one end, which reacts with amines to form amides, and a maleimide group, which reacts with thiols forming thioethers, at the other. Thus, slides prepared by condensation of EMCS with amino bearing glass surfaces were able to immobilize 5' thiol-oligonucleotides. Okamoto chose this cross-linker because of the stability of the maleimide group to neutral aqueous conditions. Chrisey (1996a, b) applied the heterobifunctional cross-linker Succinimidyl 4-[malemidophenyl]butyrate (SMPB) to trimethoxysilylpropyldiethylenetriamine (DETA) coated surfaces for attachment of thiol-oligonucleotides to aminosilane films. It should be noted that Chrisey et al. (1996a, b) avoided the use of thiol-silane modified surfaces for attachment of amine-modified oligonucleotides with heterobifunctional cross-linkers in favor of thiol specific electrophiles (i.e. maleimide and iodoactetyl) because of the concern that thiol-silane monolayer films are susceptible to reactions with ambient thiols and under certain irradiation conditions, are converted to unreactive sulfonates. Coupling via disulfide linkages was also avoided because of the potential for reductive cleavage of the linked molecules. Adessi et al. (2000) investigated coupling 5' thiol and amine modified oligonucleotides with heterobifunctionalized amino-derivatized and thio-derivatized glass slides respectively. The first method gave Adessi the highest performance, substantiating Chrisey et al.'s (1996a, b) concerns.

Comparing various methods for attaching PCR primers for solid-phase DNA amplification, Adessi et al. (2000) found the highest 5'-end specific attachment, thermal stability and reproducibility using the water soluble heterobifunctional cross-linker m-maleimidobenzoyl-N-hydroxysulfo-succinimide ester (s-MBS) to attach 5'-thiol (SH) modified oligonucleotides to amino-silanized glass slides.

Disulfide bonds

Thiol/disulfide exchange is another reaction commonly used to conjugate biomolecules and in particular proteins. This reaction is particularly appealing because side reactions involving other DNA functional groups are minimized due to the specificity of the thiol/disulfide exchange reaction. Rogers et al. (1999)

used this reaction to covalently immobilize pre-synthesized 5' disulfide DNA probes onto mercaptosilane-derivatized glass supports. This method allows direct coupling of stable 5' disulfide-modified oligonucleotides onto mercaptosilane-activated glass surfaces without pretreatment. An alternate use of this chemistry has been used to prepare silanized oligonucleotides that were printed on unmodified glass surfaces (Kumar et al., 2000). The silanized ologonucleotides were prepared as 5' thioated DNA-thiosilane mixed disulfides. The procedure is reportedly a quick method for attaching oligonucleotides and gives strong hybridization signals with negligible background (Kumar et al., 2000). Stripping and re-probing was also demonstrated. Lindroos et al. (2001) compared silanized nucleic acids to a variety of other covalent attachment strategies including isothiocyanate, aldehyde, acrydite and thiol chemistries.

Other Chemistries

Copolymerization

Oligonucleotides derivatized with the acrylic group can polymerize with free acrylamide to form a polyacrylamide-acrylic DNA copolymer (Rehman et al., 1999). Rehman and coworkers (1999), in their desire to develop a stable linkage for solid phase PCR, took advantage of this to develop a covalent-thermostable oligonucleotide attachment method. The method involves printing synthetic 5' alkylacrylamido oligonucleotides (Acrydite™ oligos) onto acrylic silane coated glass slides. The slides were post printing processed to immobilize the DNA by covering the surface with a polymerizing acrylamide solution resulting in a glass tethered acrylamide-acryloligo copolymer (Figure 18). Unlinked DNA (16-17%) was removed from the surface by electroelution. The authors speculated on the use of photochemical initiators to avoid use and inherent variability of a polymerizing slide coating solution. Kumar et al. (2000) similarly prepared

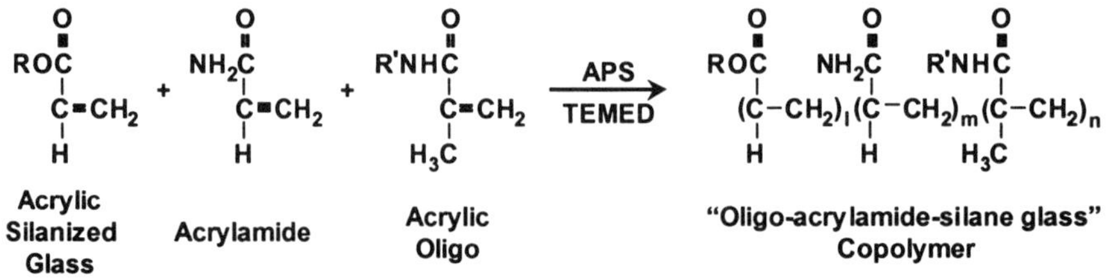

Figure 18. Acrylic oligo-acrylamide-acrylic silane copolymerization. R = modified glass surface, R' = modified DNA.

acrylicoligonucleotide-acrylic silane copolymers. As with the disulfide linked silane oligonucleotides (see disulfides above) the copolymers were printed directly onto clean glass surfaces.

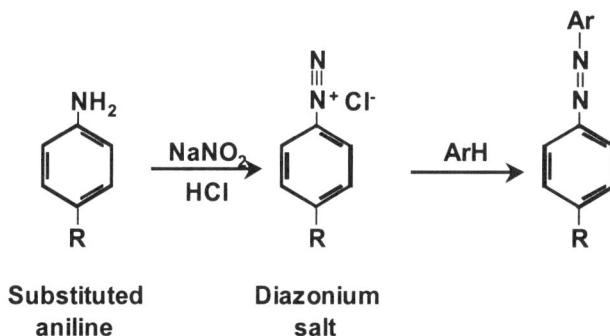

Figure 19. Conversion of substituted aniline to diazonium salt followed by reaction with an aromatic moiety (Ar). In the context of DNA immobilization, R is modified glass and Ar is presumed to be the purine nucleobases.

Diazonium Ion

Dolan et al. (2001) describe the use of diazotized glass surfaces for the covalent attachment of unmodified nucleic acids. This method entails preparation of thermally unstable diazotized slides by converting *p*-aminophenyl silane coated slides to *p*-diazophenyl silane with nitrous acid (sodium nitrite + hydrochloric acid) (Figure 19). Though the authors did not clearly speculate on the attachment chemistry, diazonium salts form C-N bonds with aromatic compounds by electrophilic attack. One would thus presume that DNA immobilization proceeds via electrophilic attack of the nucleobases by the diazonium ion, resulting in base-surface attachment. This presumption is supported by the fact that aryl diazonium salts react with dA and dG to form triazines and C8 addition products (Chin, Hung and Stock, 1981, Hung and Stock , 1982 and Gannett et. al., 1999). Microarrays prepared by this chemistry reportedly give consistent and sensitive hybridization results, have low background fluorescence, and permit stripping and re-hybridization (Dolan et al., 2001).

SURFACE ANALYSIS METHODS

Whether a slide is coated or clean, it is necessary that one characterize the prepared surface's properties. Methods often used range from the quite simple/ minimal (e.g. contact angle) to those that require sophisticated instrumentation such as for Atomic Force Microscopy and X-ray Photoelectron Spectroscopy. These methods can be used to detect and characterize coatings as well as the presence/absence of surface contaminants. In general the methods are capable of either measuring a physical property of the surface (e.g. surface energy, topology etc.) or surface molecule characterization (e.g. Raman and infrared spectroscopies). Below we briefly describe some commonly used techniques in surface analysis. Further surface characterization techniques and tutorials are readily found on the Internet. One such excellent site is the Surface Analysis Forum Surface Science Site (http://www.uksaf.org/home.html).

Contact Angle

Contact angle is one of the simplest yet most elegant methods for characterizing a surface. That is, while it does not allow identification or size of surface groups, surface topology or coating thickness, it does allow one to quickly test for the addition of a surface coating or contamination/cleanliness of a surface. Either of these processes can be quickly ascertained by noting a change in contact angle.

In the context of characterizing solid surfaces, the contact angle is the angle between a substrate surface and the tangent line at the point of contact between a liquid droplet surface. The resulting tangent line-surface plane angle and is dependent on the surface energy of the substrate and the surface tension of the liquid used (assuming the gas phase remains constant). This measurement is used to describe the hydrophilic/hydrophobic nature of surfaces. Figure 20A shows the principle of contact angle measurement with respect to surface hydrophobicity/hydrophilicity. A method for measuring contact angles entails placing a droplet of liquid on the surface and measuring the angle of a vector between the drop edge and top-center (Figure 20B, Tantec Inc. 2001). A homogenous smooth surface will have a single value across the surface.

With water, the contact angle will approach zero if complete wetting takes place. If there is only partial wetting, the contact angle will read somewhere between 0 and 180 degrees. One can estimate the contact angle by measuring the diameter of a given volume of liquid. Figure 21 shows the relationship between contact angle and drop diameter for 10 μl drops of water spotted on a

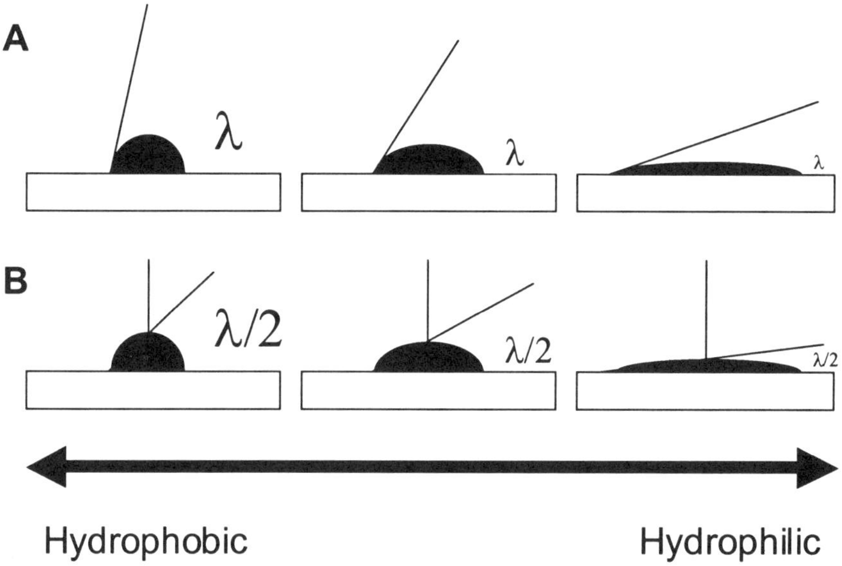

Hydrophobic **Hydrophilic**

Figure 20. A. Schematic showing drop profiles of water spotted onto surfaces of varying degrees of hydrophilicity where λ is the contact angle. *B.* Half angle method for measuring contact angles.

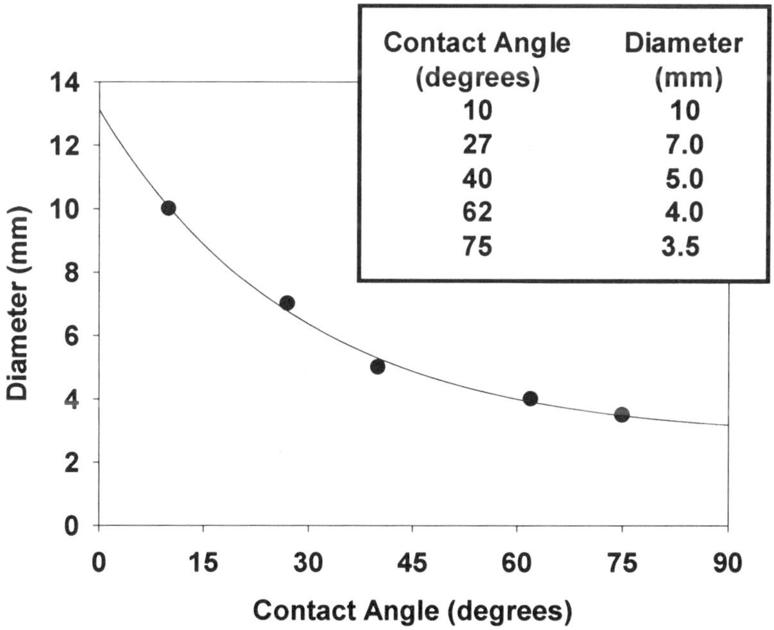

Contact Angle (degrees)	Diameter (mm)
10	10
27	7.0
40	5.0
62	4.0
75	3.5

Figure 21. Graph and table showing empirical correlation between spot diameter of 10 ml of water spotted onto slides with varying degrees of hydrophilicity. Contact angles were measured on a Tantec CAM-PLUS contact angle meter.

diameter of 10 µl water drops carefully placed on the slide surface and correlating the diameter with contact angle values obtained from the figure.

Atomic Force Microscopy (AFM)

AFM maps the topology of a surface by continually measuring the changes in probe-surface distance by passing a probe across a surface. AFM is fairly unique in that it can be performed in a dry or liquid environment. Sensitivity is dependent on probe size and AFM mode. Larger probes are less sensitive than smaller probes. There are three AFM modes: contact, tapping, and non-contact (Wright-Smith and Smith, 2001). In contact mode the probe is in continual contact with the surface with the probe being dragged across the surface to record surface topology. Tapping mode has the probe tapping up and down as it moves across the surface. This method is useful on coatings that may be sensitive to the tip, i.e. that may tear or be damaged by the passing of the tip. Noncontact scanning works by moving across the surface without touching. It measures the van der Waal's forces between the probe and surface and generates a map from these measurements. Noncontact scanning is not as sensitive as either of the other modes and is generally avoided for biological analysis. AFM can detect surface irregularities as small as 10 pm. Analysis of clean slides should yield an AFM with minimal topology. If variability beyond 2 nm exists the slide may need to be re-cleaned. AFM is expensive for day-to-day testing, but it yields a detailed surface map when developing new coatings or cleaning processes.

X-ray Photoelectron Spectroscopy (XPS)

Siegbahn developed XPS also known as ESCA (electron spectroscopy for chemical analysis) during the mid 1960's based on Einstein's photoelectric effect principle where a photon is used to eject an electron from the inner-shell orbital of an element. An electron analyzer reads the emitted photoelectron signals and presents them as a spectrum of binding energies. These binding energies (represented as peaks) are element specific. The shape and position of the binding energy peaks is dependent on the chemical state of the compound being analyzed. The strength and versatility of XPS is its effectiveness in determining changes in chemical states such as oxidation/corrosion, adsorption, and thin film growth. Studying glass with XPS is ideal because XPS can give a molecular account of the species present on the surface of the glass.

Vibrational Spectroscopy

Vibrational spectra, i.e. wavelengths and intensities of light that excite molecular vibrations, have been used for decades for molecular characterization. Infrared spectroscopy, for example, provides a fingerprint of a molecular species. Raman and/or infrared spectroscopies performed in a reflection mode, though different in mechanism, are often used to characterize the functional groups present on surfaces. Methods used to characterize surface coating/adsorbates by vibrational spectroscopy include absorption of infrared light (e.g. Fourier Transform Infra Red (FTIR), Fourier Transform Reflection Absorption Infra Red (FT RA-IR), Attenuated Total Reflection (ATR), Multiple Internal Reflection (MIR), Diffuse Reflectance Infra-red Fourier Transform (DRIFT), Reflection Absorption Infra Red Spectroscopy/ Infrared Reflection Absorption Spectroscopy (RAIRS/ IRAS)) inelastic scattered light (e.g. Surface Enhanced R aman Spectroscopy (SERS)) and by inelastic reflection of an electron beam (e.g. High Resolution Electron Energy Loss Spectroscopy (HREELS)).

Infrared (IR) absorption spectroscopy is straightforward in its process. An IR experiment consists of subjecting a sample to infrared light (2.5 - 50 µm, 4000 - 200 cm-1) and measuring the wavelength and intensity of light transmitted or in the case of surfaces, reflected. Infrared light is of appropriate energy to excite molecular vibrations and so is diagnostic of specific types of chemical bonds.

Raman spectroscopy is the wavelength and intensity measurement of inelastically scattered light. That is, a portion of an incident light's reflection is shifted by the energy of molecular vibrations in/on the reflective surface. Thus, measuring the wavelength and intensity of reflected light yields the vibrational energies of those molecules.

Bombarding a sample with a monoenergetic electron beam results in loss of beam energy by a variety of mechanisms. Examining energy losses at high resolution, as with HREELS, reveals information about the vibrations of the molecules residing on a surface.

Ellipsometry

Ellipsometry is an optical technique that is often used to characterize very thin film thickness. Films even thinner than the wavelength of the incident light to several thousand Angstroms can be characterized by this technique with a sensitivity that can detect a change of a few Angstroms in film thickness. The method impinges polarized light onto a sample and measures the polarization of light reflected parallel and perpendicular to the plane of incidence. The relative phase and amplitude change is calculated from these measurements and from this the film thickness. A film's dielectric properties can be measured if the experiment is conducted using variable wavelength light.

SUMMARY

Reproducibly preparing glass microscope slides for microarrays can be daunting. The process requires scrupulous cleaning, optimized coating methods and surface characterizations. One of the greatest lessons one learns from the exercise is how adsorbent glass is. That is, glass adsorbs molecules from the environment, thereby changing the surface in ways that in many instances results in an uncoatable/unprintable surface. It seems reasonable at this point to note too that storage containers must be chosen carefully as many plastics "off gas" monomers, plasticizers etc. that will adsorb to glass surfaces. Many groups however c hoose t o use commercially p repared s lides a llowing t hem to concentrate on obtaining bio-relevant data. In any case, Table 2 is presented to provide a ready reference to match DNA modification with slide surface chemistry. Pairing the surface and DNA modification chemistries in this way should allow one to efficiently pair a chosen slide type with an appropriate DNA modification.

REFERENCES

1   Adessi C, Matton G, Ayala G, Turcatti G, Mermod JJ, Mayer P, (2000) Solid phase DNA amplification: characterisation of primer attachment and amplification mechanisms. Nucleic Acids Res 28:e87.

2   Afanassiev V, Hanemann V, Wolfl S (2000) Preparation of DNA and protein micro arrays on glass slides coated with an agarose film. Nucleic Acids Res 28(12):e66.

3   Allemand JF, Bensimon D, Jullien L, Bensimon A, Croquette V (1997) pH-dependent specific binding and combing of DNA. Biophys J 73:2064-2070.

4   Arkles B (1977) Tailoring surfaces with silanes. Chemtech 7:766-778.

5   Awad S (1996) Ultrasonic cavitations and precision cleaning. Precision Cleaning. November, Crest Ultrasonics website (http://www.crest-ultrasonics.com/awad.html)

6   Beattie WG, Meng L, Turner SL, Varma RS, Dao DD, and Beattie KL (1995) Hybridization of DNA targets to glass-tethered oligonucleotide probes. Mol Biotechnol 4:213-225

7   Beier M, Hoheisel J (1999) Versatile derivatisation of solid support media for covalent bonding on DNA-microchips. Nucleic Acids Res 27(9):1970-1977.

8   Benters R, Niemeyer CM, Drutschmann D, Blohm D, Wöhrle D (2002) DNA microarrays with PAMAM dendritic linker systems. Nucleic Acids Res 30(2):e10.

9   Benters, R, Niemeyer CM, Wöhrle D (2001) Dendrimer-activated solid supports for nucleic acid and protein microarrays. Chembiochem 2:686-694.

10   Belosludtsev Y, Iverson B, Lemeshko S, Eggers R, Wiese R, Lee S, Powdrill T, Hogan M (2001) DNA microarrays based on noncovalent oligonucleotide attachment and hybridization in two dimensions. Anal Biochem 292:250-211

11   Birch, William R (2000) Coatings: An introduction to the cleaning procedures. Sol-Gel Gateway website (http://www.solgel.com/articles/June00/Birch/cleaning1.htm).

12   Bradshaw TK, Hutchinson DW (1977) 5-Substituted pyrimidine nucleosides and nucleotides. Chem. Soc. Rev. 6:43-62.

13   Call DR, Chandler DP, Brockman F (2001) Fabrication of DNA microarrays using unmodified oligonucleotide probes. BioTechniques 30(2):368-379.

14   Chin, A., Hung, M.H., Stock, L.M. (1981) Reactions Of Benzenediazonium Ions With Adenine And Its Derivatives. J. Org. Chem., 46(11):2203-7

15   Choi S-W, Choi W-B, Lee Y-H, and Ju B-K (2001) Effect of oxygen plasma treatment on anodic bonding. J Korean Phys Soc 38(3):207-209.

16   Chrisey LA, Lee GU, O'Ferrall CE (1996a) Covalent attachment of synthetic DNA to self-assembled monolayer films. Nucleic Acids Res 24(15): 3031-3039

17   Chrisey LA, O 'Ferrall CE, S pargo BJ, Dulcey CS, Calvert JM (1996b) Fabrication of patterned DNA surfaces. Nucleic Acids Res 24(15):3040-3047.

18   Church GM, Gilbert W (1984) Genomic sequencing. Proc. Natl. Acad. Sci. USA 81:1991-1995.

19   Diehl, F, Grahlmann S, Beier M, Hoheisel JD (2001) Manufacturing DNA microarrays of high spot homogeneity and reduced background signal. Nucleic Acids Res 29(7):e38.

20   Dolan PL, Wu Y, Ista LK, Metzenberg RL, Nelson MA, Lopez GP (2001) Robust and efficient synthetic method for forming DNA microarrays. Nucleic Acids Res 29(21):E107-7.

21   Duggan DJ, Bittner M, Chen Y, Meltzer P, Trent JM (1999) Expression profiling using cDNA microarrays. Nature Genet 21:10-14.

38

22 Eisen MB and Brown PO (1999) DNA arrays for analysis of gene expression. Methods Enzymol 303:179-205.

23 Engländer T, Wiegel D, Naji L, Arnold K (1996) Dehydration of Glass Surfaces Studied by Contact Angle Measurements. J Coll Interface Sci., 179(2):635-636

24 Erdogan F, Kirchner R, Mann W, Ropers H, Nuber U (2001) Detection of mitochondrial single nucleotide polymorphisms using a primer elongation reaction on oligonucleotide microarrays. Nucleic Acids Res 29(7):e36.

25 Gannett, P. M.; Powell, J. H.; Rao, R.; Shi, X.; Lawson, T.; Kolar, C.; Toth, B. (1999) C8-Arylguanine and C8-aryladenine formation in calf thymus DNA from arenediazonium ions. Chem. Res. Toxicol. 12(3), 297-304.

26 Ghosh SS, Musso GF (1987) Covalent attachment of oligoncucleotides to solid supports. Nucleic Acids Res. 15:5353-5372.

27 Gilham PT (1968) The synthesis of celluloses containing covalently bound nucleotides, polynucleotides, and nucleic acids. Biochemistry 7(8):2809-2813.

28 Gray DE, Case-Green SC, Fell TS, Dobson PJ, Southern EM (1997) Ellipsometric and interferometric characterization of DNA probes immobilized on a combinatorial array. Langmuir 13, 2833-2842.

29 Guiseppi-Elie A (2000) Cleaning and surface acivation of microfabricated interdigitated microsensor electrodes (IMEs), planar metal electrodes (PMEs), independently addressable microband electrodes (IAMEs), and E'Chem"Cell-On-A-Chip." Abtech Scientific Application Note 30 (website: http://www.abtechsci.com/pdfs/clean0501.pdf)

30 Guo Z, Guilfoyle RA, Thiel AJ, Wang R, Smith LM (1994) Direct fluorescence analysis of genetic polymorphisms by hybridization with oligonucleotide arrays on glass supports. Nucleic Acids Res 22(24):5456-5465.

31 Halliwell CM, Cass AE (2001) A factorial analysis of silanization conditions for the immobilization of oligonucleotides on glass surfaces. Anal Chem 73(11):2476-83.

32 Hermanson GT (1996) Bioconjugate techniques. New York: Academic Press.

33 Hung, M. H., STOCK, L.M. (1982) Reactions Of Benzenediazonium Ions With Guanine And Its Derivatives. J. Org. Chem., 47(3): 448-53

34 Joos B, Kuster H, Cone R (1997) Covalent attachment of hybridizable oligonucleotides to glass supports. Anal Biochem 247(1):96-101.

35 Kimura W, Kim G, Balick B (1994) Comparison of laser and CO_2 snow cleaning of astronomical mirror samples. SPIE 2199:1164-1171

36 Kolchinsky A, Mirzabekov A (2002) Analysis of SNPs and other genomic variations using gel-based chips. Human Mutation 19:343-360.

37   Kumar A, Larsson O, Parodi D, Liang Z (2000) Silanized nucleic acids: a general platform for DNA immobilization. *Nucleic Acids Res* **28**:E71.

38   LaForge KS, Shick V, Spangler R, Proudnikov D, Yuferov V, Lysov Y, Mirzabekov A, Kreek MJ (2000) Detection of single nucleotide polymorphisms of the human mu opioid receptor gene by hybridization or single nucleotide extension on custom oligonucleotide gelpad microchips: Potential in studies of addiction. Am J Med Genet 96:604-615.

39   Lamture JB, Beattie KL, Burke BE, Eggers MD, Ehrlich DJ, Fowler R, Hollis MA, Kosicki BB, Reich RK, Smith SR, Varma RS, Hogan ME (1994) Direct detection of nucleic acid hybridization on the surface of a charge coupled device. Nucleic Acids Res 22:2121-2125.

40   Lee P, Sawan S, Modrusan Z, Arnold L, Jr., and Reynolds M (2002) An efficient binding chemistry for glass polynucleotide microarrays. Bioconjug Chem 13(1):97-103.

41   Leonard MJ, McCredie RS, Logue MW, Cundall RL (1973) Solid state ultraviolet irradiation of 1,1'-trimethylenebisthymine and photosensitized irradiation of 1,1'-polymethylenebisthymines   J. Am. Chem. Soc. 95, 2301-2304.

42   Lindroos K, Liljedahl U, Raitio M, Syvanen AC (2001) Minisequencing on oligonucleotide microarrays: Comparison of immobilisation chemistries. Nucleic Acids Res 29:e69.

43   MacBeach G and Schreiber SL (2000) Printing proteins as microarrays for high-throughput function determination. Science 289:1760-1763.

44   Mandenius CF, Mosbach K, Welin S, Lundström I (1986) Reversible and specific interaction of dehydrogenases with a coenzyme-coated surface continuously monitored with a reflectometer. Anal. Biochem 157:283-288.

45   Maskos U, Southern EM (1992) Oligonucleotide hybridisations on glass supports: a novel linker for oligonucleotide synthesis, hybridisation properties of oligonucleotides synthesized in situ. Nucleic Acids Res 20:1679-1684.

46   Mitra RD, Church GM (1999) *In situ* localized amplification and contact replication of many individual DNA molecules. Nucleic Acids Res 27(24):e34.

47   Miyachi H, Hiratsuka A, Ikebukuro K, Yano K, Muguruma H, Karube I (2000) Application of polymer-embedded proteins to fabrication of DNA array. Biotechnol Bioeng 69(3):323-329.

48   Möller R, Csaki A, Kohler JM, Fritzsche W (2000) DNA probes on chip surfaces studied by scanning force microscopy using specific binding of colloidal gold. Nucleic Acids Res 28(20):e91.

49   Okamoto T, Suzuki T, Yamamoto N (2000) Microarray fabrication with covalent attachment of DNA using bubble jet technology. Nat Biotechnol 18:438-441.

50 Podyminogin MA, Lukhtanov EA, Reed MW (2001) Attachment of benzaldehyde-modified oligodeoxynucleotide p robes to s emicarbazide-coated glass. N ucleic Acids R es 29(24):5090-5098.

51 Proudnikov D, Timofeev E, Mirzabekov A (1998) Immobilization of DNA in polyacrylamide gel for the manufacture of DNA and DNA-oligonucleotide microchips. Anal Biochem 259:34-41.

52 Reed KC, Mann DA (1985) Rapid transfer of DNA from agarose gels to nylon membranes. Nucleic Acids Res. 20:7207-7220.

53 Rehman FN, Audeh M, Abrams E, Hammond PW, Kenney M, Boles TC (1999) Immobilization of acrylamide-modified oligonucleotides by co-polymerization. Nucleic Acids Res 27(2): 649-655.

54 Revenko I, Hansma HG (1996) Atomic force microscopy of DNA electrophoresed onto silylated mica. Scanning Microsc 10(2):323-328.

55 Rogers YH, Jiang-Baucom P, Huang ZJ, Bogdanov V, Anderson S, Boyce-Jacino MT (1999) Immobilizaton of oligonucleotides onto a glass support via disulfide bonds: A method for preparation of DNA microarrays. Anal Biochem 266:23-30.

56 Saito I, Sugiyama H, Furukawa N, Matsuura T (1981) Photochemical ring opening of thimidine and thymine in the presence of primary amines. Tett. Lett. 22:3265-3268.

57 Sanchez-Cortes S, Berenguel RM, Madejon A, Perez-Mendez M, (2002) Adsorption of Polyethyleneimine on Silver Nanoparticles and Its Interaction with a Plasmid DNA: A surface-enhanced Raman scattering study. Biomacromolecules 3:655-660.

58 Schena M, Shalon D, Davis RW, Brown PO (1995) Quantitative monitoring of gene expression patterns with a complementary DNA microarray. Science 270:467-470.

59 Schena M, Shalon D, Heller R, Chai A, Brown PO, Davis RW (1996) Parallel human genome analysis: Microarray-based expression monitoring of 1000 genes. Proc Natl Acad Sci USA 93:10614-10619.

60 Seeboth A, Hettrich W (1997) Spatial orientation of highly ordered self-assembled silane monolayers on glass surfaces. J. Adhesion Sci. Technol. 11(4):495-505.

61 Shchepinov MS, Case-Green SC, Southern EM (1997) Steric factors influencing hybridisation of nucleic acids to oligonucleotide arrays. Nucleic Acids Res 25(6):1155-1161.

62 Shirai K, Yoshida Y, Nakayama Y, Fujitani M, Shintani H, Wakasa K, Okazaki M, Snauwaert J, Van Meerbeek B (2000) Assessment of decontamination methods as pretreatment of silanization of composite glass fillers. J Biomed Mater Res 53(3):204-210.

63 Shumaker JM, Metspalu A, Caskey CT (1996) Mutation detection by solid phase primer extension. Human Mutation 7:346-354.

64 Southern E, Mir K, Shchepinov M (1999) Molecular interactions on microarrays. Nat

Genet 21(1 Suppl):5-9.

65 Stowers I (1978) Advances in cleaning metal and glass surfaces to micron-level cleanliness. J Vac Sci Technol 15:751-754.

66 Tantec Inc (2001) Tantec operating manual for contact angle meter. March.

67 Timofeev EN, Kochetkova SV, Mirzabekov AD, Florentiev VL (1996) Regioselective immobilization of short oligonucleotides to acrylic copolymer gels. Nucleic Acids Res 24(16):3142-3148.

68 Tomalia DA, Naylor AM, Goddard III WA (1990) Starburst dendrimers: molecular level control of size, shape, surface chemistry, topology, and flexibility from atoms to macroscopic matter. Angew. Chem. Intl. Ed. Eng. 29, 138-175

69 Vigs JR (1993) Chapter 6: Ultraviolet-ozone cleaning of semiconductor surfaces. In: Handbook of Semiconductor Wafer Cleaning Technology. pub: Noyes Publications (ed. Kern W) pp.233-273.

70 Williams D, Randall P (1994) Guide to cleaner technologies: Cleaning and degreasing process changes. United States Environmental Protection Agency.

71 Wright-Smith C, Smith C M (2001) Atomic force microscopy: Researchers map the topography of biological macromolecules. The Scientist 15(2):23.

72 Walsh K, Wang X, Weimer BC (2001) Optimizing the immobilization of single-stranded DNA onto glass beads. J Biochem Biophys Methods 47:221-231.

73 Yershov G, Barsky V, Belgovskiy A, Kirillov E, Kreindlin E, Ivanov I, Parinov S, Guschin D, Drobishev A, Dubiley S, Mirzabekov A (1996) DNA analysis and diagnostics on oligonucleotide microchips. Proc Natl Acad Sci USA 93: 4913-4918.

74 Yoshioka M, Mukai Y, Matsui T, Udagawa A, Funakubo H (1991) Immobilization of ultra-thin layer of monoclonal antibody on glass surface. J Chromatogr 566(2):361-368.

75 Zammatteo N, Jeanmart L, Hamels S, Courtois S, Louette P, Hevesi L, Remacle J (2000) Comparison between different strategies of covalent attachment of DNA to glass surfaces to build DNA microarrays. Anal Biochem 280:143-150.

Chapter 2

DIAGNOSTIC OLIGONUCLEOTIDE MICROARRAYS FOR MICROBIOLOGY

Levente Bodrossy, Ph.D.
Department of Biotechnology, Austrian Research Centers, Seibersdorf, Austria
www.diagnostic-arrays.com
www.arcs.ac.at/ul/ulb/bt

This chapter deals with oligonucleotide microarrays for the detection and quantification of microorganisms.

Such microarrays contain hundreds of oligonucleotide probes, each one specific for different strains/species/genera of microorganisms. They offer a fast, high-throughput alternative for the parallel detection of microbes from virtually any sample. The application potential of diagnostic microarrays covers most sectors of life sciences, including human, veterinary, food and plant diagnostics, environmental microbiology, water quality control, etc. Identification of human pathogens from clinical samples can be accomplished with a properly designed array within a matter of hours without cultivation and prior information on its approximate nature. Detection of food contaminating microbes, veterinary and plant pathogens can also be accomplished with diagnostic microarrays on a large scale. They can be used to study the effect of soil microbial diversity on soil fertility and sustainable soil quality as well as the response of microbial diversity to agricultural practices. It has only recently been acknowledged that the activity of microorganisms is essential to stabilise the biosphere. Understanding the complex interactions between the highly complex microbial communities and their environments is crucial for a sustainable development. Microbial ecology has only started to scratch the surface of this topic. High-throughput methods for the analysis of microbial diversity are badly needed for further development in this field.

The scientific and technical background, the lab protocols, the experimental design and the data mining technique for the creation and application of such microarrays will be described in detail. Actual protocols will be those successfully applied in the author's lab while references will be given to alternative approaches. Some o f the n umerous other a pplications of oligonucleotide microarrays will also be mentioned. Their detailed discussion is beyond the scope of this chapter thus the reader will be referred to more detailed literature on these topics.

INTRODUCTION

Oligonucleotide microarrays are sets of oligonucleotides (typically 10 to 100 nt in length) immobilised onto a glass carrier in a highly parallel, addressable format. Since the late 1990s when microarray technology was first described (Schena *et al.*, 1995), there has been an explosion in the application of oligonucleotide microarrays.

Applications of oligonucleotide microarrays can be classified into three major groups:
·   Diagnostic arrays
·   Transcriptional arrays
·   Arrays for sequence analysis

Oligonucleotide probes have long been applied for the detection of specific nucleic acid sequences, such as in fluorescent in situ hybridisation (detecting microorganisms in the environment, pathogens in tissues, cells within a tissue expressing a given gene, etc.) or in Southern hybridisation to identify closely related genes in different organisms. In a wider sense, PCR primers can also be considered as oligo probes. The use of specific oligonucleotide probes in PCR-based applications includes the amplification of taxon-specific 16S rRNA genes from various environments, in situ PCR with applications similar to those for fluorescent *in situ* hybridisation, and the amplification of related genes from a variety of organisms / environments. Hybridisation to a battery of such oligonucleotides immobilised onto a single microarray offers a wide range of potential applications in the detection of various organisms, genes or DNA elements from the most diverse samples. One of the most promising applications of oligonucleotide microarrays is their use as biosensors in microbiology, which will be discussed in depth throughout this chapter.

Transcriptional arrays can further be divided into two subclasses. High density, on-chip sy nthesised oligonucleotide m icroarrays (GeneChips) f rom Affymetrix contain on average 20 independent, random 25mers targeting each gene of interest. For each oligo probe these arrays also contain a negative control oligo (MM) with a single, central mismatch compared to its corresponding "perfect match" (PM) pair. In this approach probe redundancy accounts for the reliability of expression data (Lipshutz *et al.*, 1999; Alon *et al.*, 1999; de Saizieu *et al.*, 2000).

Spotted oligonucleotide microarrays are used less frequently in transcriptional profiling due to the relatively high price for the individual synthesis of single oligonucleotides and the patent (for oligonucleotide arrays immobilised onto solid supports) held by Affymetrix. They typically consist of a single, carefully selected oligonucleotide (40-80mer) per gene of interest. This type is mainly used in the form of custom made low-density (i.e. up to a few hundred different probes) microarrays for the study of only a selected group of genes. Recently, MWG-Biotech's human, rat, Escherichia coli, Helicobacter pylori and Saccharomyces cerevisiae Pan-Arrays (targeting several thousand genes by a single 40mer each) appeared on the market. Sigma-Genosys and Compugen have been producing and marketing human, mouse and rat oligo libraries ready for spotting since the middle of 2001.

The application potential of oligonucleotide microarrays in sequence analysis has not yet been explored. These include *de novo* sequencing (Yershov *et al.*, 1996), resequencing (Hacia, 1999), detection of mutations (most importantly single nucleotide polymorphisms, SNPs) (Hacia, 1999; Gunderson *et al.*, 1998; Lindroos *et al.*, 2001; Erdogan *et al.*, 2001) sequence comparison (identifying clones carrying SNPs for subsequent sequencing) (Southern, 1996), analysis of secondary structure stability and formation (Sohail *et al.*, 1999;Sohail *et al.*, 2001;Mir and Southern, 1999), and analysis of hybridisation kinetics (Fotin *et al.*, 1998). Callida Genomics, a subsidiary of Hyseq Pharmaceuticals, is developing a technique for sequencing with oligo arrays using Affymetrix GeneChip technology.

The three major approaches mentioned above require fundamentally different experimental designs. Sequence analysis applications are very diverse, and even within this type of application completely different probe design, experimental protocols, and data analysis procedures exist. The discussion of these techniques reaches beyond the scope of this chapter.

The use of oligonucleotide microarrays in transcriptional profiling is, from many points of view (like target preparation and data mining), the same as for cDNA microarrays and is thus discussed in the subsequent chapters of this book. The preparation of oligo arrays for transcriptional profiling and for detection purposes is based on the same principles and is discussed later in this chapter.

The cornerstone of oligonucleotide microarrays is the design of the oligo probes and this is the step where transcriptional profiling oligo arrays require a fundamentally different approach. The probe (or the probes, i.e. for high density, Affymetrix-type arrays) must be specific to the single targeted gene and yield minimal cross-hybridsation signal with any other DNA sequences that may be present in the sample (in most cases this means within the genome of the species investigated). As a consequence, these oligo probes are normally designed for the most variable regions of the target gene. Even though desirable, it is not necessary to have oligos of nearly identical hybridisation behaviour.

Most popular microarray analysis programs adjust the raw data output of these experiments (e.g., in spotted arrays the ratio of two differently labeled targets hybridised to the same physical spot), eliminating the effect of probe hybridisation efficiency on the results. Software for the design of such oligo sets will be discussed in some detail later in this chapter.

On the contrary, detection-type oligo arrays contain oligonucleotides targeting in most cases a set of (more or less) closely related sequences (i.e. 16S rRNA genes from related microorganisms). These probes have to be designed such that they hybridise to each member within the targeted group of sequences with the same or at least with similar efficiency. Thus, oligonucleotides of the detection-type microarrays are mainly designed for regions of medium to high sequence conservation. Further, raw data are mostly hybridisation signals with no internal normalisation. As a result, probe sets have to consist of oligos with highly similar hybridisation efficiencies to enable reliable quantification.

SCHEME OF THE EXPERIMENTAL APPROACH

Figure 1 shows a general scheme of the steps and procedures to obtain and apply a diagnostic oligonucleotide microarray. Details and alternative procedures will be discussed in the appropriate sections.

SOURCES OF VARIATION

While seemingly straightforward, the technique involves a series of steps that are susceptible to variability, from manufacturing of the array, to PCR amplification and labeling of the targets, to differences in hybridisation efficiencies of the probes. Results might reflect this introduced variability rather than actual ratios in bacterial abundance. In many cases not only quantitative but also qualitative errors may result from such variability. Therefore, one might not only mispredict the ratio of different microbial groups but also miss completely microorganisms, which in fact dominate a given sample. Also, one can apparently detect microorganisms, which are in fact absent. It is therefore important to be aware of all possible sources of error in oligonucleotide microarray-based identification and enumeration of microorganisms (see Table 1). These will be discussed in more detail in the corresponding sections.

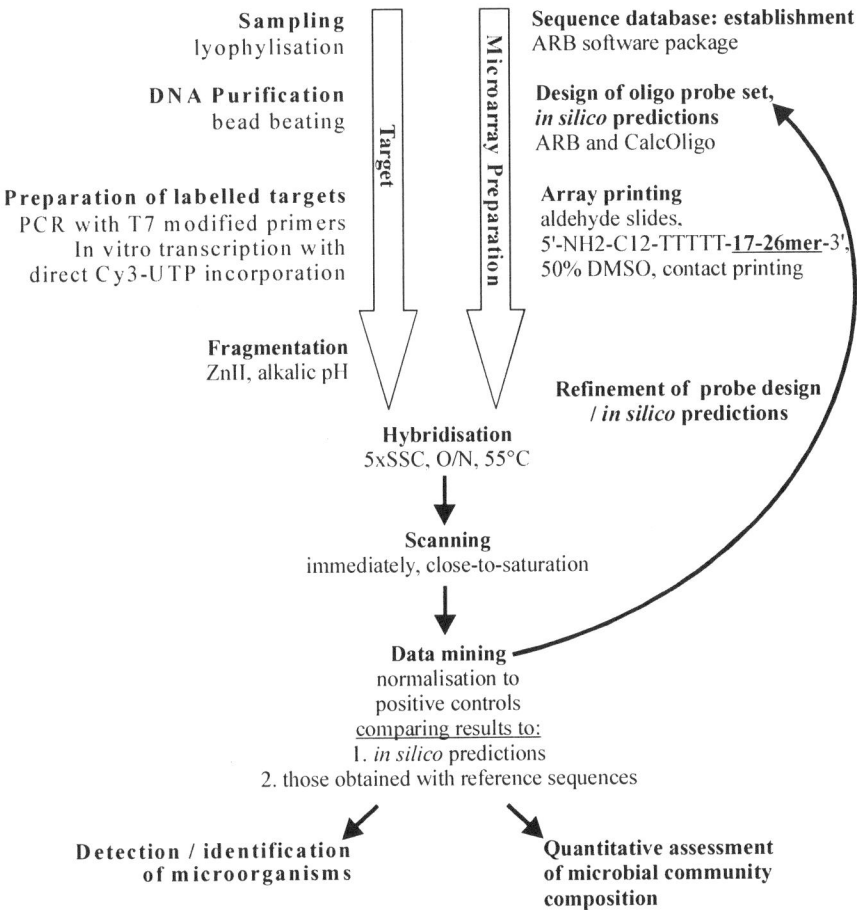

Sampling
lyophylisation

DNA Purification
bead beating

Preparation of labelled targets
PCR with T7 modified primers
In vitro transcription with
direct Cy3-UTP incorporation

Target

Microarray Preparation

Sequence database: establishment
ARB software package

Design of oligo probe set,
in silico **predictions**
ARB and CalcOligo

Array printing
aldehyde slides,
5'-NH2-C12-TTTTT-**17-26mer**-3',
50% DMSO, contact printing

Fragmentation
ZnII, alkalic pH

Refinement of probe design
/ *in silico* **predictions**

Hybridisation
5xSSC, O/N, 55°C

Scanning
immediately, close-to-saturation

Data mining
normalisation to
positive controls
comparing results to:
1. *in silico* predictions
2. those obtained with reference sequences

Detection / identification
of microorganisms

Quantitative assessment
of microbial community
composition

Figure 1. Scheme of the Experimental Approach

ESTABLISHMENT OF A SEQUENCE DATABASE

The 16S rRNA gene, encoding for one of the three RNA molecules of the ribosome, is the most widely used DNA sequence for the detection and phylogenetic analysis of microorganisms. The major advantages of the 16S rRNA gene are:

- It is generally highly conserved throughout the living world. Universal primer sets exist which can amplify the 16S rRNA gene from the overwhelming majority of Bacteria, Archea and Eucarya, respectively (Table 2).

Table 1. Sources of error in diagnostic microarray lab procedures.

Stage	Step	Possible mistake	Result
Sampling	Storage	Cells lyse during storage	Biased loss of template, skewed ratios
Nucleic acid preparation	Releasing DNA or RNA	Treatment doesn't break open all cells	
	DNA/RNA stabilisation	DNA or RNA degrades during preparation	
PCR	Conditions	Suboptimal PCR conditions (number of cycles, amount of template, temperature, degenerate probes)	Biased amplification, skewed ratios
Labelling/IVT Modified nucleotides	Incorporation	Modified nucleotides less efficiently incorporated than normal ones	Biased labelling, skewed ratios
	Hybridisation	Sequences rich in modified nucleotides may hybridise more weakly.	Biased hybridisation, skewed ratios
Array printing	Spot homogeneity	Heterogeneous spots (with high and low pixels) have a lower mean or median value than homogenous ones; high signal may be saturated.	Increasing error in the data read (mean or median intensity of spots).
	Probe density	Spots with different probe densities may behave differently during hybridisation (bind more target or - due to steric hindrance -, less).	Increasing variation across print runs.
Hybridisation	Tm	Melting temperatures can cause loss of signal if too high, and nonspecific signal if too low.	False negative / positive results; skewed ratios.
	Equilibrium	Incubation time insufficient to achieve equilibrium (especially with only diffusion aided mixing).	Uneven hybridisation; low signal.
	Bubbles	Usually in static incubation - bubbles create areas of differential hybridisation conditions.	Local differences in signal and background intensities.
Scanning	Balancing of photon yields (two-colour hybridisations)	If not addressed by adjusting PMT settings has to be accounted for later *in silico*.	Extensive *in silico* normalisation may introduce systematic error / shift into results.

Table 1 (continued).

Stage	Step	Possible mistake	Result
Scanning (cont'd)	Saturation	Saturated pixels contribute artificially little to the mean/median values of the spots	High signal mean/median values underestimated
	Stacking effect	Crowding of the fluorophores (too efficient labelling or hybridisation) causes anomalous fluorescence enhancement for Cy3 and loss for Cy5	Errors in mean/median values for some spots
	FRET (only for 2 colour hybridisation)	Frequency Resonance Energy Transfer occurs if two different fluorophores are in close proximity.	Decrease in Cy3, increase in Cy5 signal; skewed Cy3/Cy5 ratios.

Abbreviations: IVT- In vitro transcription; PMT- Photomultiplier

Table 2. Universal primer sets.

Name	Sequence (5'-3')	dir.	Comments/Reference
f27	AGAGTTTGATCMTGGCTCAG	fw	Eubacterial *Giovannoni 1991*
8f	AGAGTTTGATCCTGGCTCA	fw	Eubacterial *Edwards et al., 1989*
A8F	TCCGGTTGATCCTGCCGG	fw	archeal + eukaryotic *Kolganova et al., 2002*
E528f	CGGTAATTCCAGCTCC	fw	Eukaryotic *Sogin et al., 2002*
U514f	GTGCCAGCMGCCGCGG	fw	Universal *Sogin et al., 2002*
U1492r	ACCTTGTTACGACTT	rev	Universal *Sogin et al., 2002*
1492rpl	GGTTACCTTGTTACGACTT	rev	Eubacterial *Hershberger et al., 1996*
r1492	TACGGYTACCTTGTTACGACTT	rev	Eubacterial *Giovannoni 1991*
38r	CCGGGTTTCCCCATTCGG	rev	universal - 23S/IGS *Martin-Laurent et al., 2001*
72f	TGCGGCTGGATCTCCTT	fw	universal - 16S/IGS *Martin-Laurent et al., 2001*

- It contains more and less conserved regions allowing for the design probes specific for higher and lower taxons.
- There is an apparent lack of lateral transfer of the rRNA genes (Woese, 1987).
- During evolution, the 16S rRNA gene is believed to have changed at a fairly constant rate (Woese, 1987). Thus it can be considered as an evolutionary clock with each nucleotide difference translating to an evolutionary time unit.

Thus, the ~1500 bp sequence, amplified by universal primers carries enough information to predict the phylogeny of the host organism with high precision.

An extensive, rapidly growing database exists for this gene. As this chapter is being written, the ARB database (see later) (Strunk *et al.*, 2000; Amann and Ludwig, 2000) contains over 22,000 aligned 16S rRNA gene sequences; the RDP database (Maidak *et al.*, 1994) consist of over 45,000 sequences and a search of the NCBI database on "16S r*" returns over 67,000 hits.

Other genes (often referred to as functional, i.e. encoding for related enzymes carrying out a defined function) meeting the above conditions at least for a narrower group of microorganisms, can also be used in molecular microbial ecology. The application of functional genes narrows down the analysis to a functionally (sometimes also phylogenetically) defined group of microbes. The main advantage of this approach is that it enables the detection and analysis of microbial groups with no cultivated members. Examples are:

- *pmoA/amoA* (Bourne *et al.*, 2001; Radajewski *et al.*, 2000) - The *pmoA* gene (encoding for the active site subunit of the particulate methane monooxygenase) of methanotrophic bacteria and the evolutionarily re-lated *amoA* gene (encoding for ammonia monooxygenase) of aerobic ammonia oxidisers is widely used for the analysis of methanotrophic and/or nitrifying bacterial communities. Conserved primer pairs that enable the parallel amplification of nearly all *pmoA/amoA* and related genes from environmental samples exist. Their use has led to the dis-covery of functional groups of bacteria with novel *pmoA/amoA* related genes. Members of these groups have not yet been cultivated. They may play a role in the oxidation of atmospheric methane.
- *mmoX* (Auman *et al.*, 2000) - Encoding for one of the proteins of the soluble methane monooxygenase (present in some of the methanotrophs only). This gene has been used to complement the results of *pmoA/amoA* analyses of environmental samples.
- *mxaF* (McDonald and Murrell, 1997) - The mxaF gene encodes for the large subunit of the methanol dehydrogenase of aerobic methanol oxidising bacteria. It has been used for the analysis of methanol utilising commu-nities as well as for confirming the results of *pmoA/amoA* sequence analyses on methanotrophs.

- *nifH* (Lovell *et al.*, 2001; Widmer *et al.*, 1999) - The *nifH* gene encodes the nitrogenase iron protein. It was used for the analysis of nitrogen fixing bacterial populations in various environments.
- *norB* (Ren *et al.*, 2000) - The *norB* gene encodes for the nitric oxide reductase. Bacterial populations able to reduce nitric oxide can be analysed.
- *mcrA* (Lueders *et al.*, 2001) - Encoding for the methanogen-specific methyl-coenzyme M reductase a-subunit, this gene is used for the analysis of methanogen communities. Recently, a novel archeal lineage was revealed by analysis of mcrA sequences from rice field soil.
- *rbcL* (Wyman *et al.*, 2000) - The rbcL gene encodes the *RuBisCo* (ribulose bisphosphate carboxylase/oxidase) large subunit, present in many autotrophic bacteria. It was used to follow, and also to predict, the changes in phytoplankton communities, including phytoplankton booms.
- *ndoB* (Milcic-Terzic *et al.*, 2001) - PCR amplified fragment diversity of the ndoB (naphthalene dioxygenase) gene has been used to investigate the enrichment of aromatic hydrocarbon degrader populations in diesel-contaminated soils.
- *rpoB* (Dahllof *et al.*, 2000) - The *rpoB* gene encodes for the RNA polymerase beta subunit. It is common to all bacteria and is highly conserved. Thus, it offers an alternative to the use of the 16S rRNA gene in analysing the total bacterial diversity. It is present in only a single copy, providing a significant advantage over 16s rRNA which has up to 15 copies in Clostridium paradoxum (Rainey *et al.*, 1996).

The ARB software package (Strunk *et al.*, 2000) is the software of choice for establishing and managing phylogenetic sequence databases. For details of this step (sequence alignment, construction of phylogenetic trees) the reader is referred to information published in scientific journals (Amann *et al.*, 1995) and on the following Internet sites:

http://www.arb-home.de
http://rdp.cme.msu.edu/docs/documentation.html
http://evolution.genetics.washington.edu/phylip.html

For the basics of installing and using the ARB software the reader is referred to online documentation:

http://www.mpi-bremen.de/molecol/arb/manual/data/arb.pdf
http://www.mpi-bremen.de/molecol/arb/manual/data/ARB_install.pdf
http://www.mbl.edu/arb_tutorial/
http://groups.yahoo.com/group/arb_users/

Briefly, a database has to consist of sequences aligned in such a way that each column contains only nucleotides, which share a common ancestor. This is one of the starting assumptions in all mathematical analyses that are later applied to these sequences. Based on such an alignment, phylogenetic tree(s) reflecting evolutionary distances between the gene sequences (and thus in an optimal case, also of the host organisms) are constructed using various mathematical models. In ARB, dendrograms (phylogenetic trees) are then used to guide the design of probes for different phylogenetic taxa.

OLIGO LENGTH AND MELTING TEMPERATURE (TM); DESIGNING OLIGO SETS TUNED TO WORK TOGETHER

One of the difficulties in the design of detection-type oligo microarrays is fine-tuning the probes selected so that they can be applied at the same hybridisation temperature while not loosing specificity. One option is to use identical length oligos, and equivalent hybridisation buffers, which diminish the effect of GC content on melting temperature. The second option is to design oligos with (nearly) identical predicted melting temperatures in traditional hybridisation buffers, like 6xSSC (Sambrook *et al.*, 1989).

Hybridisation buffers containing tertiary amine salts, like tetramethyl ammonium chloride (TMACl) or tetraethyl ammonium chloride (TEACl) were successfully used to achieve GC-content independent hybridisation (Wood *et al.*, 1985; Connors *et al.*, 1997; Mir and Southern, 1999; Spiro *et al.*, 2000) on nitrocellulose or nylon membranes. The present weakness of this approach is that the thermodynamic background is missing, making it impossible to predict the effects of further factors on the behaviour of a given oligo probe.

In most cases, sets of oligonucleotide probes are fine-tuned for use in traditional, Na-containing hybridisation buffers, like 6xSSC. The nearest neighbour method (Breslauer *et al.*, 1986) is used to estimate the melting behaviour of each oligo in the set. Of the several alternative parameter sets reported, the one of Santa-Lucia and co-workers (1996) has been used most to estimate the melting temperature (Tm) of oligonucleotide duplexes. A different set for RNA/ DNA duplexes has recently been reported (Sugimoto *et al.*, 2000) enabling the use of this approach for microarrays hybridised with RNA targets. The shortcoming of this approach is that the nearest-neighbour datasets predict duplex stability with high accuracy in solution only. There are several further factors influencing hybridisation behaviour, which arise from the immobilisation of the oligo probes. To take these effects into consideration one needs to analyse the results from a statistical number of duplexes and refine the predictions based on these results, as will be described in detail later in this chapter.

Artificial nucleotide analogues like locked nucleic acids (LNA) or 2'-O-methyl-DNA oligos can substantially strengthen the hybridisation between DNA probes and RNA, especially if the modified oligos take part in the formation of AT pairs. In some cases the inclusion of such modified nucleotides may be the only solution to obtain an oligo probe of the desired specificity and Tm (Majlessi *et al.*, 1998), Petersen *et al.*, 2002).

For oligonucleotide microarrays used in transcriptional analysis, either longer (50-80mer) oligos or a larger number of randomly chosen 20-25-mers per gene are used. In both cases, problems arising from different base composition and sequence of the oligonucleotides are minimised.

Considering the flexibility in designing probes of wider specificity allowed for by shorter oligos, and the reliability of longer oligos in meeting predicted specificity, the author's choice is 18-24 mer oligos with a predicted Tm of 58-62 °C in 6xSSC.

OLIGO SET DESIGN

The design of oligo probe sets for diagnostic microarrays is a fairly laborious process. It is, however, well worth the investment of time and effort. It has to be remembered that all subsequent analyses (in a typical case this would be the analysis of several hundreds to thousands of hybridisations) will be based on a set of assumptions regarding the specificity of each single probe. If these assumptions are false, all subsequent conclusions may be wrong and misleading. There is so far no single software package, which covers the entire process. A combination of the ARB phylogenetic software package, CalcOligo and Excel is suggested for this purpose.

Oligo probe sets for transcriptional arrays have to meet different criteria. As discussed in the introduction, for such microarrays one has to choose oligonucleotides complementary to the most variable regions of the targeted gene. For custom-made transcriptional oligo arrays it is common to apply one or a maximum of two well-chosen, longer (40-80-mer) oligo(s) rather than a set of shorter oligos. Several software programs and web pages exist which can facilitate the design of such probes and probe sets; some of them are listed below

Free Web Sites

Web Primer at Stanford Genomic Resources
    http://genome-www2.stanford.edu/cgi-bin/SGD/web-primer
Primer3 at Whitehead Institute, Center for Genome Research
    http://www-genome.wi.mit.edu/cgi-bin/primer/primer3_www.cgi
PCR primer design at the Virtual Genome Center, Univ. of Minnesota
    http://alces.med.umn.edu/websub.html

Freeware

OligoArray of Jean-Marie Rouillard, Univ. of Michigan
    http://berry.engin.umich.edu/oligoarray/index.html
OligoPicker of Xiaowei Wang at Harvard
    http://genetics.mgh.harvard.edu/xwang/research/oligopicker.html
Featurama 0.7 of David Shteynberg
    http://probepicker.sourceforge.net/featurama.html

Commercial Software and Web Sites

OligoSys - microarray
    http://oligosys.firstserv.co.uk/OScr_Login.jsp
Array Designer
    http://www.premierbiosoft.com/dnamicroarray/dnamicroarray.html

Method

Create a database in ARB containing as many of the available sequences for the given gene as possible. It is extremely important to have all the available sequences for the group of microorganisms to be investigated and for their closest relatives. The more of these sequences that are provided, the better the specificity of the probe set will be. For 16S-based probes, an extensive ARB database exists and can be downloaded (http://www.mpi-bremen.de/molecol/arb). 16S rRNA gene sequences of the targeted microbial groups and closest relatives still need to be updated from a sequence databank like the NCBI. Replace all spaces with underscore ("_") characters in the full names of the entries, to enable easy formatting of the results at the end.

Align the sequences [Fig 2].

Create phylogenetic trees with bootstrap values. Bootstrap values in general indicate the validity of grouping sequences (strains) together within one branch.

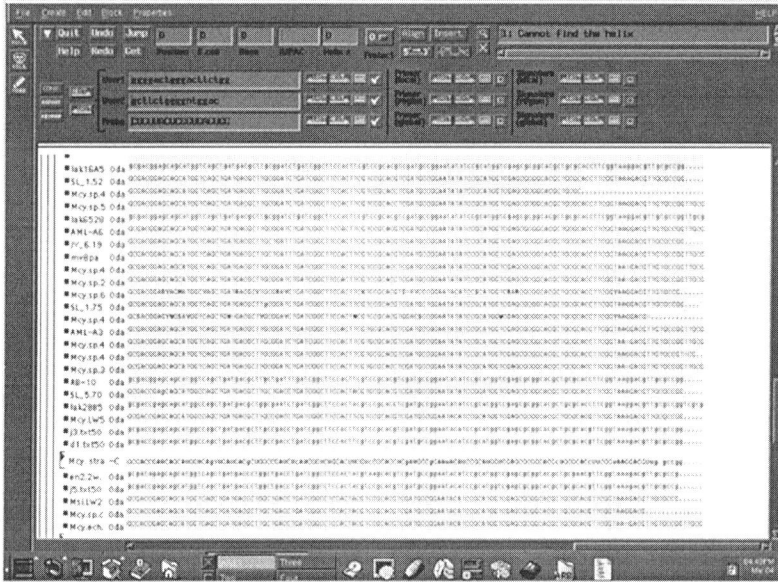

Figure 2: Sequence Alignment

Branches with high bootstrap values are likely to contain a phylogenetically coherent group for which it is relatively easier to design a general probe. In general it is a good idea to create trees using alternative calculations and mark stable branching patterns for the probe design step [Fig 3].

Identify the best tree (one with relatively high bootstrap values and branching pattern stability).

Select different groups on the tree (single strains, if such resolution is required; subspecies, species, higher groups) and use the Probe_ Design function of ARB to design probes [Fig 4].

Optimal settings for this function depend on the targeted group (how monophyletic; how closely related; closest group to be excluded; etc.). In general, the following settings may serve as a good starting point:

- Length of output - 50;
- Max. Non-group hits - 0 to 5% of the total number of sequences targeted;
- Max. hairpin bonds - 4;
- Min. group hits - 80 to 100%;

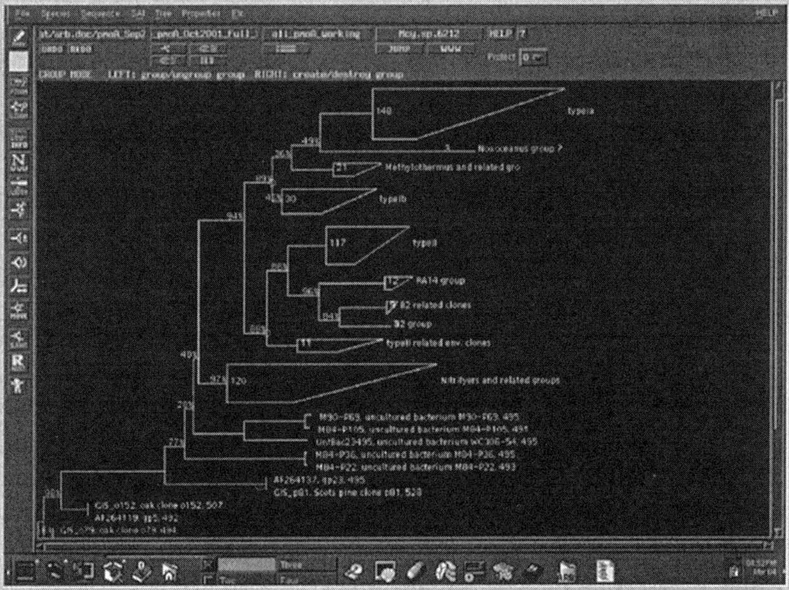

Figure 3: Picking stable branch patterns for the probe design step.

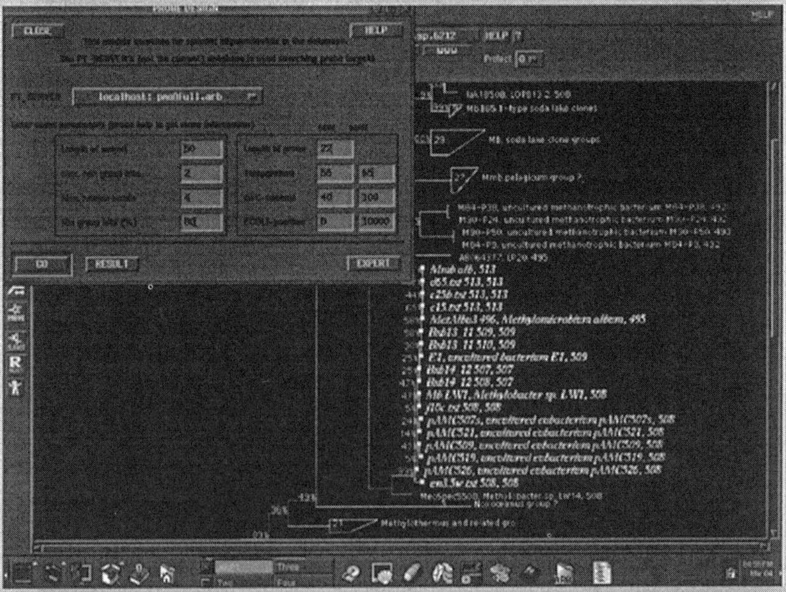

Figure 4: Using ARB's Probe_Design Function

- Length of probe - 18 to 20;
- Temperature - 50 to 70 °C; G+C content - 35 to 75%;
- ECOLI position - 0 to 10000 (thus, probe design process will never exclude regions).

The output of the Probe_Design window [Fig 5] can be used in the next step as input for the Probe_Match function [Fig 6]. Use this function to visualise the predicted specificity of the potential probes listed in the Probe_Design result window [Fig 7].

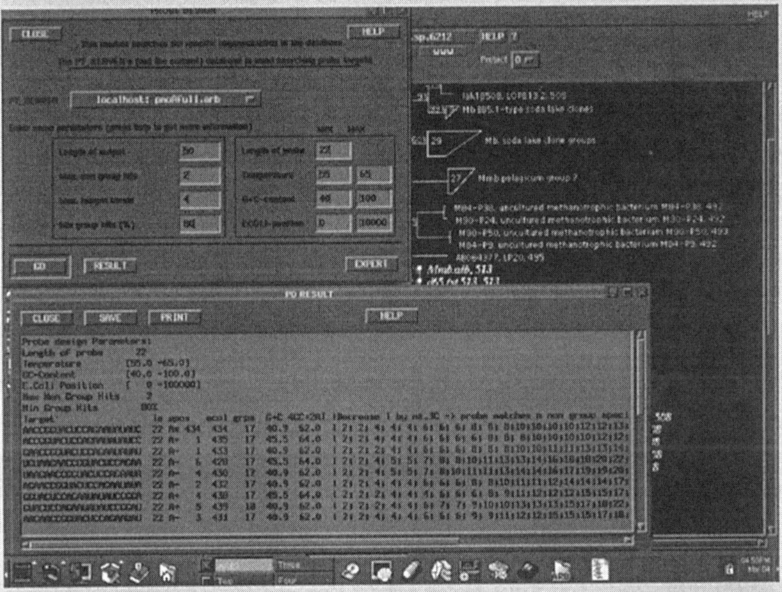

Figure 5: Probe_Design output.

Probes with acceptable specificities are then checked for melting temperature predicted by the nearest neighbor model. Web sites of most oligo synthesis companies (i.e.: http://www.idtdna.com/program/oligocalc/oligocalc.asp),as well as most molecular biological software, offer this function. When the predicted Tm does not fall within the range of 58-62°C, repeat Probe_Design with longer/shorter setting for probe length.

Select a number (3-4) of alternative probes with Tm of 58 - 62 °C (if available) and run Probe_Match, allowing for 5 mismatches. Save results as a text file [Fig 8] (print to ASCII file).

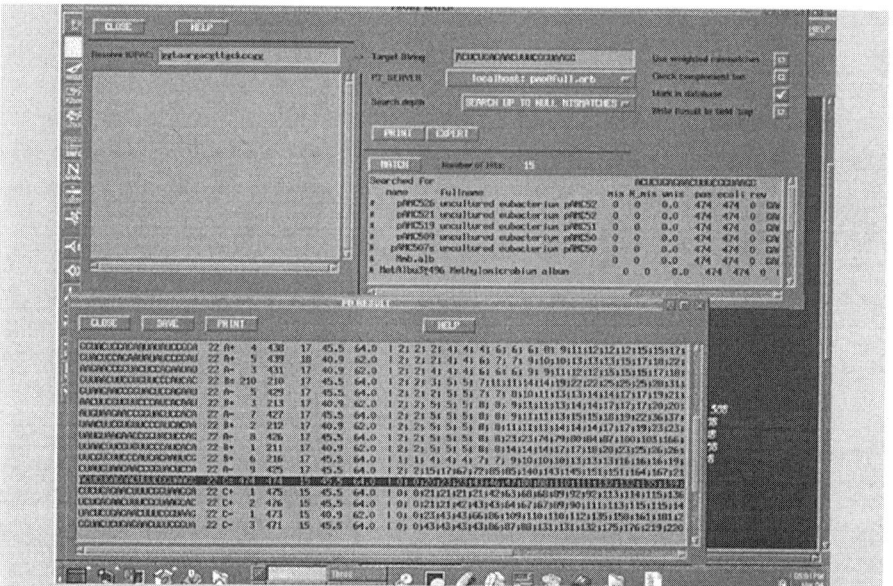

Figure 6: Probe_Match function based on Probe_Design output.

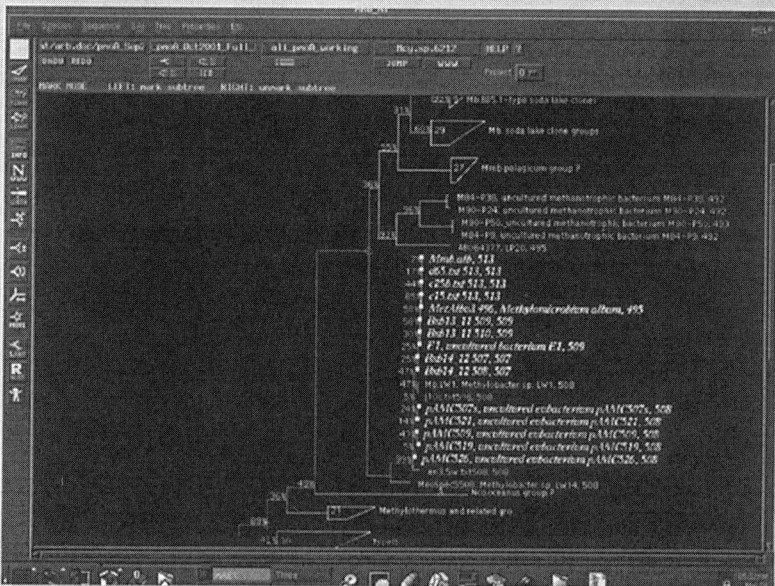

Figure 7: Probe_Design results help visualize predicted specificity.

Using the saved Probe_Match files as input files, run CalcOligo.exe (http:\\www.diagnostic-microarrays.com\calcoligo\). CalcOligo is software for the design and optimisation of oligonucleotide probe sets for diagnostic

Searched for "gccgaccgagcaatatggc"; RA542

name	fullname	mis	N_mis	wmis	pos	ecoli	rev	'GCCGACCGAGCAAUAUGGC'
RA26	uncultured_methanotroph_Rold_	0	0	0	371	371	0	AUUCCAUCA-================-ACGCUGAUG
RA14	uncultured_methanotroph_Rold_	0	0	0	371	371	0	AUUCCAUCA-================-ACGCUGAUG
G2	uncultured_bacterium_G2	0	0	0	371	371	0	AUUCCAUCA-================-ACGCUGAUG
Maine8	uncultured_methanotroph_'Main	0	0	0	371	371	0	AUACCAUCA-================-ACGCUGAUG
Maine6	uncultured_methanotroph_'Main	0	0	0	371	371	0	AUACCAUCA-================-ACGCUGAUG
Rold1	uncultured_methanotroph_'Rold	0	0	0	371	371	0	AUACCAUCA-================-ACGCUGAUG
Rold_3	uncultured_methanotroph_'Rold	0	0	0	371	371	0	AUUCCAUCA-================-ACGCUGAUG
G1	uncultured_bacterium_G1	0	0	0	371	371	0	AUUCCAUCA-================-ACGCUGAUG
RA9	uncultured_methanotroph_Rold_	0	0	0	371	371	0	AUCACAUCA-================-ACGCUGAUG
G3	uncultured_bacterium_G3	0	0	0	371	371	0	AUUCCAUCA-================-ACGCUGAUG
MR5	uncultured_bacterium_MR5	0	0	0	371	371	0	AUUCCAUCA-================-ACGCUGAUG
Rold_5	uncultured_methanotroph_'Rold	1	0	1.1	371	371	0	AUUCCAUCA-===============-G-ACGCUGAUG
MR4	uncultured_bacterium_MR4	1	0	1.5	371	371	0	AUUCCAUCA-===C===========-ACGCUGAUG
SL_5.70	soda_lake_5-70	3	0	2.4	371	371	0	GUUCCACCA-=G==========gC====-CAGCUGAUG
lak28B	LOPA13.5	3	0	2.4	371	371	0	GUUCCACCA-=G==========gC====-CAGCUGAUG
j3.txt	j3	3	0	2.4	371	371	0	GUUCCACCA-=G==========gC====-CAGCUGAUG
Mcy.LW5	Methylocystis_sp._LW5	3	0	2.4	371	371	0	GUUCCACCA-=G==========gC====-CAGCUGAUG
d1.txt	d1	3	0	2.4	371	371	0	GUUCCACCA-=G==========gC====-CAGCUGAUG
Msi.tri.l41	Methylosinus_trichosporium_sp	3	0	2.4	371	371	0	UCUGCAUCA-=G==========gC====-CAGCUGAUG
Msi.tri.SC6	Methylosinus_trichosporium_sp	3	0	2.4	371	371	0	UCUGCAUCA-=G==========gC====-CAGCUGAUG
Msi.tri.l34p	Methylosinus_trichosporium_sp	3	0	2.4	371	371	0	UCUGCAUCA-=G==========gC====-CAGCUGAUG
Sp-55	Sp-55	3	0	3.2	371	371	0	UUACCAUCA-=G=U==u====-UUGCUCAUG
JY_6.48	japan_yumoto_6-48	3	0	3.2	371	371	0	UUACCAUCA-=G=U==u====-UUGCUCAUG
RB-60	RB-60	3	0	3.2	371	371	0	UUACCAUCA-=G=U==u====-UUGCUCAUG
lak28A	LOPB13.4	3	0	3.2	371	371	0	UUACCAUCA-=G=U==u====-UUGCUCAUG
lak14	LOPA12.6	3	0	3.2	371	371	0	UUACCAUCA-=G=U==u====-UUGCUCAUG
SL_4.51	soda_lake_4-51	3	0	3.2	371	371	0	UUACCAUCA-=G=U==u====-UUGCUCAUG
SL_5.66	soda_lake_5-66	3	0	3.2	371	371	0	UUACCAUCA-=G=U==u====-UUGCUCAUG
peat1-7	peat1-7	3	0	3.2	371	371	0	UUACCAUCA-=G=U==u====-UUGCUCAUG
peat11-4	peat11-4	3	0	3.2	371	371	0	UUACCAUCA-=G=U==u====-UUGCUCAUG
UntBac56	uncultured_bacterium_Bakchar3	3	0	3.7	371	371	0	UUAUCAUCA-=G====G==A=====-CUUCUCAUG
UntBac57	uncultured_bacterium_Bakchar7	3	0	3.7	371	371	0	UUAUCAUCA-=G====G==A=====-CUUCUCAUG
MhhB2	methanotroph_B2	3	0	3.7	371	371	0	UUAUCAUCA-=G====G==A=====-CUUCUCAUG
Msi.spo.SC8	Methylosinus_sporium_sp._SC8	3	1	1.9	371	371	0	NUUCCACCA-=G==========gN====-u-CAGCUGACG

Figure 8: Alternative Probe_Match results.

microarrays. It creates a table (saved as tab delimited text file, importable into Excel) with predicted hybridisation behaviour of all the designed probes versus all the sequences considered during Probe_Design.

Use the Conditional Formatting function of Excel to highlight probe-target pairs of 0, 1 or 2 mismatches [Fig 9]. Organise the rows (sequences) of the Excel table according to the order of sequences on the phylogenetic tree used during Probe_Design. Organise columns (probes) into a similar phylogenetic order [Fig 10].

Remove unneeded probes (if possible, keep 2 to 3 alternative probes for each targeted group / species / strain).

Create a subtable with reference strains representing the entire known diversity of the targeted group of microorganisms and obtain the references strains for testing the probes [Fig 11].

Order probe set as described in the next section.

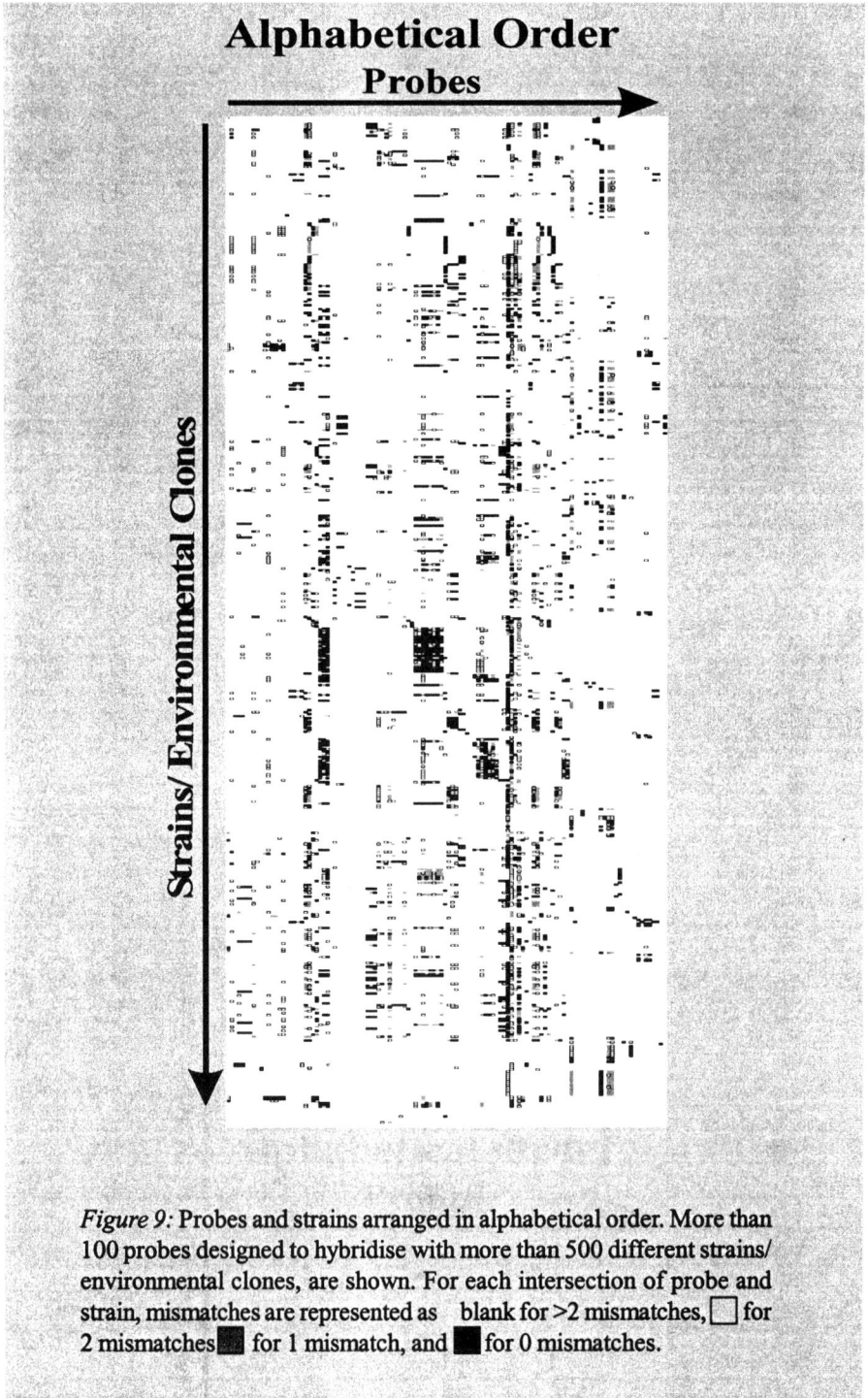

Figure 9: Probes and strains arranged in alphabetical order. More than 100 probes designed to hybridise with more than 500 different strains/ environmental clones, are shown. For each intersection of probe and strain, mismatches are represented as   blank for >2 mismatches, ☐ for 2 mismatches ▓ for 1 mismatch, and ■ for 0 mismatches.

Phylogenetic Order

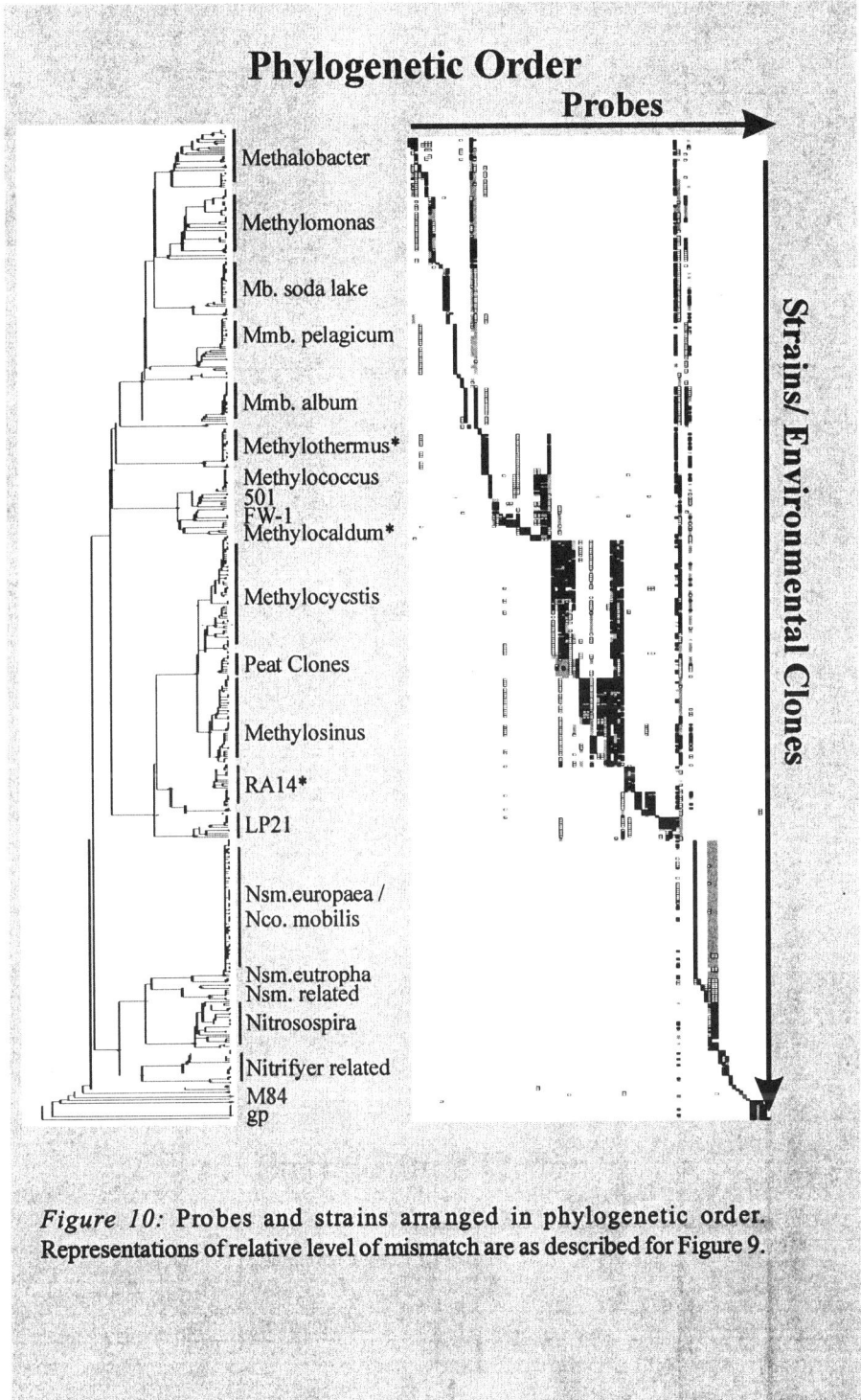

Figure 10: Probes and strains arranged in phylogenetic order. Representations of relative level of mismatch are as described for Figure 9.

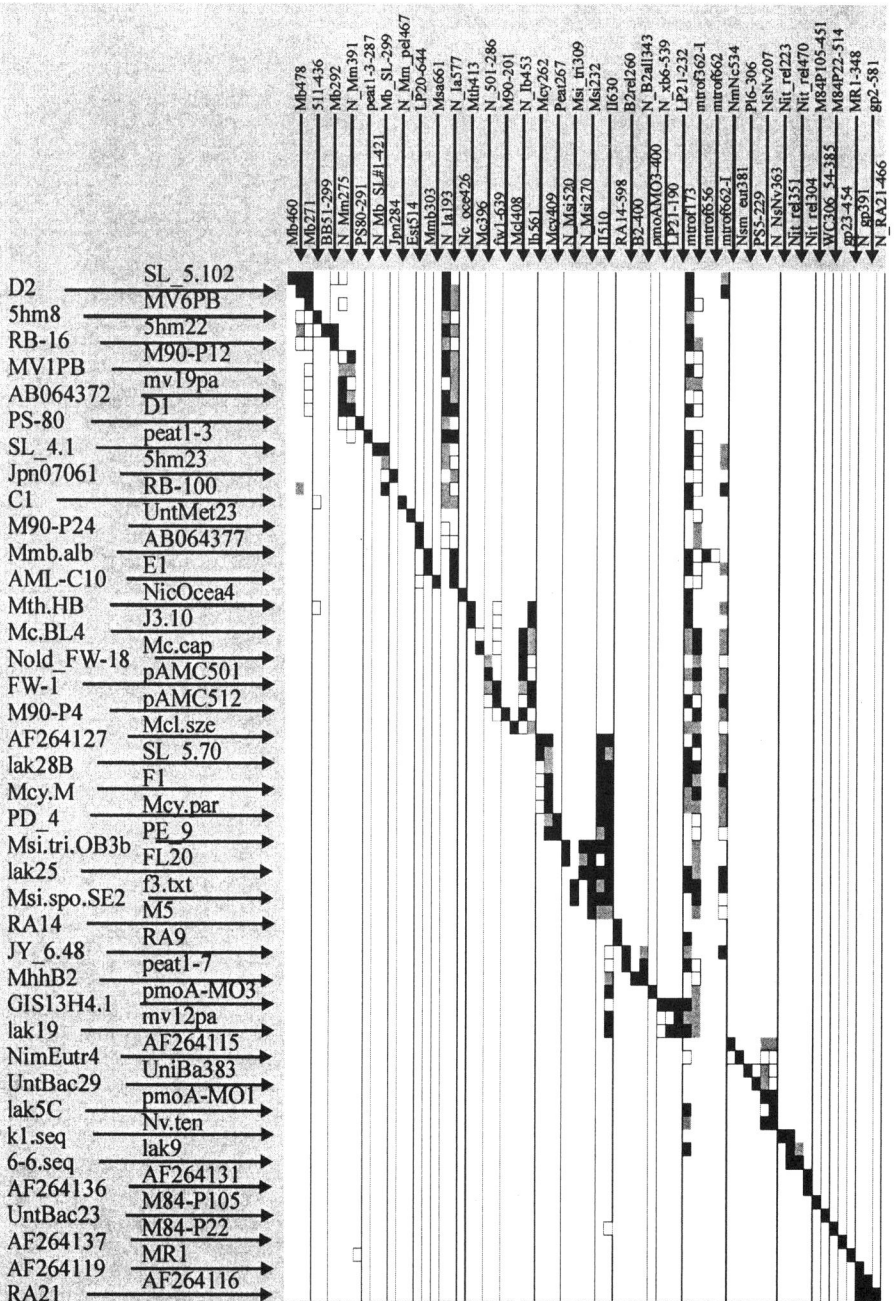

Figure 11. Subtable of clones and probes reflecting the detectable diversity of the genomes of interest. Clones and probes are in phylogenetic order, with mismatch counts coded as in figures K and L. Minor divisions in Probe phylogeny are represented by thin lines, major divisions are represented by thick black lines.

CHOICE OF OLIGO/SURFACE BINDING CHEMISTRY

In traditional Southern blotting and for most microarrays comprised of DNA fragments, the immobilised DNA is coupled to the surface by crosslinking the DNA strand at a random (mostly internal) location to the surface. The immobilised DNA fragment is thereby efficiently cut into two halves (or more parts) as far as its hybridisation potential is considered. While this is not a problem for long (i.e. several hundreds of nucleotides in length) nucleic acids, it can seriously affect the hybridisation potential of oligonucleotides. Thus, in most cases, oligonucleotides are covalently attached to glass supports via a reactive terminal group, leaving the entire molecule available for hybridisation. Other methods use non-covalent immobilisation via passive adsorption or affinity binding. Recently it has been reported that unmodified oligonucleotide probes can also be efficiently immobilised without compromising hybridisation efficiency (Call *et al.*, 2001).

Steric effects can seriously decrease the hybridisation capacity of the immobilised oligos. These effects are steric interference of the solid support on the hybridisation properties of the immobilised oligos, and steric hindrance resulting from the crowding of immobilised oligos. These effects are successfully mitigated by the application of spacer molecules (Guo *et al.*, 1994; Shchepinov *et al.*, 1997). The length, flexibility and hydrophobicity of the spacers all influence hybridisation behaviour. The effect of spacer molecules depends on the exact experimental set-up, most importantly on the: composition of the hybridisation buffer, nature of the solid surface (hydrophobicity, distance of the active groups from the glass/matrix surface, density of the active groups, porosity), hybridisation temperature applied, and length of the oligonucleotides used. There is an optimal spacer length, above which hybridisation efficiency rapidly decreases. This value was found to be 40-60 atoms for amine-functionalised polypropylene support (Shchepinov *et al.*, 1997), but it does vary between different experimental set-ups.

Oligonucleotides are thus normally covalently coupled to the solid surface via a well-defined chemical reaction usually by their 5' or 3' ends. More commonly used coupling chemistries for oligonucleotide microarrays (surface-oligo) include: Amino-aldehyde, amino-epoxy, amino-isothiocyanate, biotin-streptavidin and thiol-thiol. For a detailed description, see Chapters 1 and 3 in this book.

The most widely used spacers to mitigate steric effects are alkane spacers of 6 or 12 C atoms (C6 and C12). Alternatively or in combination with them 5 extra thymidines at the attached end of the oligo are frequently applied.

CSS aldehyde slides from Cel Associates are routinely used for economical reasons. If a lower detection limit is needed, more expensive aldehyde or epoxy slides with lower autofluorescence and/or higher immobilisation capacity should be considered.

Oligonucleotides for immobilisation are synthesised with a 5' NH2 group, followed by a C12 spacer and 5 thymidines residues preceding the probe sequence. Under these experimental conditions, extending the spacer further has no clear positive effect. Order probe sets at a 0.2 µM scale (Table 3). This is normally enough for a minimum of 10 source plates, each source plate enabling the spotting of up to 3000 slides in triplicate.

Table 3. Commercial slides suitable for oligo arrays.

Slide name	Chemistry	Company/Web Address
SuperAldehyde	Amino-Aldehyde	Arrayit/Telechem *http://arrayit.com/Products/Substrates/substrates.html*
SuperEpoxy	Amino-Epoxy	
CSS	Amino-Aldehyde	Cell Associates *http://arrayit.com/Products/CEL_Associates/cel_associates.html*
ALS	Amino-Aldehyde	
EPO	Amino-Epoxy	
Vivid Membrane	Cross-linking	Pall Life Sciences *http://www.pall.com/laboratory/*
Immobilizer	Amino-???	Exiqon *Http://www.exiqon.com/*
PicoRapid GmbH	Amino-???	Picoslide *http://www.picorapid.de/produkte/slides.html*
Code-Link	Amino-???	Motorola (earlier: 3D-Link by SurModics) *http://www.motorola.com/lifesciences/codelink/slides.html*
SuperChip	Amino-Epoxy	Erie Scientific Company *http://www.eriesci.com/*
Genorama SAL	Amino-isothioncyanate	Asper Biotech *http://www.asperbio.com/coated_glass.htm*
QMT Epoxy	Amino-Epoxy	Quantifoil *http://www.quantifoil.com/*
QMT Aldehyde	Amino-Aldehyde	
Bioslide Aldehyde	Amino-Aldehyde	Bioslide Technologies *http://www.bioslide.com/bioslidema.html*
Xenoslide D	Amino-Aldehyde	Xenopore *http://www.xenopore.com/*
Xenoslide E	Amino-Epoxy	
Xenoslide S	Biotin-Streptavidin	
Xenoslide U	Thiol-Thiol	
Epoxy Covalent Binding Slide	Amino-Epoxy	NoAb BioDiscoveries *http://www.noabdiagnostics.com/index.htm*
Aldehyde Covalent Binding Slide	Amino-Aldehyde	
Hydrogel Epoxy Covalent Binding Slide	Amino-Epoxy	
Hydrogel Aldehyde Covalent Binding Slide	Amino-Aldehyde	
ARChip Epoxy	Amino-Epoxy	ARC Seibersdorf research GmbH *http://www.arcs.ac.at/ul/ul/ulb/bt/p1*
ARChip UV	Amino-???	

ARRAY PRINTING

For a detailed discussion of the different options in spotting (spotting buffer, pins, humidity, etc.), please see Chapter 3 of this book.

50% DMSO used as a printing buffer has the following advantages:

- It doesn't dry during long spotting rounds (a routine spotting of 100 oligos onto 100 slides takes about 6 hours) unlike aqueous solutions, such as 3xSSC or phosphate buffers.
- It provides uniform spots on the slides applied. Standard deviation in signal intensities between replicate spots is 10-15% as opposed to 20-30% for 3xSSC (Table 4).
- 384 well plates are used because of the smaller evaporation rate and smaller volume required.
- 1 pin is applied to avoid variations inherent in spotting with multiple pins.
- 50% humidity and 22°C provides optimal conditions in our hands to yield uniform, homogenous spots from 50% DMSO.
- Reduction of the free aldehydes makes prehybridisation with BSA or other aminated compounds (which then serve to block free aldehyde groups) unnecessary.

Table 4: Comparison of spotting solutions.

| | | F635 Median: | | | | F635 SD: | |
| | | Variation within blocks (same pin) | | Variation between pins | | | |
	Average	St.Dev.	St.Dev.%	St.Dev.	St.Dev.%	Average	% of Mean
50%DMSO	4200	505	**13%**	1279	**30%**	2437	60%
3xSSC	6066	2116	**25%**	3517	**58%**	15612	123%

Notes: F635 Median- Median value of pixel intensities within spots; average and standard deviation values for 16 x 20 spots (spotted with 16 pins), F635 SD- Standard deviation of pixel intensities within spots; average and standard deviation values for 16 x 20 spots (spotted with 16 pins)

Method:

Prepare a 384 well flat bottom plate with 30 µl of 50 µM oligonucleotide solutions in 50% DMSO.

Spot samples with an OmniGrid spotter (1 TeleChem SMP3 pin) at 50% relative humidity, 22°C. This pin takes 250 nl sample per run and deposits 0.6 nl of it per spot.

Incubate spotted slides overnight at room temperature at <30% relative humidity. The formation of the Schiff base between the aldehyde and amino groups yields water. Low humidity levels help this reaction.

Rinse slides twice in 0.2% (w/v) SDS for 2 min at room temperature (20-25°C), with vigorous agitation to remove the unbound DNA.

Rinse slides twice in dH2O for 2 min at room temperature, with vigorous agitation.

Transfer slides into dH2O at 95-100°C for 2 min to denature the DNA.

Allow slides to cool at room temperature (~5 min).

Treat slides in a freshly (right before use) prepared sodium borohydride solution for 5 min at room temperature to reduce free aldehydes. Sodium borohydride solution: Dissolve 0.5 g NaBH4 in 150 ml phosphate buffered saline (PBS; 8g NaCl, 0.2g KCl, 1.44g Na2HPO4, 0.24g KH2PO4, in 1000 ml H2O, pH 7.4, autoclaved), then add 44 ml of 100% ethanol to reduce bubbling.

Rinse slides three times in 0.2% (w/v) SDS for 1 min each at room temperature.

Rinse slides once in dH2O for 1 min at room temperature.

Dry slides, one by one, using an air gun fitted with a cottonwool filter inside (to keep oil microdroplets away from the slide surface). Apply a modest stream of air first to the area containing the array, blowing the drops down on the slide, rather than drying them onto it. Dried slides can be stored at room temperature in the dark for several months.

TARGET PREPARATION

Target preparation for diagnostic arrays normally has to start with a PCR amplification step. For applications where the aim is only the detection of microorganisms, PCR amplification is necessary to achieve satisfactory levels of sensitivity. For environmental applications where the relative abundance of different microorganisms should also be precisely assessed, the bias inherent in PCR amplifications with degenerate primers can be a problem. There are efforts to use RNA (isolated from environmental samples) directly as labeled target (Small *et al.*, 2001). In most cases, however, the amplification of target molecules by PCR is unavoidable.

A growing literature exists on the bias inherent in PCR using conserved, degenerate primers. Many potential solutions have been proposed and tested to minimise this bias. These included the use of high concentrations of template in combination with minimal cycle numbers (Polz and Cavanaugh, 1998); the addition of acetamide (Reysenbach *et al.*, 1992); and decreasing annealing temperature below optimal (Ishii and Fukui, 2001).

There are many alternative procedures to produce fluorescently labeled nucleic acid targets for microarray hybridisation. The basic decisions to be made are: Should the target be:

- end-labeled or internally labeled
- double or single stranded
- DNA or RNA
- full length or fragmented (if fragmented: randomly or not)

The most widely used fluorescent dyes in microarray technology are Cy3 and Cy5 made by Amersham Biosciences and their very closely related alternatives cyanine 3 and cyanine 5 by NEN. The most important requirements for the chosen dyes are:

- High and similar quantum yields
- Minimal and similar photobleaching and decay under routine storage conditions of the hybridised slide
- Dyes should have a minimal Fluorescence Resonance Energy Transfer (FRET). FRET is a process that shifts energy from an electronically excited molecule (the donor fluorophore) to a neighbouring molecule (the acceptor or quencher), returning the donor molecule to its ground state without emission of light. The basic requirement for FRET is that the emission spectrum of the donor and the excitation spectrum of the acceptor molecule overlap. A close proximity of the donor and acceptor molecules is required, but no physical contact is needed. Two-colour microarray hybridisations do suffer from FRET; some of the excitation energy of Cy3 can be transferred to nearby Cy5 molecules, which then emit this as light. The result is a shift of Cy3/Cy5 ratios in the higher intensity ranges. Even though it has not been shown with other dyes being introduced to the microarray arena, FRET is likely to affect most dye pairs. Further information on FRET is available at: http://www.cci.virginia.edu/fret_spectra.html
- Little or no fluorescence enhancement or quenching upon stacking of the same dye (Gruber et al., 2000).
- Dyes should interfere with the labelling and hybridisation reactions to a minimal extent, and possibly to the same extent. There is now a widen-

ing range of methods to label nucleic acids with fluorescent dyes. In some cases an enzyme (i.e. Taq polymerase, reverse transcriptase, RNA polymerase) has to incorporate labeled nucleotides. This step is always hindered by the inclusion of labeled nucleotides; it is important that the nucleotides can still be handled by the enzyme and that the two dyes represent a similar challenge to the enzyme. To avoid discrepancies, more and more applications use chemical coupling of the fluorescent dyes to either unlabeled nucleic acids, or to nucleic acids that contain aminoallyl-modified nucleotides. This step, however, can again differ in efficiency with regard to the two dyes used.

For microarrays with oligomers in the range of 15-30 nt, it is important to apply randomly fragmented nucleic acid targets. This increases signal and facilitates subsequent quantification of the results.

Finally, it has to be noted that all protocols below use DNA as the starting material. This way one can obtain information on the presence, diversity and distribution of microorganisms in the sample or environment investigated. However there will be no differentiation between active and inactive populations. All the methods below can be applied for experiments where the starting material is RNA by carrying out RT-PCR instead of conventional PCR. In this case, results predominantly reflect the active populations only, and, in the case of quantification, their level of activity as well as their relative abundance.

Method: Generation of Fragmented, Cy Labeled RNA Target

I. PCR Amplification of DNA Sequence for In Vitro Transcription

Design PCR primers to amplify the gene of interest. The primer of the strand to be labeled has to contain the T7 promoter site as well: 5'-TAATACGACTCACTATAG – actual primer-3'. In our case, this is always the reverse primer. Consequently, oligonucleotide probes are of the same sequence as the positive strand of the gene.

For each target start 3 PCR reactions of 50 µl each. Each 50 µl reaction contains: 5 µl 10x PCR buffer, 4 µl dNTP mixture (2.5 mM for each dNTP), 1.5 µl 50 mM MgCl2, 1-1 µl of both primers (15 pmol/µl » 100 ng/µl), 1U Taq polymerase (Gibco Life Sciences/Invitrogen). Leaving space for template DNA (10 ng for environmental DNA, 1 ng for genomicDNA (gDNA) or 0.1 ng for plasmid DNA), add ultrapure water to 50 µl.

95°C, 5 min. Pause @95°C. Add template DNA ("hot start" to minimise mispriming). 32 cycles of 1 min @ 95°C; 1 min at the annealing temperature; 1

min @72°C for every 1000 bp to be amplified. A final 10 mins @72°C to allow the completion of all ongoing or unfinished amplification reactions.

Pool parallel PCR products (3x50µl) and purify with a commercial PCR purification kit according to manufacturer's instructions (we use the HighPure PCR purification kit from Macherey-Nagel). Dissolve or elute purified DNA in ultrapure water keeping in mind that the final concentration has to be adjusted to 50 ng/µl.

Measure the concentration of purified DNA by spectrophotometry (Concentration of dsDNA = (A260 x Dilution rate)/0.02 [ng/µl]). For example: Dilute 1µl DNA in 49µl water, measure absorbance at 260nm in a 50µl cuvette with 10mm light path. An A260 of 0.1 refers to (0.1x50)/0.02=250 ng/µl dsDNA concentration. Adjust concentration to 50ng/µl with ultrapure water. Store at -20° C.

II. In Vitro Transcription

Work under RNAse-free conditions.

This is achieved by observing the following basic rules:

1. Always wear clean gloves when working with RNA

2. Use filter tips from separate boxes used for RNA work only

3. Use plasticware autoclaved at 121°C for 60 minutes

4. Use RNAse free solutions

Note: Most solutions (and also glassware) can be made RNAse free by DEPC (diethylpyrocarbonate)-treatment (add 0.01% DEPC, shake, incubate at 37°C overnight, autoclave at 121°C for 60 minutes). Tris and other amine-containing buffers cannot be treated with DEPC; it is best to buy them RNAse free from major suppliers. Warning: DEPC is toxic; Autoclaving is needed to inactivate it!!!

Use chemicals dedicated to RNA work; try to measure them straight into RNAse free glassware without using spatulas. If spatulas have to be used, sterilise them by ethanol flaming.

Into an RNAse-free Eppendorf tube add:

1. 8 µl 50 ng/µl purified PCR product;

2. 4 µl 5x T7 RNA polymerase buffer;

3. 2 µl 100mM DTT;

4. 0.5 µl 40 U/µl RNasin (Promega);

5. 1 µl each of 10mM ATP, CTP, GTP;

6. 0.5 µl 10mM UTP;

7. 1 µl 40U/µl T7 RNA polymerase (Gibco BRL);

8. 1 µl 5mM Cy3 or Cy5-UTP

Incubate @37°C, for 4 hours.
Purify labeled RNA immediately.

III. RNA Purification

We use the Quiagen RNeasy kit .
This step removes unincorporated nucleotides, DNA template, T7 polymerase and salts.
Work under RNAse free conditions.
1. Add 80μl DEPC treated water to the IVT mix.
2. Add 350μl RLT solution (provided with the kit); mix thoroughly.
3. Add 250μl EtOH; mix thoroughly.
4. Sample (700μl) into an RNeasy mini column. 15 secs @>10,000 rpm.
5. Transfer column into a new 2ml collection tube.
6. Add 500μl RPE solution (provided with the kit), 15 secs @>10,000 rpm.
7. Add 500μl RPE, 2 minutes @>10.000 rpm.
8. Transfer column into a 1.5 ml collection tube.
9. Add 50μl RNAse free water.
10. 1 minute @>10,000 rpm.
11. Transfer the eluate into a new 1.5 ml Eppendorf tube.
12. Proceed to fragmentation.

IV. Zn2+ fragmentation of RNA

This procedure fragments RNA into pieces with an average length of 50 nt.
Work under RNAse free conditions.
1. To 50 μl purified RNA (in a 1.5 ml Eppendorf tube, from the previous step) add:
    1.43 μl, 1M Tris.Cl pH 7.4
    5.71 μl, 100mM ZnSO4
2. Mix; incubate at 60°C, 30 minutes.
Note: Use dry block and do not mix during the incubation because the condensation on the lid of the tube is also included in the optimisation of the protocol (it causes gradual concentration of the reaction, thus influencing efficiency of fragmentation).
3. Add 1.43 μl 500mM EDTA pH 8.0 to stop the reaction (by chelating Zn2+).
4. Put on ice for 1 min; add 1 μl 40 U/μl RNasin.
Fragmented, labeled RNA target can now be stored at -20°C for several months.

Alternative Method: Generation of End-labeled, Double Stranded DNA Target

Design PCR primers to amplify the gene of interest. The primer of the strand to be labeled (i.e. complementary to the probes; in our case this is always the reverse strand) has to be synthesised with a 5' Cy3 / Cy5 modification.

Per target start 1 PCR reaction of 50 μl volume. Each 50 μl reaction contains: 5 μl 10x PCR buffer, 4 μl dNTP mixture (2.5 mM for each dNTP), 1.5 μl 50 mM MgCl2, 1-1 μl of both primers (15 pmol/μl » 100 ng/μl), 1U Taq polymerase (Gibco Life Sciences / Invitrogen). Leaving space for template DNA (10 ng for environmental DNA, 1 ng for genomic DNA or 0.1 ng for plasmid DNA) add ultrapure water to 50 μl.

95°C, 5 min. Pause @ 95° C. Add template DNA ("hot start" to minimise mispriming). 32 cycles of 1 min @ 95° C; 1 min at the annealing temperature; 1 min @ 72° C for every 1000 bp to be amplified. A final 10 mins @ 72° C to allow the completion of all ongoing or unfinished amplifications.

Purify PCR products with a commercial PCR purification kit according to manufacturer's instructions (we use the HighPure PCR purification kit from Macherey-Nagel). Dissolve or elute purified DNA in ultrapure water.

Labeled DNA can now be stored at -20° C for several months.

Alternative Method: Generation of Internally Labeled, Double Stranded DNA Target

Design PCR primers to amplify the gene of interest.

Per target start 1 PCR reaction of 50μl volume. Each 50μl reaction contains: 5 μl 10x PCR buffer, 1 μl each of 10 mM dATP, dGTP and dTTP, 4 μl of 10mM dCTP, 0.6 μl of 5 mM Cy3-dCTP, 1.5 μl 50 mM MgCl2, 1-1 μl of both primers (15pmol/μl » 100ng/μl), 1U Taq polymerase (Gibco Life Sciences / Invitrogen). Leaving space for template DNA (10 ng for environmental DNA, 1 ng for genomic DNA or 0.1 ng for plasmid DNA) add ultrapure water to 50 μl.

95° C, 5 min. Pause @ 95° C. Add template DNA ("hot start" to minimise mispriming). 35 cycles of 1 min @ 95°C; 1 min at the annealing temperature; 2 min @ 65° C and 2 min @ 72° C for every 1000 bp to be amplified. A final 10 mins @ 72° to allow the completion of all ongoing or unfinished amplifications.

Purify PCR products with a commercial PCR purification kit according to manufacturer's instructions (we use the HighPure PCR purification kit from Macherey-Nagel). Dissolve or elute purified DNA in ultrapure water.

Labeled DNA can now be stored at -20° C for several months.

Alternative Method: Generation of Internally Labeled, Single Stranded DNA Target

Design PCR primers to amplify the gene of interest. The primer of the strand to be removed (in our case this is always the positive, forward strand) has to be synthesised with a 5' biotin modification.

Per target start 1 PCR reaction of 50μl volume. Each 50μl reaction contains: 5μl 10x PCR buffer, 1μl each of 10 mM dATP, dGTP and dTTP, 4 μl of 10mM dCTP, 0.6 μl of 5 mM Cy3-dCTP, 1.5 μl 50 mM MgCl2, 1-1 μl of both primers (15 pmol/μl » 100 ng/μl), 1U Taq polymerase (Gibco Life Sciences / Invitrogen), leave space for template DNA (10 ng for environmental DNA, 1 ng for genomicDNA or 0.1 ng for plasmid DNA), ultrapure water to 50 μl.

95° C, 5 min. Pause @95° C. Add template DNA ("hot start" to minimise mispriming). 35 cycles of 1 min @ 95° C; 1 min at the annealing temperature; 2 min @ 65° C and 2 min @ 72° C for every 1000 bp to be amplified. A final 10 min @ 72° C to allow the completion of all ongoing or unfinished amplifications.

Purify PCR products with a commercial PCR purification kit according to manufacturer's instructions (we use the HighPure PCR purification kit from Macherey-Nagel). Dissolve or elute purified DNA in ultrapure water.

Strand separation:
1. Wash 20 μl 10 μg/μl Dynabeads (Roche) in 200μl 1xTS solution (100 mM Tris.Cl, 150 mM NaCl, pH 7.6) each
2. Dissolve beads in 6 μl 6x TS solution (600 mM Tris.Cl, 900 mM NaCl, pH 7.6)
3. Add 40 μl purified PCR product
4. 30 minutes incubation @ 37° C with low speed shaking
5. Wash 2x with 200 μl 1x TS
6. Add 20 μl 0.2M NaOH; incubate at room temperature for 5 min; spin and take supernatant. Repeat once more.
7. Precipitate pooled supernatant (»40ml) with 120 μl 90% EtOH / 0.3 M sodium acetate (NaOAc) pH 5.5; 60 mins @ -20° C.
8. Pellet DNA by centrifugation for 30 mins, @ 13,000 rpm, 4°C.
9. Wash with ice-cold 70% EtOH: spin as above for 10 mins.
10. Air-dry; dissolve in 20 μl TE.

Note: the binding capacity of 20μl (200μg) Dynabeads is 2 pmoles of ds DNA. This is equivalent to 660 μg of a 500 bp fragment.

Labeled DNA can now be stored at -20°C for several months.

Alternative method: Fragmentation of internally labeled DNA by partial DNAseI treatment

Method taken from (de Saizieu et al., 2000).
1. Into a 1.5 ml Eppendorf tube add:
   DNA to be fragmented (500-2000 ng)
   10 µl 10x RQ1 DNAse buffer (Promega)
   10 µl 0.01 U/µl RQ1 DNAseI (Promega)
2. ultrapure water to 100µl Incubate at 37° C for 5 mins.
3. Add 10 µl of DNAse stop solution to stop the reaction.
4. Incubate at 65°C for 30 mins to inactivate DNAseI.
5. Add 200 µl 100% EtOH and 20 µl 3 M NaOAc pH 5.5; 60 mins @-20° C.
6. Pellet DNA by centrifugation for 30 mins, @13,000 rpm, 4° C.
7. Wash with ice-cold 70% EtOH: spin as above for 10 mins.
8. Air-dry; dissolve in 20 µl TE.
   Labeled, fragmented DNA can now be stored at -20° C for several months.

HYBRIDISATION

Hybridisation should be carried out in the dark whenever possible to minimise photobleaching of the fluorescent dyes. Differential photobleaching can make this problem more serious when two-colour hybridisation is applied.

As hybridisation under a cover slip or under a disposable hybridisation chamber is a diffusion-limited process, it is recommended to carry out hybridisations overnight (14-18hr).

To ensure good control over hybridisation temperature, a custom tailored aluminium block is used with holes of 11mm in depth for 14 slides. The block is used as an insert for a Belly Dancer (Stovall Life Sciences, Greensboro, NC, USA). 1-2 mm spacing between the block and the wall of the water bath chamber allows for the addition of about 20 ml water, which is required for reliable heat transfer to the aluminium block.

Method:

1. Preheat hybridisation block to 55° C. Allow for at least 30 minutes for the temperature to stabilise.
2. Preheat an Eppendorf incubator (dry block) to 66° C.

3. Apply 200 μl HybriWell (Grace BioLabs) chambers onto the slides containing the arrays. Preheat assembled slides on top of the hybridisation
   block. The BellyDancer should be set to maximum bending.
4. Per hybridisation add into a 1.5 ml Eppendorf tube:
   124 μl DEPC-water
   2 μl 10% SDS
   4 μl 50x Denhardt's reagent (Sigma)
   60 μl 20x SSC
   10 μl target RNA
   Incubate at 65° C for 1 - 15 mins.
5. Apply preheated hybridisation mixtures onto assembled slides via the port in
   the lower positions (to minimise risk of air bubbles being trapped within the
   chamber). Seal chambers with seal spots.
6. Incubate overnight in Belly Dancer (30-40 rpm circulation at maximum
   bending).
7. Take slides one by one, remove sticky chamber and put them immediately
   into 2xSSC, 0.1% SDS at RT.
8. Wash slides by shaking at RT for:
   5 min in 2xSSC, 0.1% SDS;
   2x5 min in 0.2x SSC;
   5 min in 0.1x SSC
9. Dry slides one by one using an airgun with a cottonwool filter inside (to
   keep oil microdroplets away from the slide surface). Apply a modest
   stream to the area containing the array first to blow the drops down on the
   slide, rather than drying them onto it.
10. Scan slides the same day.

SCANNING

There is a relatively large choice of microarray scanners on the market. While
complex, high-tech instruments, their most important characteristics from an
average user's point of view can be summarised as follows:

Resolution Range

Spatial resolution is normally expressed as pixel size. Highest resolutions are
between 1 and 10 μm for most scanners. The smaller the spots are the higher
the resolution has to be. One spot has to be resolved into a minimum of 50-100

pixels in order to be able to correctly calculate and consider edges, edge effects and background. For spots of 100-150 μm in diameter, a resolution of 10 μm is sufficient; in most cases, further increase in resolution only results in increased memory demand for scanned image storage.

Detection System

Commercial scanners apply either photomultipliers (PMTs) or, in a few cases, CCD (charged coupled device) cameras. There is little information available on the comparison of the two detection systems, however it is believed that PMT scanners perform better than CCD ones.

Sensitivity

Sensitivity is usually expressed as 'minimal number of fluorophores per μm² detectable by the scanner. A scanner's sensitivity changes with the wavelength. Typical values are 0.02 to 0.1 fluor/μm², usually measured with Cy3 and/or Cy5.

Maximal Field Size

Microarray slides are in the size of "standard" microscope slides (i.e., either 25.0 x 76.0 mm- metric standard, or 25.4 x 76.2 mm- US standard, 1 x 3"). The useful surface limits the size of the array that can be spotted. Scan sizes range from 22 x 60 to 25 x 75 mm.

Throughput (Scan Rate)

A typical 20x60 mm image is scanned at 5-10 μm resolution in 4-15 min by most commercial scanners. CCD scanners have the potential for a significantly faster throughput, however, at the price of higher detection limits (i.e., less sensitive detection). Simultaneous data acquisition mode can dramatically increase throughput; however, it may also increase cross talk between channels. Throughput also depends on automation options. Most recently developed scanners are able to take and scan multiple slides without human supervision.

Number of Channels; Excitation and Detected Wavelengths; Choice of Fluorophores

The vast majority of microarray labs are currently using only two fluorophores, in most cases, Cy3 and Cy5, thus requiring only two channels on their scanners. However, throughput and performance of microarray techniques can be drastically improved by the use of multiple fluorophores in a single experiment. Most new scanners have 3, 4, 5 or even 8 channels.

As the number of channels increase, so does the chance of cross talk between them. Care must be taken, when choosing a multiple-channel scanner to check the technical details of the instrument. Each fluorophore has an optimal excitation and detection wavelength, strongly influencing the sensitivity of the instrument. If emission spectra of two dyes used in separate channels overlap, significant cross talk may occur between them. The scanner should have a built-in technical solution for such problems.

Confocal vs. Non-confocal

Most scanners are equipped with confocal optical systems. According to a recent report (Ramdas *et al.*, 2001), performance of the two systems (confocal and non-confocal) are comparable and neither of them is superior over the other.

Method:

In our hands, an Axon 4000A scanner (driven by a GenePix Pro software) at 10µm resolution (spots are 110-130µm in diameter) performs perfectly.

Prescan the whole slide without averaging. Start with a photomultiplier voltage likely to be near but still below saturation. When prescan is started, finish to the end of the spotted area as each round of scanning causes photobleaching of the spots. Partial scanning of the images may thus introduce variation. Find the optimal photomultiplier voltage that yields the highest signals possible without saturation.

Scan the spotted area at 3 lines to average, 10mm resolution. Save as multi-layer Tiff image (required to enable later analysis of the scanned image).

DATA ANALYSIS

Once designed and spotted, a probe set first has to be evaluated by hybridising with reference sequences. In order to make results comparable and to allow for any quantification it is recommended to compare all results to a (set of) reference probe(s). It also helps in refining the probe set by providing information on the binding efficiency of each probe. A reference probe can be a highly conserved probe hybridising to all or almost all of the possible targets. Alternatively, probes hybridising to the PCR primers can be used. A third alternative is to spike the IVT reaction with known amounts of a foreign gene fragment and then use a reference probe that targets this gene fragment.

There is no specialised software available for the analysis of diagnostic oligo arrays. Results can be analysed by general-purpose statistical software, like Excel or the R package (http://www.r-project.org).

Raw data further analysed are usually mean or median fluorescence values of the spots minus background fluorescence. The use of median values is recommended, as very high and very low values outside the linear range of the scanner are unreliable, and have less effect on the median than on the mean value.

The hybridisation behaviour of oligonucleotide probes immobilised onto a solid surface depends on several factors, most of which can be accounted for in CalcOligo:

1. Length, GC content and exact sequence of the probe. These together are considered when predicting Tm for the oligos by using the nearest neighbour method. Note: the predicted Tm applies for free oligos in solution.
2. Position of GC and AT pairs. The middle of the probe is more important in stabilising hybridisation. A probe where the middle contains all Gs and Cs binds its target much stronger than another one with homogenous GC distribution (but with identical length and GC content).
3. Secondary structures of the probe and of the corresponding target. When any of these two are of significant strength compared to the strength of hybridisation between the probe and the target, a significant drop in hybridisation efficiency is expected.
4. The exact nature of the overhanging nucleotides on the target. This comes from the nearest neighbour model, but isn't normally accounted for, because the overhangs of the target sequence are not considered.
5. Number and type of mismatches. The destabilizing effect of mismatches varies; some have little effect, while others have very strong destabilising effects.
6. Position of mismatches. Mismatches in the middle are more destabilizing than mismatches in end positions.

7. Factors arising from the immobilised nature of the probes. Steric effects can hinder the formation of hybrids between the target and the bound probe. This effect is much stronger for the immobilised end of the probe. Thus, the bound end of the probe plays a lesser role in the hybridisation than the free end. This applies for points 2 and 6.

8. Hybridisation between DNA oligos and RNA fragments, as in our case, has slightly different thermodynamics than that of DNA-DNA hybridisation (Hung *et al.*, 1994; Sugimoto *et al.*, 2000)

Once the hybridisation behaviour (specificity, intensity) of the probe set is determined, further probes can be designed if necessary (if some of the probes in the set do not work as predicted). Once a final probe set is designed and spotted, real experiments can commence. CalcOligo allows for the fitting of the predictions to the actual results. By doing this, one can generate an improved parameter set for CalcOligo, enabling the better prediction of hybridisation behaviour and minimising the number of faulty probes designed and ordered.

Results are always referred back to a table with the hybridisation behaviour of the probe set. When the aim is only to detect microorganisms from a sample, results are used in a simple plus or minus matter (qualitative). For quantification of the results (i.e. when community structures, relative amounts, differences in biodiversity, etc. are questioned) one has to relate the intensities of the spots to the intensities of the same spots when hybridised with a single target. This is a task which becomes difficult with the increasing number of different probes used, and so far there is no automation strategy worked out for it. For up-to-date information the reader is referred to the frequently updated homepage of the author:

    http://www.diagnostic-arrays.com

An alternative approach for the quantitative analysis of results is the use of a reference target set labeled with another dye (Cy5 in our case). In this case, one works with the ratios of the median intensities for each spot. The results are first normalised to the positive control spots and then the ratios of the target molecules (à of the different microorganisms in the analysed sample) are calculated from the intensity ratios of the corresponding probe spots.

APPLICATIONS IN MICROBIAL IDENTIFICATION

The number of publications in this field has not expanded yet at the rate seen for transcriptional profiling using microarrays. The full potential of microarray technology in microbial identification has clearly not been explored yet.

The concept of oligo chips for microbial detection was first described and demonstrated by Andrei Mirzabekov's group (Guschin *et al.*, 1997). They developed a special microarray format consisting of individual polyacrylamide gel micropads with immobilized oligonucleotides. Nine oligonucleotide probes applied onto such glass immobilised gel elements were successfully used to discriminate various phylogenetic groups of ammonia oxidising bacteria. In a more recent paper (Bavykin *et al.*, 2001), they describe a portable system enabling probe preparation, hybridisation and data imaging on site within less than an hour. The special platform enabled not only hybridisation but also on-chip PCR and on-chip ligase detection reactions. All three approaches were succesfully demonstrated to detect and differentiate between rifampicin-resistant *Mycobacterium tuberculosis* strains (Mikhailovich *et al.*, 2001). A chip with a set of 15 probes targeting the *crm*B genes was developed for the species specific identification of orthopoxviruses (Lapa *et al.*, 2002).

David Stahl's group is using the same platform to develop diagnostic microarrays for environmental microbiology. A set of nine probes, custom-designed for the microbial community investigated, was used for the characterization of aromatic hydrocarbon degrading consortia (Koizumi *et al.*, 2002). A new software under development aims at reliably predicting the hybridisation behaviour of oligonucleotide probes, considering array-specific effects like position of mismatch (Urakawa *et al.*, 2002).

French and co-workers (Anthony *et al.*, 2000) used a 30 oligonucleotide set targeting variable regions of the 23S rRNA gene. Their oligo set, immobilised onto a nylon membrane ("macroarray"), was used to detect and discriminate 30 common strains causing bacteremia directly from infected blood samples. Their oligo set was first evaluated with pure cultures. Misidentification occurred in only 6% of the 125 blood samples.

Rudi and co-workers (Rudi *et al.*, 2000; Rudi *et al.*, 2002) developed sets of 9 and 7 16S rRNA probes respectively targeting cyanobacteria and microbial communities in ready-to-eat vegetable salads. Community structures of marine cyanobacterial communities and of vegetable salads packed under modified atmospheres were analysed with the probe sets developed. Sequence-specific labeling of PCR products (amplified with a universal 16S rRNA primer set) was employed to secure specificity. Quantitative analysis of the results was carried out as outlined in the above section (1-colour hybridisation).

A set of 51 oligonucleotide probes targeting the 16S and 23S rRNA genes of the genus Enterococcus has been developed and succesfully applied in the detection of *Enterococcus* strains from drinking water samples as well as from waste water treatment plants (Behr *et al.*, 2000). The oligonucleotide probes were designed to target the *Enterococcus* group at different levels, from 'species' through 'species groups' to the whole genus. Several probes had partially overlapping specificities. This approach is described as the application of a nested set of probes.

Six oligo probes targeting the femA gene (encoding a precursor of the peptydoglucan, highly conserved amongst Staphylococci) were used to differentiate between five clinically relevant strains of Staphylocci (*S. aureus*, *S. epidermidis*, *S. haemolyticus*, *S. hominis* and *S. saprophyticus*) (Hamels *et al.*, 2001). No false positive or false negative results were obtained when tested with 16 reference strains of Staphylococci.

Raskin and co-workers (Hansen et al., 1999) used five oligonucleotide probes to quantify and to determine the community structure of syntrophic fatty acid-ß-oxidising bacteria in methanogenic environments. They used traditional slot blot hybridisation by immobilising rRNA was onto nylon membranes and then hybridising these membranes with the oligonucleotide probes.

A high-density Affymetrix GeneChip containing over 30,000 16S rRNA targeting oligo probes was used to identify culture collection species and subsequently to characterize populations of airborne bacteria at the level of higher phylogenetic taxa (Wilson *et al.*, 2002)
All (20) tested human group A rotaviruses were succesfully differentiated at the genotype level by a set of 47 probes targeting a gene, VP7, conserved among rotaviruses (Chizhikov *et al.*, 2002).

Targeting multiple genes on a single array is limited by problems associated with multiplex PCR. In these applications, a single PCR reaction is expected to yield specific amplification products from multiple genes (there is a specific PCR primer pair included for each target gene). The increase in the number of primers quickly leads to an inhibition of some of the PCR reactions due to complex interactions between the primers. In a recent work (Chizhikov *et al.*, 2001), six microbial virulence genes were successfully amplified in a multiplex PCR by employing relaxed annealing conditions. The resulting non-specific products did not interfere with the hybridisation to the specific probes, thus enabling the specific detection of all six virulence factors.

The oligonucleotide probe database (ODP, http://www.cme.msu.edu/OPD/; Zheng *et al.*, 1996) is a very useful initiative to collect and make available already tested phylogenetic oligonucleotide probes. Unfortunately the database is not updated, however many authors follow the naming convention of the ODP.

Alternatives for Microbial Identification

Tiedje and co-workers proposed the use of whole genome DNA-DNA hybridisation in the detection and community analysis of microorganisms (Cho and Tiedje, 2001; Murray *et al.*, 2001). The forerunner of this approach is described as reverse sample genome probing by Voordouw and co-workers (1991).

Gene fragment based DNA microarrays offer the potential of quantifying the relative abundance of microbial groups carrying out certain functions (e.g. nitrogen metabolism) without considering their phylogenetic relatedness and deeper community structure. Such an array was developed and shown to differentiate between genes of less than 80-85% sequence identity (Wu *et al.*, 2001). A simple and elegant method to quantify specific microbial genes from a sample has recently been published (Cho and Tiedje, 2002). In this method gene fragments (500-900bp in length) were applied as probes. Each spot consisted of a mixture of an individual probe and a common reference gene fragment. Hybridisation was done with a Cy3 labeled environmental mixture and a Cy5 labeled reference DNA. Quantification was based on the Cy3/Cy5 ratios. Unfortunately, the same principle cannot be applied to the quantification of oligonucleotide chip results due to the inherent differences in hybridisation efficiency between oligo probes. The drawback of this approach is that the resolution power does not allow for the differentiation within closely related species. On the other hand, this handicap becomes a bonus if one wants to look at global processes mediated by microbes, i.e. nitrogen or methane cycling, where the aim is to detect and quantify bacteria carrying out the same function, irrespective of their phylogenetic relatedness.

Recent Advancements in Oligonucleotide Microarray Technology

Small and co-workers (Small *et al.*, 2001) used a novel system for the direct species-specific detection of 16S rRNA in environmental samples. The combination of two methods was required to overcome the limitations associated with the strong secondary structure of RNA. Environmental RNA preparations were chemically fragmented prior to hybridisation in order to reduce the strength of RNA secondary structures. Visualisation of the RNA was achieved by the application of labeled detector probes binding to the bound RNA adjacent to the capture probes immobilised on the microarray. It was hypothesised that the capture and detector probes enhanced each other's binding to the target RNA molecule via a chaperone effect.

A special format applying bacterial magnetic particles (BMPs) was developed by Matsunaga *et al.* (2001). 50-100 nm sized BMPs were conjugated to oligonucleotide probes. Hybridisation was done in 96 well microtiter plates. Particles were washed and concentrated using a magnetic-separation robot and then spotted into microwells on a prefabricated silicon wafer. The requirement for highly sophisticated equipment was compensated by a very short (5 minutes) hybridisation and the possibility to reuse the probes and to purify and analyse the captured target. Another advantage of the approach is that by concentrating the probe-target mixtures after hybridisation, a high sensitivity is theoretically possible. In practice however current limitations in sensitivity are more due to background signal arising from weak nonspecific hybridisations.

Microelectronic chips offer the potential for parallel amplification of various genes on the same platform, followed by specific detection of the products (Westin *et al.*, 2001). Such chips enable the parallel targeting of multiple genes on a single array by avoiding primer interactions inherent in multiplex PCR reactions.

ACKNOWLEDGEMENTS

The author would like to thank J. Colin Murrell and Fodor Szilvia for critical reading of the manuscript and for their useful comments and suggestions. Sándor Bottka, József Csontos, Angela Sessitsch and many active members of the Yahoo microarray / Gene-Arrays listserv discussion group (groups.yahoo.com/ group/microarray/ and gene-arrays@itssrv1.ucsf.edu) are gratefully acknowledged for their help and suggestions concerning various aspects of the work presented here. I would also like to thank Eric Blalock for not only doing an excellent job as editor but also providing a lot of help in preparing the figures for this chapter.

LINKS RELATED TO DIAGNOSTIC MICROBIAL MICROARRAYS

Academic

Group of Andrei Mirzabekov, Engelhardt Inst. of Molecular Biology, Russian Acad. Sci.
   http://www.eimb.relarn.ru/eimb/mirzabekov.htm
Group of David Stahl, University of Washington
   http://stahl.ce.washington.edu/index.html
Group of Michael Wagner, TU Munich:
   http://www.microbial-ecology.de/research.html
Homepage of James Tiedje:
   http://www.msu.edu/unit/mic/facpages/tiedje.html

MIDI-CHIP (Microbial Diversity Chip):
  http://155.253.6.98/index.htm
Phylogenic Microchip Project (funded by DARPA):
  http://www.darpa.mil/spo/programs/componenttechnologies.htm
The identification DNA array platform (maintained at the University of Wageningen, the Netherlands)
  http://www.ftns.wau.nl/id-array/index.htm

Software

ARB - a phylogenetic software for the establishment of databases, computation of phylogenetic trees and taxon-specific probe design
  http://www.arb-home.de
CalcOligo - a software predicting hybridisation behaviour of a whole probe set against all the sequences available in an ARB file; also enabling user-defined accounting for microarray-specific effects.
  http://www.diagnostic-arrays.com/calcoligo/index.htm
Neural Network at the Stahl group - to predict probe hybridisation behaviour; enables algorithm optimisation to match results
  http://stahl.ce.washington.edu/programpage.jsp
Primrose - a user friendly platform for 16S rRNA based probe design based on RDP databases
  http://www.cardiff.ac.uk/biosi/research/biosoft/
DEODAS - efforts to create a software for the design and analysis of degenerate oligonucleotide probes
  http://deodas.sourceforge.net/
ProbeSelect - Algorythm to select optimal oligos and to predict their hybridisation behaviour (does not account for array-specific effects)
  available upon request: Fugen Li, lif@ural.wustl.edu
LNA Tm prediction - predicts Tm of DNA and DNA/LNA combined oligos
  http://lna-tm.com/

Companies

OGT - Oxford Gene Technologies - Technology
  http://www.ogt.co.uk/technology.html
GeneScan Europe, NutriChip
  http://www.genescan.com

Advanced Array Technologies, StaphyChip
  http://www.aat-array.com/staphy.htm
Lambda GmbH, Parodontosis Chip
  http://www.lambda.at/e/e_parocheck.shtml
Argonne National Laboratories, *Mycobacterium tuberculosis* chip
  http://www.anl.gov/OPA/vtour/biochip.htm
Dr. Chip - various microbial diagnostic "macroarrays"
  http://www.bio-drchip.com.tw/eweb/
iQserv - offering integrated probe design, hybridisation behaviour prediction and
data analysis solutions
  http://www.isenseit.com

Fluroescence Resonance Energy Transfer

Integrated DNA Technologies - Technical Bulletins on FRET:
  http://www.idtdna.com/program/techbulletins/Fluorescent_Dye_Labeled
  _Oligonucleotides_Printer.asp
W.M. Keck Center For Cellular Imaging - FRET:
  http://www.cci.virginia.edu/frethead5.html

REFERENCE LIST

1. Alon U, Barkai N, Notterman DA, Gish K, Ybarra S, Mack D, Levine AJ (1999) Broad patterns of gene expression revealed by clustering analysis of tumor and normal colon tissues probed by oligonucleotide arrays. Proc Natl Acad Sci U S A 96: 6745-6750.

2. Amann RI, Ludwig W, Schleifer KH (1995) Phylogenic Identification and in situ detection of individual microbial cells without cultivation. Microbiol Rev 59: 143-169.

3. Amann R, Ludwig W (2000) Ribosomal RNA-targeted nucleic acid probes for studies in microbial ecology. FEMS Microbiol Rev 24: 555-565.

4. Anthony RM, Brown TJ, French GL (2000) Rapid diagnosis of bacteremia by universal amplification of 23S ribosomal DNA followed by hybridization to an oligonucleotide array. J Clin Microbiol 38: 781-788.

5. Auman AJ, Stolyar S, Costello AM, Lidstrom ME (2000) Molecular characterization of methanotrophic isolates from freshwater lake sediment. Appl Environ Microbiol 66: 5259-5266.

6. Bavykin SG, Akowski JP, Zakhariev VM, Barsky VE, Perov AN, Mirzabekov AD (2001) Portable system for microbial sample preparation and oligonucleotide microarray analysis. Appl Environ Microbiol 67: 922-928.

7. Behr T, Koob C, Schedl M, Mehlen A, Meier H, Knopp D, Frahm E, Obst U, Schleifer K, Niessner R, Ludwig W (2000) A nested array of rRNA targeted probes for the detection and identification of enterococci by reverse hybridization. Syst Appl Microbiol 23: 563-572.

8. Bourne DG, McDonald IR, Murrell JC (2001) Comparison of pmoA PCR primer sets as tools for investigating methanotroph diversity in three Danish soils. Appl Environ Microbiol 67: 3802-3809.

9. Breslauer KJ, Frank R, Blocker H, Marky LA (1986) Predicting DNA duplex stability from the base sequence. Proc Natl Acad Sci U S A 83: 3746-3750.

10. Call DR, Chandler DP, Brockman F (2001) Fabrication of DNA microarrays using unmodified oligonucleotide probes. Biotechniques 30: 368-72, 374, 376.

11. Chizhikov V, Rasooly A, Chumakov K, Levy DD (2001) Microarray analysis of microbial virulence factors. Appl Environ Microbiol 67: 3258-3263.

12. Chizhikov, V., M. Wagner, A. Ivshina, Y. Hoshino, A. Z. Kapikian, and K. Chumakov. 2002. Detection and Genotyping of Human Group A Rotaviruses by Oligonucleotide Microarray Hybridization. J. Clin. Microbiol 40:2398-2407.

13. Cho JC, Tiedje JM (2001) Bacterial species determination from dna-dna hybridization by using genome fragments and dna microarrays. Appl Environ Microbiol 67: 3677-3682.

14. Cho JC, Tiedje JM (2002) Quantitative detection of microbial genes by using DNA microarrays. Appl Environ Microbiol 68: 1425-1430.

15. Connors TD, Burn TC, VanRaay T, Germino GG, Klinger KW, Landes GM (1997) Evaluation of DNA sequencing ambiguities using tetramethylammonium chloride hybridization conditions. Biotechniques 22: 1088-1090.

16. Dahllof I, Baillie H, Kjelleberg S (2000) rpoB-based microbial community analysis avoids limitations inherent in 16S rRNA gene intraspecies heterogeneity. Appl Environ Microbiol 66: 3376-3380.

17. de Saizieu A, Gardes C, Flint N, Wagner C, Kamber M, Mitchell TJ, Keck W, Amrein KE, Lange R (2000) Microarray-based identification of a novel Streptococcus pneumoniae regulon controlled by an autoinduced peptide. J Bacteriol 182: 4696-4703.

18. Edgcomb, V. P., D. T. Kysela, A. Teske, G. A. de Vera, and M. L. Sogin. 2002. Benthic eukaryotic diversity in the Guaymas Basin hydrothermal vent environment. Proc. Natl. Acad. Sci. U. S. A 99:7658-7662.

19. Edwards, U., T. Rogall, H. Blocker, M. Emde, and E. C. Bottger. 1989. Isolation and direct complete nucleotide determination of entire genes. Characterization of a gene coding for 16S ribosomal RNA. Nucleic Acids Res. 17:7843-7853.

20. Erdogan F, Kirchner R, Mann W, Ropers HH, Nuber UA (2001) Detection of mitochondrial single nucleotide polymorphisms using a primer elongation reaction on oligonucleotide microarrays. Nucleic Acids Res 29: E36.

21. Fotin AV, Drobyshev AL, Proudnikov DY, Perov AN, Mirzabekov AD (1998) Parallel thermodynamic analysis of duplexes on oligodeoxyribonucleotide microchips. Nucleic Acids Res 26: 1515-1521.

22. Giovannoni, S. J. 1991. The polymerase chain reaction, p. 177-203. In E. Stackebrandt and M. Goodfellow (eds.), Nucleic acid techniques in bacterial systematics. Wiley and Sons., Chichester.

23. Gruber HJ, Hahn CD, Kada G, Riener CK, Harms GS, Ahrer W, Dax TG, Knaus HG (2000) Anomalous fluorescence enhancement of Cy3 and cy3.5 versus anomalous fluorescence loss of Cy5 and Cy7 upon covalent linking to IgG and noncovalent binding to avidin. Bioconjug Chem 11: 696-704.

24. Gunderson KL, Huang XC, Morris MS, Lipshutz RJ, Lockhart DJ, Chee MS (1998) Mutation detection by ligation to complete n-mer DNA arrays. Genome Res 8: 1142-1153.

25. Guo Z, Guilfoyle RA, Thiel AJ, Wang R, Smith LM (1994) Direct fluorescence analysis of genetic polymorphisms by hybridization with oligonucleotide arrays on glass supports. Nucleic Acids Res 22: 5456-5465.

26. Guschin DY, Mobarry BK, Proudnikov D, Stahl DA, Rittmann BE, Mirzabekov AD (1997) Oligonucleotide microchips as genosensors for determinative and environmental studies in microbiology. Appl Environ Microbiol 63: 2397-2402.

27. Hacia JG (1999) Resequencing and mutational analysis using oligonucleotide microarrays. Nat Genet 21: 42-47.

28. Hamels S, Gala JL, Dufour S, Vannuffel P, Zammatteo N, Remacle J (2001) Consensus PCR and microarray for diagnosis of the genus Staphylococcus, species, and methicillin resistance. Biotechniques 31: 1364-2.

29. Hansen KH, Ahring BK, Raskin L (1999) Quantification of syntrophic fatty acid-beta-oxidizing bacteria in a mesophilic biogas reactor by oligonucleotide probe hybridization. Appl Environ Microbiol 65: 4767-4774.

30. Hershberger, K. L., S. M. Barns, A. L. Reysenbach, S. C. Dawson, and N. R. Pace. 1996. Wide diversity of Crenarchaeota. Nature 384:420.

31. Hung, S. H., Yu, Q., Gray, D. M., and Ratliff, R. L. Evidence from CD spectra that d(purine).r(pyrimidine) and r(purine).d(pyrimidine) hybrids are in different structural classes. Nucleic Acids Res. 22[20], 4326-4334. 11-10-1994.

32. Ishii K, Fukui M (2001) Optimization of annealing temperature to reduce bias caused by a primer mismatch in multitemplate pcr. Appl Environ Microbiol 67: 3753-3755.

33. Koizumi, Y., J. J. Kelly, T. Nakagawa, H. Urakawa, S. El Fantroussi, S. Al Muzaini, M. Fukui, Y. Urushigawa, and D. A. Stahl. 2002. Parallel characterization of anaerobic toluene- and ethylbenzene-degrading microbial consortia by PCR-denaturing gradient gel electrophoresis, RNA-DNA membrane hybridization and DNA microarray technology. Appl Environ Microbiol 68:3215-3225.

34. Kolganova, T. V., B. B. Kuznetsov, and T. P. Turova. 2002. Designing and testing oligonucleotide primers for amplification and sequencing of archaeal 16S rRNA genes. Mikrobiologiia 71:283-286.

35. Lapa, S., M. Mikheev, S. Shchelkunov, V. Mikhailovich, A. Sobolev, V. Blinov, I. Babkin, A. Guskov, E. Sokunova, A. Zasedatelev, L. Sandakhchiev, and A. Mirzabekov. 2002. Species-level identification of orthopoxviruses with an oligonucleotide microchip. J. Clin. Microbiol 40:753-757.

36. Lindroos K, Liljedahl U, Raitio M, Syvanen AC (2001) Minisequencing on oligonucleotide microarrays: comparison of immobilisation chemistries. Nucleic Acids Res 29: E69-E69.

37. Lipshutz RJ, Fodor SP, Gingeras TR, Lockhart DJ (1999) High density synthetic oligonucleotide arrays. Nat Genet 21: 20-24.

38. Lovell CR, Friez MJ, Longshore JW, Bagwell CE (2001) Recovery and phylogenetic analysis of nifH sequences from diazotrophic bacteria associated with dead aboveground biomass of Spartina alterniflora. Appl Environ Microbiol 67: 5308-5314.

39. Lueders T, Chin KJ, Conrad R, Friedrich M (2001) Molecular analyses of methyl-coenzyme M reductase alpha-subunit (mcrA) genes in rice field soil and enrichment cultures reveal the methanogenic phenotype of a novel archaeal lineage. Environ Microbiol 3: 194-204.

Chapter 2

40. Maidak BL, Larsen N, McCaughey MJ, Overbeek R, Olsen GJ, Fogel K, Blandy J, Woese CR (1994) The ribosomal database project. Nucl Acid Res 22: 3485-3487.

41. Majlessi, M., N. C. Nelson, and M. M. Becker. 1998. Advantages of 2'-O-methyl oligoribonucleotide probes for detecting RNA targets. Nucleic Acids Res. 26:2224-2229.

42. Martin-Laurent, F., L. Philippot, S. Hallet, R. Chaussod, J. C. Germon, G. Soulas, and G. Catroux. 2001. Dna extraction from soils: old bias for new microbial diversity analysis methods. Appl. Environ. Microbiol. 67:2354-2359.

43. Matsunaga, T., H. Nakayama, M. Okochi, and H. Takeyama. 2002. Fluorescent detection of cyanobacterial DNA using bacterial magnetic particles on a MAG-microarray. Biotechnol. Bioeng. 73:400-405.

44. McDonald IR, Murrell JC (1997) The methanol dehydrogenase structural gene mxaF and its use as a functional gene probe for methanotrophs and methylotrophs. Appl Environ Microbiol 63: 3218-3224.

45. Mikhailovich, V., S. Lapa, D. Gryadunov, A. Sobolev, B. Strizhkov, N. Chernyh, O. Skotnikova, O. Irtuganova, A. Moroz, V. Litvinov, M. Vladimirskii, M. Perelman, L. Chernousova, V. Erokhin, A. Zasedatelev, and A. Mirzabekov. 2001. Identification of rifampicin-resistant Mycobacterium tuberculosis strains by hybridization, PCR, and ligase detection reaction on oligonucleotide microchips. J. Clin. Microbiol 39:2531-2540.

46. Milcic-Terzic J, Lopez-Vidal Y, Vrvic MM, Saval S (2001) Detection of catabolic genes in indigenous microbial consortia isolated from a diesel-contaminated soil. Bioresources Technol 78: 47-54.

47. Mir KU, Southern EM (1999) Determining the influence of structure on hybridization using oligonucleotide arrays. Nat Biotechnol 17: 788-792.

48. Murray AE, Lies D, Li G, Nealson K, Zhou J, Tiedje JM (2001) DNA/DNA hybridization to microarrays reveals gene-specific differences between closely related microbial genomes. Proc Natl Acad Sci U S A 98: 9853-9858.

49. Petersen, M., K. Bondensgaard, J. Wengel, and J. P. Jacobsen. 2002. Locked nucleic acid (LNA) recognition of RNA: NMR solution structures of LNA:RNA hybrids. J. Am Chem. Soc. 124:5974-5982.

50. Polz FM, Cavanaugh CM (1998) Bias in template-to-product ratios in multitemplate PCR. Appl Environ Microbiol 64: 3724-3730.

51. Radajewski S, Ineson P, Parekh NR, Murrell JC (2000) Stable-isotope probing as a tool in microbial ecology. Nature 403: 646-649.

52. Rainey FA, Ward-Rainey NL, Janssen PH, Hippe H, Stackebrandt E (1996) Clostridium paradoxum DSM 7308T contains multiple 16S rRNA genes with heterogeneous intervening sequences. Microbiology 142 (Pt 8): 2087-2095.

53. Ramdas L, Coombes KR, Baggerly K, Abruzzo L, Highsmith WE, Krogmann T, Hamilton SR, Zhang W (2001) Sources of nonlinearity in cDNA microarray expression measurements. Genome Biol 2: RESEARCH0047.

54. Ren T, Roy R, Knowles R (2000) Production and consumption of nitric oxide by three methanotrophic bacteria. Appl Environ Microbiol 66: 3891-3897.

55. Reysenbachal, Giver LJ, Wickham GS, Pace NR (1992) Differential amplification of rRNA genes by polymerase chain reaction. Appl Environ Microbiol 58: 3417-3418.

56. Rudi K, Flateland SL, Hanssen JF, Bengtsson G, Nissen H (2002) Development and Evaluation of a 16S Ribosomal DNA Array-Based Approach for Describing Complex Microbial Communities in Ready-To-Eat Vegetable Salads Packed in a Modified Atmosphere. Appl Environ Microbiol 68: 1146-1156.

57. Rudi K, Skulberg OM, Skulberg R, Jakobsen KS (2000) Application of sequence-specific labeled 16S rRNA gene oligonucleotide probes for genetic profiling of cyanobacterial abundance and diversity by array hybridization [In Process Citation]. Appl Environ Microbiol 66: 4004-4011.

58. Sambrook J, Fritsch EF, Maniatis T (1989) Molecular cloning: A laboratory manual. New York: Cold Spring Harbor Laboratory, Cold Spring Harbor.

59. SantaLucia J, Jr., Allawi HT, Seneviratne PA (1996) Improved nearest-neighbor parameters for predicting DNA duplex stability. Biochemistry 35: 3555-3562.

60. Schena M, Shalon D, Davis RW, Brown PO (1995) Quantitative monitoring of gene expression patterns with a complementary DNA microarray [see comments]. Science 270: 467-470.

61. Shchepinov MS, Case-Green SC, Southern EM (1997) Steric factors influencing hybridisation of nucleic acids to oligonucleotide arrays. Nucleic Acids Res 25: 1155-1161.

62. Small J, Call DR, Brockman FJ, Straub TM, Chandler DP (2001) Direct Detection of 16S rRNA in Soil Extracts by Using Oligonucleotide Microarrays. Appl Environ Microbiol 67: 4708-4716.

63. Sohail M, Akhtar S, Southern EM (1999) The folding of large RNAs studied by hybridization to arrays of complementary oligonucleotides. RNA 5: 646-655.

64. Sohail M, Hochegger H, Klotzbucher A, Guellec RL, Hunt T, Southern EM (2001) Antisense oligonucleotides selected by hybridisation to scanning arrays are effective reagents in vivo. Nucleic Acids Res 29: 2041-2051.

65. Southern EM (1996) DNA chips: analysing sequence by hybridization to oligonucleotides on a large scale. Trends Genet 12: 110-115.

66. Spiro A, Lowe M, Brown D (2000) A Bead-Based Method for Multiplexed Identification and Quantitation of DNA Sequences Using Flow Cytometry. Appl Environ Microbiol 66: 4258-4265.

67. Sugimoto N, Nakano M, Nakano S (2000) Thermodynamics-structure relationship of single mismatches in RNA/DNA duplexes. Biochemistry 39: 11270-11281.

68. Strunk, O., Gross, O., Reichel, B., May, M., Hermann, S., Stuckman, N., Nonhoff, B., Lenke, M., Ginhart, A., Vilbig, A., Ludwig, T., Bode, A., Schleifer, K.-H., and Ludwig, W. ARB: A software environment for for sequence data. 2000. www.mikro.biologie.tu-muenchen.de.

69. Urakawa, H., P. A. Noble, S. El Fantroussi, J. J. Kelly, and D. A. Stahl. 2002. Single-base-pair discrimination of terminal mismatches by using oligonucleotide microarrays and neural network analyses. Appl. Environ. Microbiol. 68:235-244.

70. Voordouw G, Voordouw JK, Karkhoff-Schweizer RR, Fedorak PM, Westlake DWS (1991) Reverse sample genome probing, a new technique for identification of bacteria in environmental samples by DNA hybridization and its application to the identification of sulfate-reducing bacteria in oil field samples. Appl Environ Microbiol 57: 3070-3078.

71. Westin L, Miller C, Vollmer D, Canter D, Radtkey R, Nerenberg M, O'Connell JP (2001) Antimicrobial resistance and bacterial identification utilizing a microelectronic chip array. J Clin Microbiol 39: 1097-1104.

72. Widmer F, Shaffer BT, Porteous LA, Seidler RJ (1999) Analysis of nifH gene pool complexity in soil and litter at a Douglas fir forest site in the Oregon cascade mountain range. Appl Environ Microbiol 65: 374-380.

73. Wilson, K. H., W. J. Wilson, J. L. Radosevich, T. Z. DeSantis, V. S. Viswanathan, T. A. Kuczmarski, and G. L. Andersen. 2002. High-density microarray of small-subunit ribosomal DNA probes. Appl Environ Microbiol 68:2535-2541.

74, Woese CR (1987) Bacterial evolution. Microbiol Rev 51: 221-271.

75. Wood WI, Gitschier J, Lasky LA, Lawn RM (1985) Base composition-independent hybrid-
     ization in tetramethylammonium chloride: a method for oligonucleotide screening of highly
     complex gene libraries. Proc Natl Acad Sci U S A 82: 1585-1588.

76. Wyman M, Davies JT, Crawford DW, Purdie D A (2000) M olecular and physiological
     responses of two classes of marine chromophytic phytoplankton (Diatoms and
     prymnesiophytes) during the development of nutrient-stimulated blooms. Appl Environ
     Microbiol 66: 2349-2357.

77. Yershov G, Barsky V, Belgovskiy A, Kirillov E, Kreindlin E, Ivanov I, Parinov S, Guschin D,
     Drobishev A, Dubiley S, Mirzabekov A (1996) DNA analysis and diagnostics on oligonucle-
     otide microchips. Proc Natl Acad Sci U S A 93: 4913-4918.

78. Zheng D, Alm EW, Stahl DA, Raskin L (1996) Characterization of universal small-subunit
     rRNA h ybridization probes for quantitative molecular microbial ecology studies. Appl
     Environ Microbiol 62: 4504-4513.

Chapter 3

PRINTING TECHNOLOGIES AND MICROARRAY MANUFACTURING TECHNIQUES: MAKING THE PERFECT MICROARRAY

Todd Martinsky
Vice President, TeleChem International, Sunnyvale, CA

INTRODUCTION

Much has been published regarding the power of microarray technology for use in Genomic and Proteomic applications. What was not clear to the life science industry in 1995, when Mark Schena and his colleagues at Stanford wrote the first paper demonstrating the usefulness of microarray technology, is clear now...microarray technology is here to stay.

One situation that speaks to the arrival of microarray technology is that it has caught the attention of the governmental regulating bodies like the FDA. Currently FDA regulators are trying to decide if microarray data should be incorporated and/or required in the drug evaluation process. Clearly microarray technology is revolutionizing toxicology testing, but when can industry start trusting microarray data? When should the FDA start using microarray data to make critical decisions about the safety and efficacy of new drugs? When will DNA microarray-based diagnostics go mainstream? There are several factors holding up this decision; a major one is the reproducibility and reliability of microarray-generated data. Regardless of the application, generating high quality microarray data requires reliable manufacturing and processing techniques. This chapter will focus on the methods and techniques used to spot perfect microarrays, the kind the FDA and the rest of us can hang our hats on.

MICROARRAY MANUFACTURING

It is important to note that mechanically identical mechanisms perform identical tasks. Precise engineering must be employed in the tools and devices used to produce microarrays. Most microarray manufacturing is done using computer controlled X, Y, Z motion control systems equipped with micro fluidics handling technology. Samples are picked up out of 96 or 384 well source plates and delivered to microarray substrates in columns and rows of samples. The sample delivery technology can be either contact or non-contact. This chapter will focus on the printing technologies used to spot microarrays and will not cover *in-situ* synthesis methods employed to manufacture microarrays. Additionally, please keep in mind that printing technology represents only 1/5 of the variables that need to be controlled. If there is one golden rule to follow in the microarray experimental lifecycle it is as follows: "if there is a variable in your system, control it." The other key microarray manufacturing components that must be tightly regulated are **robotics, surface chemistry, sample preparation and environment**. If any one of these components is poorly managed, inconsistent microarray printing will be the result.

Robotics

Robotic systems used to print microarrays are commonly called microarrayers. Microarrayers must have the following characteristics.

- Accuracy and repeatability on the micron level
- Computer controlled graphical user interface that produces proper movements and allows for sample tracking
- Wash and dry station to eliminate the cross-contamination of samples
- Humidity and t emperature control i n a c losed, positive p ressure "cleanroom quality" chamber

Accuracy and Repeatability

Moving a printing mechanism on a microarray robot can be achieved with linear drives or various types of linear actuators. Linear drives offer the advantages of smooth motion, high spatial resolution, low maintenance, and compact design. Smooth motion is important since vibration must be kept to a minimum, high spatial resolution is critical since 100 micron size spots are being placed

~30 microns apart. Reliability and durability are key because microarrayers can be run for days or even weeks at a time without stopping. Bench space is always at a premium and arrayers should be built to accommodate modest size environments. When shopping for a microarrayer, keep in mind that features added to a machine's complexity make achieving accuracy and repeatability more difficult.

Computer Software

Since most of us are not computer programmers, an easy to use graphical user interface that supplies a high degree of flexibility in array design is necessary for successful microarray printing. The minimum parameters that should be configurable in software to run a microarrayer are:

- Number of sample delivery mechanisms and the center-to-center spacing of said mechanisms (4.5mm or 9mm centers to pickup samples from 96 or 384 well microplates, respectively).
- The total number of samples to be printed
- Offsets relative to the substrate
- Number of replicates of each sample
- Center-to-center distance between spots
- Number of columns and rows
- Number of substrates/slides to be printed
- Wash/dry parameters for the printing mechanisms between printing cycles.

In addition to the above parameters, a sample tracking system that maps each sample's location(s) on the substrate is critical. The samples from the source plates rarely map directly to the spots on the microarray since sample delivery mechanisms are typically set at 4.5 mm center-to-center spacing, and are printed at centers in the hundred-micron range. Tracking the columns and rows in each subgrid (set of printed elements delivered by a printing implement) is difficult to perform manually (Figure 1).

Avoiding Cross Contamination of Samples

Preventing cross contamination of samples is the job of the wash/dry station of a microarraying system. There are many variations to the standard configuration of a wash/dry station, but most designs include a circulating water bath and vacuum dry station. In Figure 2 there is an orifice for each tip of the printing mechanism for drying by vacuum suction. There are many variations on this

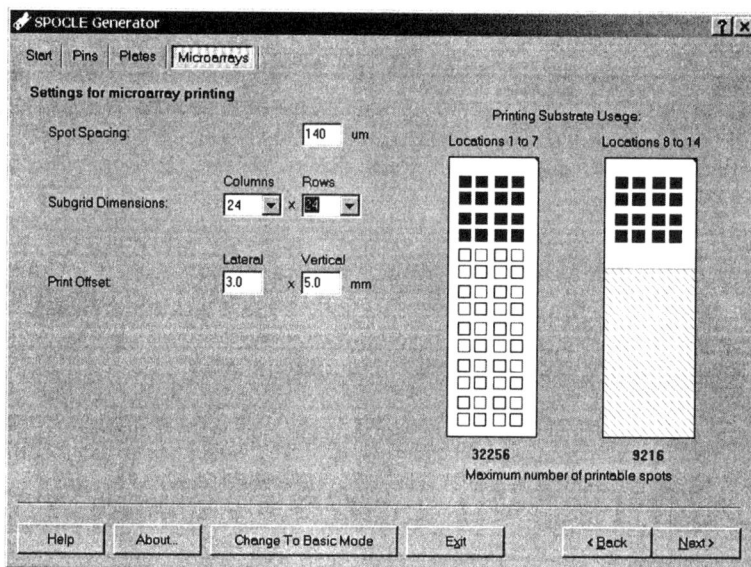

Figure 1. An example of a typical programming dialog box of a graphical user interface. In this example the center-to-center distance of the spots (spot spacing), number of columns and rows, and print offset are defined without complex programming skills. Additional screen shots are available electronically (http://arrayit.com/Products/Printing/Spotbot/spotbot.html).

standard set up, such as multiple baths with unique wash buffers and ultrasonics. Vacuum drying has become a standard in industry. It is a smart way to both dry and move potential contaminate away from printing tips, regardless of the type of printing technology.

The keys to this setup working properly are (1) a clean and continuous distilled water source (2) sufficient vacuum airflow to dry the printing mechanisms, and (3) proper programming. Wash water should be continually replenished during a print run to avoid build up of residual sample. A build up of sample could cause serious cross contamination of samples. A vacuum dry station does not rely on vacuum pressure to dry printing tips, it relies on a large amount of air passing over the tips to facilitate drying. Multiple wash/dry cycles are necessary to make sure that all residual samples are washed away. A common error in programming a wash/dry cycle is drying the tips of the printing mechanism too long after the first wash cycle. Only dry completely after the last wash cycle. If any residual sample is on the tip and it is dried there, it is much harder to wash away. The result is poor printing performance and/or cross contamination of sample.

Figure 2. A typical wash dry station for a pin printing technology is composed of a circulating water bath and a vacuum dry station.

Surface chemistry

For DNA and protein microarrays, samples are typically deposited onto specially derivatized glass substrates. In order for the same amount of sample to be immobilized at each array location on the glass, the surface chemistry must have capture reactive groups evenly distributed across the entire printing surface. In other words, the chemical treatment used to attach genetic material to the glass must be completely homogenous in the printing area. A microarray capable of generating reliable data has the same amount of sample at each array location on the printing substrate. If the substrate is not homogenous, meaning that it does not have the same amount of reactive groups at each array location on the substrate, then different amounts of sample will attach to the surface.

Common problems with Popular Surface Chemistries

Coating microscope slide glass with poly-L-lysine (PLL) has been used extensively for the manufacture of DNA microarrays. Although some satisfactory performance has been achieved with this surface chemistry, the fact that the chemistry is non-covalently bonded to low quality glass causes key problems. Firstly, the PLL applied to the glass has a tendency to age quickly and come off the glass. This makes the window of opportunity to generate quality data short. The absence of covalent linkage, and the presence of heterogeneous glass, makes it impossible to create a homogenous surface. Regardless of the source, for best results it is recommended that microarrays be printed on optical quality surfaces with low auto-fluorescence, stable surface chemistries covalently bonded to a rigid surface manufactured in a clean room setting.

Choosing the right surface chemistry is one of the most important decisions one can make before proceeding to the manufacturing step. When deciding to make homemade slides or buy them from a vendor, there are important scientific as well as economic considerations. Keep in mind that covalent linkage will stand up to harsher processing conditions, but not all covalent linkage approaches are stable for long periods of time in ambient storage printing conditions.

A basic rule of thumb to immobilize DNA is that short oligos require end termination and thus should be covalently attached through specific means. Generally oligos are amino modified and immobilized onto aldehyde surfaces (Figure 2B) or isothiocyanate reactive surfaces. In this case, aldehyde reactive groups are favored over isothiocyanate type chemistries for stability reasons. The sulfur contained in chemical bonds of the isothiocyanate chemistry loses electrons easily when exposed to air (oxidation). Long oligos and PCR products, because of their length, are more suited to non-specific attachment chemistries such as amine and epoxy (Figure 3A and C). Again, if the surface is not homogenous, array data will be greatly compromised. For any new user, it is a good idea to experiment with the different chemistries and see what works best in your own hands and for your specific application.

Sample Preparation

Sample preparation is often the most overlooked variable in the microarray experimental life cycle. The key is to have the right concentration of high purity sample re-suspended in the right spotting buffer.

Many printing problems can be attributed to poor sample preparation. Micro assays like DNA microarrays need a high level of sample purity to work well.

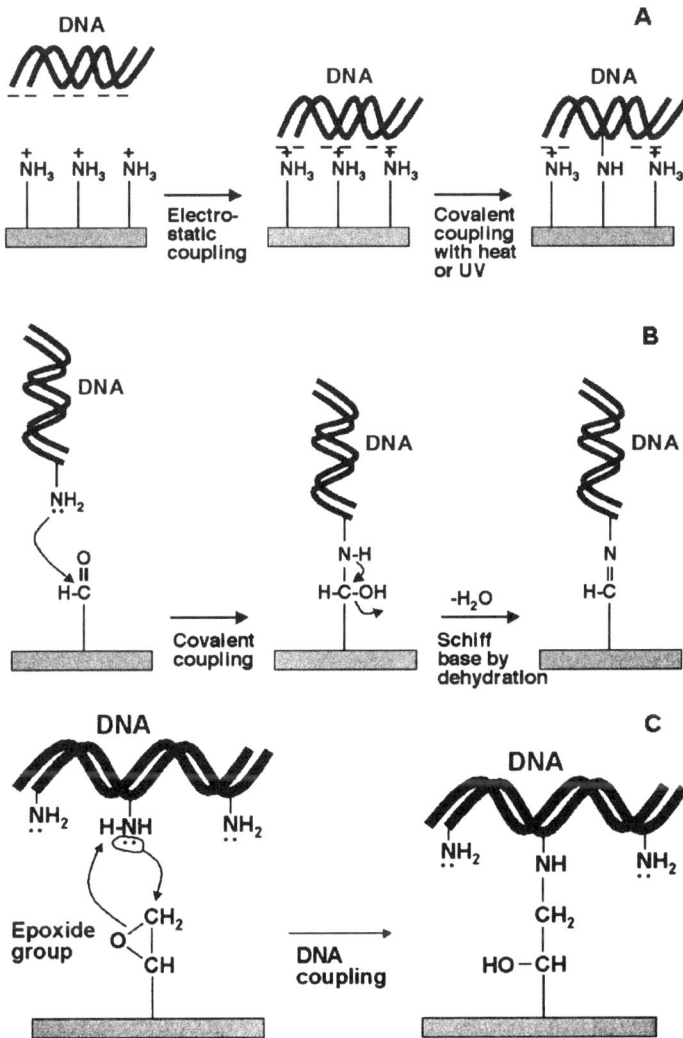

Figure 3: Common reactive surface chemistry. (A) amine, (B) aldehyde (C) epoxy.

Contaminates can inhibit the sample/surface chemistry coupling reactions, inhibit hybridization kinetics and can clog the small capillaries of pin and piezo ink jet nozzles of printing mechanisms. When solvents used to precipitate or synthesize DNA get downstream into the microarray printing mechanism, they compromise the surface tension properties of a sample, making it difficult or impossible to spot consistently. If one sample is contaminated with a solvent, and the sample next to it is not, this virtually guarantees that those spots will be different sizes. We have experienced this frustrating phenomenon first hand and it is devastating. Microarrayers are programmed to spot at specific center-to-center distances and spots that are merged together ruin data analysis.

Spotting Buffer

Once samples are prepared, they need to be added to a suitable printing buffer at the right concentration. An optimized printing buffer accomplishes many different tasks.
- Print even, small, round spots
- Disperse the sample evenly within the spot
- Promote sample binding to the array surface
- Retard evaporation within the source plates
- Dry evenly once spotted
- Wash away easily in processing
- Optimize attachment to the microarray surface
- Stabilize sample for prolonged storage
- Stabilize printed sample
- Visible after printing for easy quality control

Depending on the sample and application, all of these requirements may not be necessary, and other characteristics may be desirable. Our lab has found that using a printing buffer that is visible after printing makes checking spot quality under an inexpensive microscope or magnifier quick, easy and effective (e.g. Micro Spotting Plus). Some DNA microarray scientists prefer to use a denaturing agent in their printing buffer to avoid the boiling, denaturing step of printed cDNA prior to hybridization (e.g. DMSO). However a denaturing agent in the printing buffer would be the demise of an antibody-based microarray. The hydrophilic natures of denaturing based printing buffers require strict control of temperature and humidity, or inconsistent spot morphology will result.

Be mindful of your surface chemistry when formulating and/or choosing a printing buffer. Some chemistries, such as epoxy immobilization chemistry for proteins, do not require the sample to be dry for coupling to the surface reactive groups. Aldehyde surfaces on the other hand require a dehydration reaction to form the Schiff's base amine-aldehyde covalent bond. PCR products and oligos will not cross-link to amino-silane surfaces when wet. Additionally, spots can move if the microarrays are manipulated before spots have a chance to dry. What you use to spot is as important as the sample...it has to be right.

Microplates and Sample Preparation

Putting your sample in the right microplate is as important as having the right concentration and purity. As we'll discuss later in this chapter, efficient use of sample is important in all phases of the microarray manufacturing process. Printing technologies that pick up samples from the lowest volume are best. Plates with 384 wells are preferred for best low volume handling, but not all printing technologies are compatible with 384 well plates. Plates with round shaped wells, and conical or round bottoms are best (Figure 4). Plates with flat bottoms and/ or plates made from polystyrene (clear hard plastic) should be avoided. Samples will wick into the corners of flat bottom plates, and naked polystyrene binds DNA and protein.

PCR Products vs. Oligonucleotides

The two main templates for DNA microarrays are PCR products and oligonucleotides. For most DNA microarray surface chemistries, PCR products can be printed at a final concentration of 0.2 to 1 mg/ml and oligos at 30-60 mM. It is best to optimize these concentrations with your particular surface chemistry,

Microplates and Samples

- 384 round wells
- Rigid polypropylene construction
- V or U bottom shaped wells
- 5-15 microliters of sample per well

Polypropylene   **Polystyrene**

Figure 4: The effect of microplate material on sample handling. Polypropylene is hydrophobic and repels the samples so that it "bubbles up" in the source plate. Polystyrene material is hydrophilic, so samples tend to stick to the sides of the plate. Polypropylene plates work with much lower working volumes (3-15 µl), whereas the minimum working volume in polystyrene plates starts at or near 15 µl.

printing technology and environment.

When preparing PCR products be mindful that PCR dropout and differential amplification can affect results. Contaminated stocks and poorly tracked samples will ruin an experiment. Once these hurdles are overcome, PCR products need to be highly purified and, although ethanol precipitation can be made to work, it is not recommended. If precipitates get down stream, they can greatly affect spotting, sample attachment and hybridization results (Hedge et. al. 2000). One key advantage to using PCR products is that sequence information is not needed to prepare spot ready samples. This means that important gene function information can be found without prior knowledge of any sequence information (Schena et. al 1996).

The dropping cost of oligonucleotide synthesis and the amount of sequence information becoming more readily available is making the use of oligonucleotides (oligos) in gene expression microarrays more wide spread. The advantages of oligos are:

- Oligos designed to genes of choice – if sequence is available
- Able to QC oligos before spotting: *50mers by mass spec*
- Know sequence on microarray
- Highly purified and easy to spot
- Controlled amount spotted (Equimolar amounts)
- Single stranded templates
- An individual 100 nmole synthesis = 10,000+ microarrays

However, oligos are not void of sample preparation problems. In our own lab at TeleChem we have observed oligos contaminated with CPG. When buying oligos for microarrays, ask if the oligos are mass spec quality checked prior to shipping. Only then can you be sure that the oligos correspond to full-length, printable samples.

Environment

The key elements of environmental control are cleanliness, temperature and humidity. There are two environments to control, one inside the chamber of the microarrayer and one outside the chamber. Cleanliness in and around the microarrayer is important for obvious reasons. Clean room level environments are essential for commercial level manufacturing, but are not essential for research. Many microarrayers are equipped to filter the air and provide positive pressure, generating clean room level quality inside the arraying chamber. Temperature and humidity are important to keep samples from evaporating during a print run. Samples must stay hydrated in the source plate as well as in the printing mechanism while they are being spotting. Microarrayers equipped with plate handling equipment must come with plate lid lifters to avoid sample evaporation. For many DNA samples and surface chemistries, 50% humidity and 22° C have proven to be the best conditions for array manufacturing. Some unique microarray surfaces require long term storage in high humidity conditions to establish coupling of the sample to the substrate.

Controlling the Environment

Not every lab can afford to perform microarray manufacturing in a clean room setting. However, there are numerous things that every lab can do to improve their environment. Placing filtration on incoming air conditioning vents will help keep the room clean. Stand-alone hepa air filter units can be installed and are not too expensive for any laboratory to use. Old ceiling tiles are a common source of contamination and can be replaced or covered with materials that do not leach fibers. Environments subject to high humidity conditions can take advantage of typical dehumidifiers found at the local hardware store. In practice it is much easier to remove humidity from the room, and add it to the arraying chamber as opposed to trying the control the humidity in an entire room. Additionally, room wide environmental control typically has too much air turbulence for an arrayer to work without an enclosure.

Printing Technology

Of course the most critical component of high quality microarray production is the micro fluidics handling technology used to deliver the samples. Many of the current technologies used by industry and research meet most of the key characteristics to some degree. Some of them completely fail in critical areas, making the generation of reliable microarray data difficult. Making small spots with a high degree of uniformity and consistency is not trivial. We should appreciate the fact that 1 picoliter is to 1 liter as 1 cm is to 13 round-trips to the moon! (Source HP web site http://www.hp-go-supplies.com/english/did_you_know/thermal.htm). Accurate dispensing of nanoliter and picoliter volumes requires extremely sophisticated printing technology.

The Methodology of Printing Technologies

The methodology of printing technologies brings to light an important conceptual distinction between the tools used to print microarrays (properly "methods") and the principles that determine how such tools are used and interpreted. The most sophisticated microarray manufacturing technologies used in a research setting will adhere to the following 12 rules, but not all are necessary to have a functional printing mechanism.

1. Print uniform spots measured in microns
2. Print individual spots in regular array patterns that can be tracked by computer
3. Easy to implement
4. Cost effective/affordable
5. Print without damaging the sample or surface chemistry
6. Saturate the immobilization surface chemistry at each spot location
7. Amenable to high and low density
8. Change spot sizes and sample volumes easily
9. Load and deliver a specific amount of sample each time
10. Easy to fix and maintain, with no special tooling or tech visits required
11. Compatible with a variety of scientific applications
12. Print multiple samples, multiple times on multiple substrates with one low volume loading of sample

Rationale of the 12 Rules

Rule 1) Print Uniform Spots Measured in Microns

One of the largest benefits of microarray technology is using very small amounts of precious sample to generate tremendous amounts of valuable data. By definition, a microarray contains array elements that are microscopic (impossible to be studied by the naked eye). Any printing technology that makes spots measured in millimeters is not suitable for microarrays, since miniaturization and high throughput would be lost.

Rule 2) Print Regular Array Patterns that Can be Tracked by Computer

In mathematics, an array is a rectangular arrangement of quantities in rows and columns. Samples must be printed in order and kept in order. A spotting technology that does not put samples in ordered rows and columns is useless, since it would be impossible to track results.

Rule 3) Easy to Implement

For a microarray printing technology to be used, it must work in a traditional research setting with a minimal amount of training. Although technological development in any scientific process is interesting work, scientists using microarray technology are interested in the data that can be generated. How the data is generated is relevant to the point that it assures good data integrity. Any technology that requires atypical conditions or specialized training will not be widely used. The technologies that are easily implemented will be widely used. Widely used technologies allow labs all over the world to share methods, which will ultimately lead to standards. Microarray technology has no official governing body, therefore the most widely used methods and technologies become *de facto* standards.

Rule 4) Cost Effective/affordable

The limited budgets of research institutions require a technology to be cost effective and affordable. A technology must be durable, easy to maintain, and scalable. Most labs start out slow, and build up to high-density microarrays. Therefore, having a sample delivery technology that is scalable increases its cost effectiveness and utility. An example of this is Cartesian Technologies' synQuad (see below) system, where a user can start with one synQUAD delivery mechanism and add as many as 8 without a new investment in robotics or controller software. The Stealth Micro Spotting Device from TeleChem has the flexibility to print with 1 to 64 delivery mechanisms (see below).

Rule 5) Print without Damage to the Sample or Surface Chemistry

Sample preparation is often the most laborious and expensive part of any microarray assay. Isolating proteins, amplifying DNA by PCR, and synthesizing oligos are expensive procedures. Any technology that damages the sample during deposition should be avoided. DNA's superior stability has made DNA microarrays very robust, however fragile complex biomolecules may require special handling. Not every sample is going to survive being shot out of a piezo ink jet mechanism, stamped out by a pin, or heated by a thermal ink jet printer. Some contact printing technologies have been documented to damage glass and nylon based surfaces. The type of samples and surface chemistries being implemented in a microarray assay is an important consideration in selecting the proper microarray manufacturing technology.

Rule 6) When Spotting, Saturate the Surface Chemistry at each Location with the Proper Amount of Sample.

Most microarray surface chemistries at this time are derivatized glass slides. These microarray substrates have a maximum binding or holding capacity for the DNA and proteins that are spotted onto them. What does not stick to a saturated binding site washes away in processing. However, if insufficient biological material is deposited in the first place, data from those spot locations deficient of sample will be compromised. The printed sample on the substrate must be in excess to the labeled material reacted to the samples on the microarray. Not only must a spotting technology saturate each spot location, but also the surface chemistry must be homogenous to capture the same amount of sample at each spot location. Only then can true quantitative data be generated.

Rule 7) Amenable to High and Low Density

The inherent diversity of biology creates the need for a variety of spotted patterns on microarrays. Some genomes have large sets of genes and others have small sets. Additionally, a genome wide hybridization is not necessary to answer many important biological questions. Focused microarrays studying particular biological questions and small genome microarrays require a low number of spots. In order to get good hybridization efficiency on DNA microarrays, the spots should always be close together, but not merged. This reduces the amount of fluorescent label required for the microarray hybridization and increases the ratio of labeled material to array elements. A good microarray printing technology should allow the user to adjust the number of delivery mechanisms to print a variety of high and low density arrays in the desired array patterns.

Rule 8) Flexibility to change spot sizes and sample volumes easily

Again, the diversity of biology and scientific research leads us to appreciate flexible technologies. Additionally the personal preferences of scientists need to be met. Different applications may require larger sample volumes, higher concentrations of sample and possibly larger delivery volumes and concentrations to saturate surface chemistry binding sites, or make hard-to-see reactions in each spot detectable. A spotting technology that allows the user to easily change spot size and delivery volumes is more valuable in a typical research and development environment.

Rule 9) Load and deliver a specific amount of sample each time.

Having defined loading and sample delivery volumes allows users to fundamentally know how many times they can spot with one uptake of sample, and how many times that they can re-use a sample source plate. Without defined uptake and delivery parameters, the daily activity of microarray manufacturing is impossible to plan. Furthermore, an ordered array requires the same amount of sample to be spotted at each array location on the substrate. Different delivery volumes in spot locations can ruin data integrity.

Rule 10) Easy to fix and maintain, with no special tooling or tech visits required

All laboratory equipment requires some maintenance. Typically the more sophisticated the technology, the more difficult it is to maintain and repair. Microarray printing technologies with high-level performance, easy maintenance and repair, with minimal effort and training are best.

Rule 11) Compatible with a variety of scientific applications

Microarray technology has its roots in DNA research, but it is not limited to genomics. A variety sample types are being printed in the fields of proteomics, glycomics (ref), and even material science (ref). The disciplines taking advantage of microarray technology are endless. The best microarray printing technologies are compatible with many fields of study and capable of printing many types of samples.

Rule 12) Print multiple samples, multiple times on multiple substrates with one efficient loading of sample

The ability for microarray printing technologies to meet this rule is ultimately responsible for the paradigm shift in genetic research from a gene science to a genome science. Miniaturization is more valuable when accompanied with high throughput and *vice versa*. Both of these aspects must remain intact during all the handling steps involved in manufacturing microarrays. This includes the working volumes in the source plates where many samples are picked up at one time, in addition to the volumes of sample the printing technology uses to make a set number of microarrays. Sample preparation is tedious, expensive and primarily done in 96 and 384 well formats, which makes the handling of samples very important. Loss of sample at any step of microarray manufacturing, storage, and/or processing should be minimized as much as possible. Until sample preparation steps are miniaturized, a microarray printing technology will pick up samples from multi-well plates. Printing multiple times from one loading of sample is critical for high throughput; however, this goal should not be achieved at the expense of the amount of sample used. The best technologies make the most spots, over the highest number of substrates, from the lowest single sample volume.

COMPARING PRINTING TECHNOLOGIES

Microarray manufacturing technologies can be broken up into two basic groups, contact and non-contact. Contact printing t echnologies for manufacturing microarrays currently include, Solid Pins, Solid Pins and Rings, Split Pins, Quills and Tweezers, Capillary Tubes, and Micro Spotting Pins. Non-contact technologies include, thermal, piezo and solenoid devices. For a more complete description of non-contact technologies please read, Theriault et al., Application of inkjet printing technology for the manufacture of molecular arrays (Theriault et. al 2000). For the sake of this discussion we will stick to the three main categories, Thermal, Solenoid and Piezoelectric.

Contact or Non-Contact?

There are many considerations in choosing the type of technology to use for any given application. Budget constraints, throughput needs, and density requirements are just a few. One critical decision is choosing between contact and non-contact printing systems. Very few systems offer the flexibility to use both on the same robotics equipment, and having both is an expense most labs cannot afford. In general, non-contact delivery methods are best when the number of samples to spot is low and the number of times to spot those same samples is high. Non-contact also becomes a smarter decision if the surface being printed on is unusually fragile. Since most microarray users have many samples and need a high number of microarrays, contact printing has been the technology of choice.

Table 1. The factors that define spots size and delivery volume with solid pins.

- Diameter of the end of the tip of the pin.
- Surface tension of the liquid being printed.
- Surface tension of the tip of the pin.
- Speed of the removal of pin from sample source.
- Force of the pin tapping the printed substrate.
- Depth to which the pin is submerged in source plate.
- Dwell time of the pin on the printing substrate.
- Surface tension of the printing substrate and dwell time.
- Temperature and humidity of the printing conditions.

Assessing the Technologies According to the Rules

I believe it is important to point out that is very easy to be smart in retrospect. Comparing technologies and pointing out what is good and bad with some of the current technologies being implemented to manufacture microarrays is much easier than actually producing and delivering a product that works in the field. This information is based on my knowledge of the field and troubleshooting microarray-manufacturing problems on a daily basis over the last 5 years.

Solid Pins

Out of all the printing technologies, solid pins work by the simplest mechanism. Sets of 96 and 384 solid pins have been used to pick up samples out of the 96 and 384 well source plates. Traditionally these "replicators" have been used to manufacture macro style membrane arrays (Nizetic et al. 1991). A solid pin is dipped into the well of a source plate and whatever volume sticks to the end of the tip is touched off onto the printing surface. Spot volume and size are determined by the factors listed in Table 1. MiraiBio has implemented a somewhat sophisticated solid pin. They have borrowed technology from the semiconductor industry and use a solid pin that has tiny fingers or prongs on the end of the tip. This increases the surface tension on the end of the tip for a larger and more consistent delivery of sample to a solid support. These pins have multiple points of contact at the end of the tip, increasing the reliability of the delivery volume. Without the extra surface tension provided by the tiny fingers, solid pins can struggle to deliver a saturating amount of sample to hydrophobic surfaces like glass slides. Standard solid pins have only one point of contact, thus reduced efficiency. Solid pins can be made to work and meet many of the 12 rules. One big advantage to solid pins is since there is no capillary; they never clog and are very easy to clean. Generally this technology is considered to be too slow to be effective since only one sample delivery per sample uptake can be made. This makes the arraying process very slow. Another draw back is the volume in the source plate defines how much sample sticks to the end of the tip, thus changing the sample load and delivery volume.

Solid Pin and Ring

Affymetrix purchased an ingenious adaptation to solid pins from Genetic Micro Systems in 1999. Designed by Dr. Stanley Rose, the technology uses complex control devices to move a solid pin through an open ring set parallel to the printing substrate. The ring loads by capillary action and holds a predetermined volume of sample by surface tension. As a solid pin is moved through the ring of sample by mechanical control device, sample sticks to the end of the solid pin and is delivered to the printing surface. The amount of sample that is delivered by the solid pin-ring mechanism is defined by many of the same factors as standard solid pins (See the list in figure 1). This technology has several advantages and meets many of the 12 rules for microarray manufacturing. One key advantage is the ability of each solid pin and ring mechanism to print multiple samples, multiple times over multiple substrates with one loading of sample. While other technologies use much lower volumes to make just as many spots, the solid pin and ring technology is consistent and never misses a spot when all other key elements are in place (robotics, surface chemistry, sample preparation, environment). Unfortunately the current design of this technology is limited to 4 solid pins and rings set in a square pattern at 9mm center-to-center spacing. This makes the technology only amenable to low throughput applications. Currently designs are underway to increase the number of delivery mechanisms and provide some flexibility in terms of spot size. It will be interesting to see if this technology can be adapted to allow a user to define the number of pin-ring mechanisms they wish to use in individual print runs. This technology also many of the limitations of standard solid pins, as they have been known to struggle to deliver saturating amounts of sample to current DNA immobilization surface chemistries. Luckily this has been overcome by printing multiple times in the same spot location, but this adds time to the arraying process. Flexibility is limited since the user cannot change out the solid pin used by the pin-ring mechanism, nor can they define the number of pin-ring mechanisms they wish to use. The future of this technology is unknown since currently there are no known

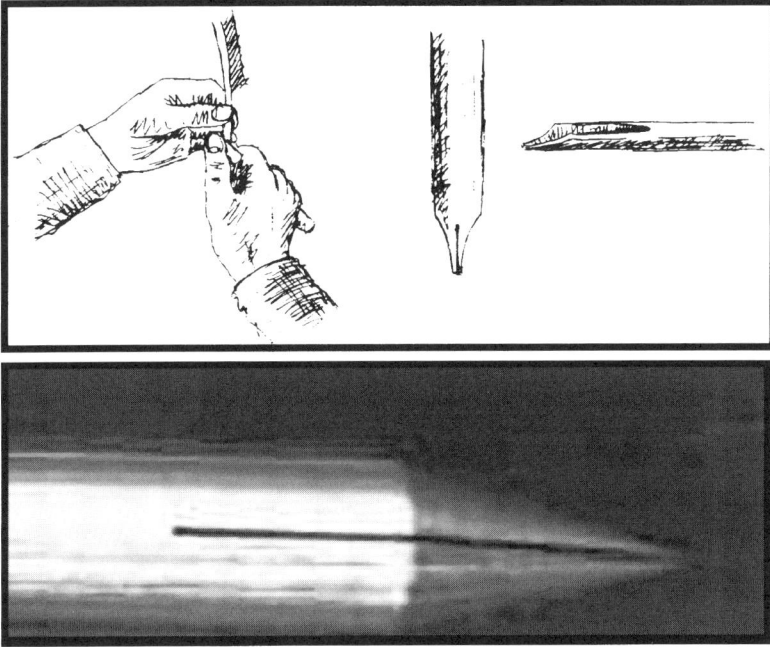

Figure 5: A quill pen being carved from a feather next to a MicroQuill Pin made by

references to the product on Affymetrix's website.

Split Pins, Tweezers and/or Quills

Historically sharpening and splitting the point or nib of the stock of a feather makes a quill pen. The Declaration of Independence and the U.S. Constitution were written and signed with a quill pen. Good or bad, the term "quill" was used to describe some of the early microarray manufacturing technologies used. There really was not a better word around to describe a fine tipped steel rod with a split down the middle, regardless of how the technology worked (Figure 5).

Tidhar Dari Shalon and Patrick O. Brown at Stanford University were the first to implement this type of technology. They describe their technology very accurately in US Patent Number 5,807,522 as "…a tweezers-like, open-capillary dispenser tip". Some of the advantages as described in the patent are:

- The open channel of the tip facilitates rapid, efficient washing and

drying before reloading with new reagent
- Passive capillary action can load the sample directly from the standard micro well plate with sufficient sample in the open capillary reservoir to print numerous arrays
- Open capillaries are less prone to clogging than closed capillaries
- Open capillaries do not require a perfectly faced bottom surface for fluid delivery

Sample is loaded into an open uptake channel by capillary action and dispensed by tapping the tip of the tweezers-like tip on a substrate with sufficient force to expel a sample onto a substrate by breaking the meniscus. Tap them harder to deliver larger volumes of sample and lighter to deliver less. Many types of different tip designs can be seen at:

http://cmgm.stanford.edu/pbrown/mguide/tips.html

This technology meets many of the 12 rules. It has some flexibility since it is possible to easily change the numbers of pins. Unfortunately changing spot size is not as straight forward since the leading manufacturer of these pins, Majer Precision, has limited options. Another drawback is that the pins must be spring loaded to be made operable by the tapping forces used to expel sample. Any pin mechanism that is spring loaded has more difficultly making up for small surface variations than a pin technology wherein each pin floats independently. The tapping force, along with the increased force of the springs, puts too much stress on the ends of the tips and quickly wears them out. The tapping force has also been known to damage some types of surface chemistries. All this being said, many users find they are able to carefully optimize the conditions for these pins and generate data.

Capillary Tubes

A capillary tube is simply a thin hollow tube. They can be made out of glass, plastic, metal, or other material. Primarily, capillary tubes used in microarray manufacture are made from stainless steel (hypodermic needles). The company with the most successful application of this technology is Amersham Pharmacia. One of the things that make capillary tubes so useful is of course "capillary action", or the proclivity of sample to wick up into the tube. Once there it can be tapped out in a similar fashion as a split pin- or tweezers-based printing technology. Capillary tubes themselves are rather inexpensive, but according to Amersham Pharmacia's website site, it is necessary to match a set in order to get good consistency. Once this is done, the performance has reportedly been very good and the capillaries can be moved at high speed for rapid

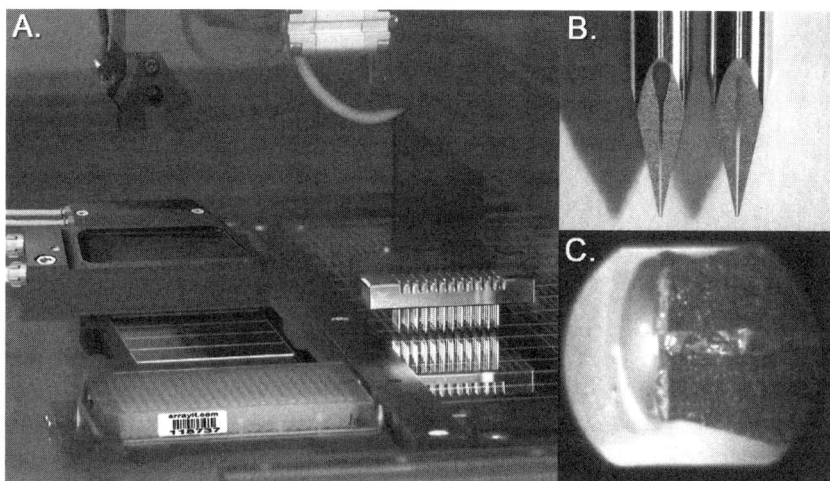

Figure 6: A. Inside a typical microarray robot, showing a source plate, wash/dry station, substrates on the platen, and Stealth Micro Spotting Device. The arrow is pointing to the tip of a Micro Spotting Pin. B. Magnified images of the tips of Micro Spotting Pins showing two different sample uptake channels with identical size tips. C. High magnification photo of a sample being pulled off the horizontally level tip by surface tension.

printing. Its current commercial implementation is somewhat limited since 12 capillary tubes are set in a row of 4.5 mm centers and cannot be changed, limiting array design possibilities. Also, the user cannot replace tubes, since each set must be matched. This means if one goes bad, they all need to be replaced. Since the tubes must be spring loaded in order to be tapped and expel sample, they have difficulty making up for surface variations. Tubes are closed capillaries and are more difficult to clean between samples than technologies that have "exterior" capillaries.

Micro Spotting Pins

A novel, patented approach for printing is currently being provided by TeleChem International, Inc. The parts that make up the Stealth Micro Spotting Device are named Micro Spotting Pins and Printheads. The key elements that make this pin technology work are two-fold: a flat (horizontally level) tip and a defined uptake channel (Figure 6). This unique flat surface at the end of the pin allows samples to load by capillary action, but unlike other capillary spotting technologies the delivery mechanism is surface tension rather than tapping. Once this type of pin is loaded, a thin layer of sample forms at the end of the tip (Figure 6C). When the thin layer of sample comes in contact with a substrate, the surface tension from the substrate pulls the thin layer of sample off the end of the tip. In fact, physical contact between the printing surface and the pin is not necessary to facilitate sample delivery though most users employ light contact to calibrate uneven printing platens. Spot size and delivery volume is controlled by the size of the end of the tip, and many tips sizes are available.

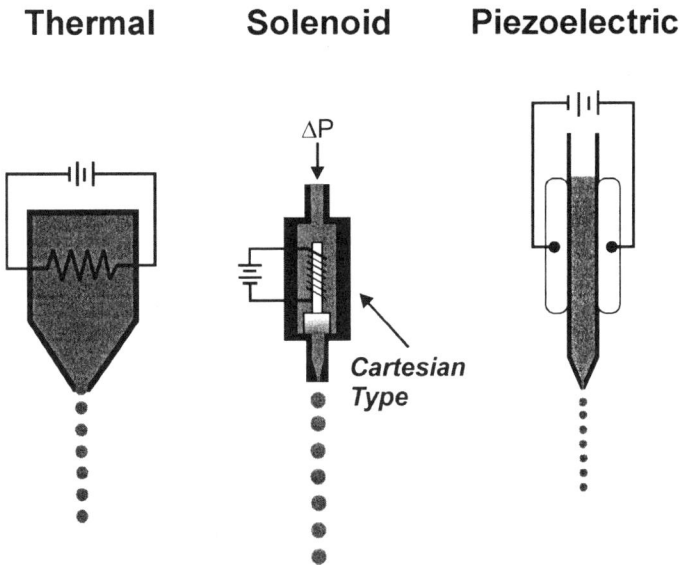

Figure 7. The 3 main types of non-contact printing technologies. This image is courtesy of Cartesian Technologies Inc.

Many design elements are included to meet user demands and adhere to the 12 rules. The printhead that holds the pins is designed to allow the removal or replacement of pins without the disassembly of any parts or special tooling. Pins can be replaced manually in less than 30 seconds, allowing them to be changed quickly during print runs. Pins are equipped with collars to keep them moving in the printhead by gravity only. This keeps the force on the ends of the tips as low as possible, drastically improving the lifetime of the tip. The collars also prevent the pins from rotating during the printing process, thereby increasing the spotting accuracy. Additionally, since there is no tension holding the pins in a rigid position (each pin floats independently), they can easily make up for minor surface variations such as an uneven slide nest. Spring-loaded technologies and non-contact technologies cannot make up for surface variations because of their fixed position. Since pins are set in the printhead at 4.5 mm centers in four rows of 8, 12 or 16 and have the ability to be removed or replaced easily, the technology has the flexibility to pick up samples out of 96 and 384 well plates. This also means that between one to sixty-four pins can be used. Two uptake volumes are available (0.25 and 0.6 ml) and a variety of delivery volumes are available (0.4 nl to 2 nl) to facilitate the efficient use of sample. Actual number of printed spots depends on the pin configuration, viscosity of the sample, hydrophobicity of the substrate and environmental conditions. It is routine for users to print between 100-2000 spots with one uptake, using virtually the entire load of sample picked up. Cleaning between samples is easy since the technology has an exterior uptake channel (capillary) that is easy to clean. Closed capillaries used in other contact and non-contact spotting technologies are much harder to clean if they become clogged. Stealth has many different pin options, all digitally quality checked, to create any spot size and delivery volume desired. A minor drawback to this, or any other technology that uses pins, is the fragile nature of the ends of the tips outside Z-axis movement.

Non-Contact

Non-contact printing technologies include piezoelectric capillary, piezoelectric cavity, thermal, acoustic and continuous flow. For the sake of this discussion we break these down into 3 categories, Thermal, Solenoid and Piezoelectric (Figure 7).

Thermal Inkjet Printers

Thermal inkjet printers use heat as part of the delivery mechanism. The ink (sample) is repeatedly heated to create bubbles, which force drops of sample from the nozzles to the printing substrate. Two years ago a team of Japanese researchers, led by Nobuko Yamamoto, adapted a Canon brand bubble-jet printing head for printing spots of DNA onto glass slides to create microarrays. However, to date there is no product available to researchers using this method. A few of the technical hurdles this group had to over come were the heat (200° C) and shearing stress (10 m s^{-1}) on the DNA sample. DNA is a very stable molecule and survived the printing process; however, it is difficult to determine how much of the DNA was damaged in the heated reservoir. We have to wonder if the concentration of damaged sample would increase over time. No short oligos were found after HPLC purifications; therefore shearing was not a major problem. However, it is likely that the heat generated in this system would disrupt protein structure to some extent. For this technology to go mainstream, there two key technical hurdles to overcome. First, how to efficiently pick up samples from micro plates, and secondly how to change samples in the thermal reservoir while avoiding cross contamination (Okamoto et. al. 2000).

Piezoelectric Printing Technology

Piezoelectricity is the generation of electricity or of electric polarity in dielectric (non conductive) crystals subjected to mechanical stress, or the generation of stress in such crystals when voltage is applied. A piezoelectric printing mechanism uses a small dielectric crystal closely apposed to a fluid reservoir. Inkjet printers from Epson are based on this technology. The advantage that Epson has over the microarray community is complete and precise control over the ink (sample) and the paper (surface) they are printing on. These elements have been highly optimized with the hardware to print photo quality documents. Microarray scientists implementing this technology must go through a similar optimization of sample preparation, surface chemistry and hardware to get optimal results. For example, samples at the wrong viscosity will produce satellite spots (small-unwanted spots around the printed spots).

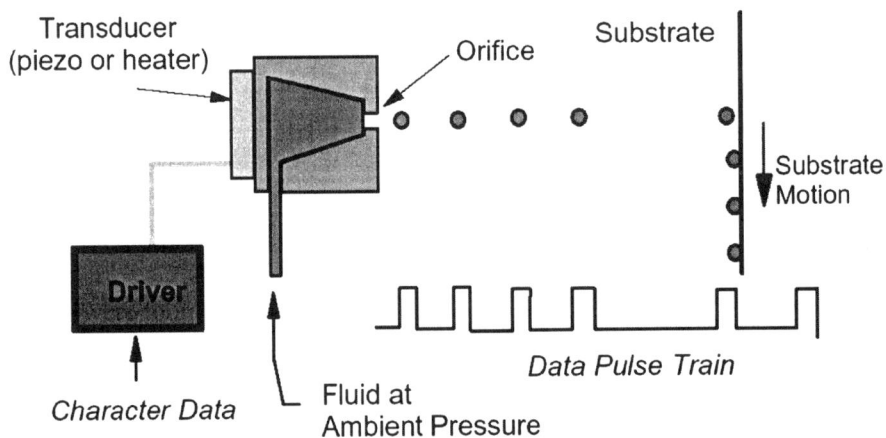

Figure 8. A schematic showing the typical configuration of a piezo or thermal

In a typical configuration for microarray printing, a crystal is in contact with a glass capillary that holds the sample fluid. The sample is drawn up into the glass capillary and voltage is applied to deform the crystal and squeeze the glass capillary to eject a small amount of fluid from the tip. Because of the ability to apply computer control and fast response time of the piezoelectric mechanism, this technology has been called "drop-on-demand". The small deflection of the crystal results in drop volumes on the order of hundreds of picoliters. To date, piezoelectric dispensing technology has been shown to work for making small numbers of gene expression microarrays (Schena et. al. 1998). The main draw-back of non-contact printing is speed. Current manifestations of this technology have a maximum of 8 delivery tips, while pin technologies commonly have 48. The efficient use of sample has been greatly improved by separating the system liquid and the sample by an air-gap or a user-defined buffer. Although cumbersome and time consuming, up to 90% of un-dispensed sample in the tip can be put back into a sample plate. (http://www.packardbioscience.com/reference_matl/ 258.asp). That means an uptake of 10 ml loses 1 ml of sample for every sample load. Not nearly as efficient as the most simplified pin technologies.

Piezo delivery mechanisms allow delivery volumes to be increased by "firing" multiple times in the same spot. Since the piezo electric crystal can be fired from hundreds to many thousands of times of times per second, increasing delivery volumes without changing tips is easy (http://www.microfab.com/papers/ mfabri97/index.htm). The current technology is perfectly suited to print a small number of samples multiple times over multiple substrates, but until 48 or more delivery mechanisms are implemented into a printhead, it will not compete with pin printing in terms of throughput (Figure 8).

A unique combination of a piezo technology and micro-fluidic cartridge is being implemented by IMTEK, Freidburg Germany. The printhead consists of 96 micro-fluidic-channels. Each channel feeds 96 nozzles by capillary action set in a 4 x 24 or 8 x 12 pattern. The nozzles are set at 500 um spacing and sample is pushed out of each nozzle by a pneumatic pump activated by a piezo actuator. All 96 samples are delivered simultaneously in nanoliter volumes, but the large spacing of a low number of samples is sub-optimal for DNA hybridization reactions. The reservoirs in the printhead hold 3-5 ml of sample for making thousands of spots from one load. This technology is suitable for low density arrays and, like all non-contact printing technologies, is best suited for printing a low number of samples (hundreds) a high number of times over many substrates.

What the company considers a distinct advantage is likely considered a disadvantage by most of the microarray community. The micro-fluidic channels on the printhead are loaded at the top of the printhead with a manual pipette or pipetting robot (minimum load volume 3 ml). Since just about every biological sample that is prepared for microarray spotting is done in plates, this makes changing samples in the printhead cumbersome. If more than 96 samples are printed, multiple 96-nozzle printheads are implemented (as many as 10). Unfortunately the configuration of this 10-printhead arrayer is 3300 mm x 1000 mm in size.... too large for most labs. This is clearly a great setup for industrial level manufacturing, but not a research lab. Nor will it be suitable for an application that requires a high degree of sample flexibility.

Syringe-Solenoid Printing Technology

Syringe-solenoid technology takes advantage of a syringe pump with a solenoid assembly that operates as a fast opening and closing valve to facilitate the accurate dispensing of liquid in the low nanoliter range. For a great description on what solenoid valves are and how they work please read
http://www.detroitcoil.com/whatis.htm.

The printing mechanism of a syringe-solenoid can be described as a tube filled with water that is connected to a micro-solenoid valve. Sample is loaded into the mechanism by withdrawing the syringe causing the sample to move upward into the tip just like a syringe needle. In order to expel a sample from the tip, the water in the tube is pressured by a syringe (there is an air bubble between the water and the sample). The opening and closing of the micro-solenoid valve lets the sample out. This technology, commercialized by Cartesian Technologies under the trade name synQuad, is capable of dispensing volumes in the range of 5-20 nl. The technology struggles to meet the delivery volumes necessary to be considered "micro", but the positive displacement nature of the dispensing mechanisms is reliable. Like all non-contact systems, they struggle to effectively meet the needs of rule 12.

Troubleshooting in the Field

A commercial microarray lab in the Midwest recently had me in as a consultant to help troubleshoot their printing problems. According to the lab, they could not get any consistent printing with their custom-built system. We started by making a list of the 5 key parameters and started to check out their system in this order

and in detail.
1. Printing Mechanism
2. Robotics
3. Sample Preparation
4. Surface chemistry
5. Environment

The first thing we did was check their Stealth Micro Spotting Device (TeleChem / arrayit.com Sunnyvale, CA). We properly cleaned their Micro Spotting Pins and examined them for mechanical damage. The Pins were slightly damaged from improper Z-axis speeds, but only some very slight blunting was observed on the very tips of the pins (microscope examination). The cleaning procedures of the Pins and Printhead prior to loading them onto the arrayer was checked and found to be good. The determination was that the printing mechanism was not the source of the problem.

Next we checked the motion control system. There were several problems here. It moved very accurately, however, the Z-axis speed was set too fast. The force of the pins touching the glass was much too strong. We already expected this due to the physical damage on the ends of the Pins, however, watching the machine run showed that we were not getting the soft contact necessary for good surface tension delivery. The motion was quick and abrupt and needed to be slowed way down. We adjusted the parameters near the following settings.

- Z axis acceleration: 30 cm/sec2
- Z axis velocity: 10-20 mm/sec

Programming these parameters into the robotics does not exactly translate into these exact speeds and every robot handles this differently. Setting the right speed is empirical and visually seeing the robot move and touch the surface softly is actually better than knowing the numbers to program into the machine.

Once we had the right motion control parameters set, we checked the wash/ dry station setup. The wash/dry station hardware installed on this custom arrayer was kindly provided by GeneMachines (San Carlos, CA) and follows the traditional setup described earlier in this chapter (water bath & vacuum dry). The users were running 100% ethanol through the bath instead of distilled water. 100% ethanol is not a good selection for a wash buffer, since it precipitates DNA and the chemicals used to drive off the water in 100% ethanol manufacturing change the surface tension properties of Micro Spotting Pins. We immediately changed the wash buffer for the bath to distilled water. After that we needed to calibrate the wash/dry station. GeneMachines has an ingenious design to their dry station, which makes it very easy to calibrate. They have designed a hole for each pin so that the Pin can be inserted and touched to the bottom of the dry station. Once there, it is then slightly backed off to put the Pin in the optimal drying position. Their design is so effective; a Pin can be dried in about ½ second. Since we do not want to dry a Pin until the last wash cycle is complete, we calibrated the arrayer to move the Pins to a non-optimal drying position for the first 3 dry cycles of the 4-cycle wash/dry setup we wanted to implement. Remember that it is important to make sure all residual sample is washed from a printing mechanism prior to completely drying it. We tested many different heights by quickly removing a Pin from the Printhead and checking to see that it was only partially dry. We tried about 6 different positions before we found the right setting. Once this was done, we felt confident that the arrayer was ready for a test print run.

~0.25-mg/ml of purified genomic DNA was re-suspended in 3XSSC and printed onto SuperAmine Substrates (TeleChem / ArrayIt.com Sunnyvale, CA). These were good choices, since we could use them with a high degree of confidence that they would not cause any problems (clean sample in a suitable buffer on a homogenous substrate). 3XSSC is not considered an optimal printing buffer (*see sample preparation section of this chapter*). However this simple formulation is suitable for checking to see if arrayer is working properly since the printed spots can be easily checked for quality under an inexpensive stereomicroscope.

The user has a terrific environment to print in, a class 1000 cleanroom set a 22 deg. C and 50% humidity. No adjustments to the environment were needed. By this time we were 3 hours into the work of optimization and we were keeping our fingers crossed that we would get a good print run. It worked perfectly the first time, reinforcing the fact microarray manufacturing really is a science. The user now knows that they have a functioning microarrayer and will no longer struggle to print. The challenges left to them are preparing their samples to spot, processing/hybridization conditions and analysis.

CONCLUSIONS

Printing high quality microarrays is not difficult, if you follow the rules. The most simple rule states "If there is a variable in your system, control it." The 5 places to find problems have been discussed at length in this chapter and they are, **printing technology, robot, sample preparation, surface chemistry, and environment**. Microarray manufacturing technologies can be be broken up into two main groups, contact and non-contact. Understanding the principles that determine how microarray printing technologies are deployed and interpreted leads to a better understanding of the practical considerations to accomplish high quality microarray manufacturing. The fact is, detailed knowledge of *what* you need to accomplish and *why* will always lead to the best *how*. It is my hope that this chapter serves scientists to avoid many common problems assoicated with generating high quality microarray data.

ACKNOWLEDGMENTS

Special thanks go to Mark Schena for teaching me most of what I know about DNA microarray technology and to Robin Stears for dealing with my daily technical questions on microarray technology. Also, to Dr. E.F. Codd (Inventor of the Relational Model) and Dr. Clynch Salley (Inventor of OLAP) for teaching me the value of methodological critical thinking.

REFERENCE LIST

1.   Chen, B., Parker, G. II, Han, J., Meyyappan, M., Cassell, A.M. Heterogeneous Single-Walled Carbon Nanotube Catalyst Discovery and Optimization. Chem. Mater. 14, 1891-1896, 2002.

2.   Hegde, P., Qi, R., Abernathy, K., Gay, C., Dharap, S., Gaspard, R., Hughes, J.E., Snesrud, E., Lee, N., Quackenbush, J. A concise guide to cDNA microarray analysis. Biotechniques. 29:548-550, 2000.

3.   Nizetic, D., Zehetner, G., Monaco, A.P., Gellen, L., Young, B.D., Lehrach, H. Construction, arraying, and high-density screening of large insert libraries of human chromosomes X and 21: their potential use as reference libraries. Proc. Natl. Acad. Sci. U.S.A. 88:3233-3237, 1991.

4.   Okamoto, T., Suzuki, T., Yamamoto, N. Microarray fabrication with covalent attachment of DNA using bubble jet technology. Nat Biotechnol 18:438-441, 2000.

5.   Rose, S.D. Spotted Arrays: Technology Overview. Microarrays and Cancer Research, J. Warrington, R. Todd & D. Wong (editors) Eaton Publishing, Westborough, MA 2002, pp. 3-14.

6.   Rose D. Microfluidic Technologies and Instrumentation for Printing DNA Microarrays. In Microarray Biochip Technology, M Schena (editor) BioTechniques Books, Westborough, MA 2000, pp 19-38

7.   Schena, M., R.A. Heller, T.P. Theriault, K. Konrad, E. Lachenmeier, And R.W. Davis 1998. Microarrays; biotechnology's discovery platform for functional genomics. Trends Biotechnol. 16: 301-306.

8.   Schena, M., Shalon, D., Heller, R., Chai, A., Brown, P.O., Davis, R.W. Parallel human genome analysis: microarray-based expression monitoring of 1000 genes. Proc Natl Acad Sci U S A 93:10614-10619, 1996.

9.   Theriault, T.P., Winder, S.C., Gamble, R.C. Application of Ink-Jet Printing Technology to the Manufacture of Molecular Arrays. In DNA Microarrays: A Practical Approach, M. Schena (editor), 2nd Edition, Oxford University Press, Oxford, UK, pp. 101-120, 2000.

10. Wang, D., Liu, S., Trummer, B.J., Deng, C., Wang, A Carbohydrate microarrays for the recognition of cross-reactive molecular markers of microbes and host cells. Nat Biotechnol 20:275-281, 2002.

Chapter 4

ARRAYS FOR THE MASSES – SETTING UP A MICROARRAY CORE FACILITY

Robert P. Searles, Ph.D.

Manager – HEDCO / Oregon Cancer Institute Spotted Microarray Core,* Vaccine and Gene Therapy Institute, Oregon Health and Science University, Beaverton, Oregon

PREFACE

Early in the fall of 2001, the Oregon Health and Science University (OHSU) Spotted Microarray Core (SMC) was about to open for business when we noticed something odd about our cDNA arrays. We were printing arrays with 5700 genes, which took about 15 hours to complete. Probe DNA printed early in the print run stained brightly with the dye Syto61 and hybridized well to targets derived from human heart RNA. DNA printed after 7 hours was progressively fainter by both assays and any DNA printed after about 12 hours was completely absent from the slide. We ran numerous tests and came to the conclusion that the slides themselves were the problem. After a month of working with the slide manufacturer to solve the problem, which was supposedly unique, I posted a message to the microarray listserv about the problem. No one had any solutions, but I was inundated with e-mail from other cores having the same problem and not knowing what to do. Eventually, through a series of tests in our core, we learned two things: 1) most amino-silane slides had the same problem, but not all facilities experienced it, and 2) a new generation of slides coming onto the market no longer exhibited this problem.

Step forward to June, 2002. Work on this chapter flagged significantly as the SMC became overwhelmed with samples and as I began to wonder if I really

* *The HEDCO / Oregon Cancer Institute Gene Microarray Shared Resource at Oregon Health and Science University was established through a generous donation from the HEDCO Foundation.*

had anything to pass along to new cores (some may come to the conclusion that I actually didn't.) Nine months after our slide problems were solved, a core operator from another university visited our affiliated Affymetrix core. She noticed our operation and commented that their spotted core might close because of all the problems they had – most notably their slides seemed to die over time and they couldn't figure out what to do about it. Obviously there were problems we had overcome that were still causing serious damage to other cores. Maybe our core experiences would actually be useful to others.

Because setting up a core is less documented in the literature and more practical in nature than data analysis, this chapter will be different than the other chapters in this book. The number of cores starting at any given time is small compared to the number of people actually using arrays, so the target audience is not very big. However, given that each core should be providing arrays to dozens of research laboratories and possibly hundreds of individual researchers, the impact of problems that continue to plague microarray facilities is very large. Because there aren't a lot of papers available on putting together a core, the reference list for this chapter is small. Most of the information is anecdotal. This is more of a tips-of-the-trade chapter, designed to help in the planning and set-up of a core facility. I address some of the common problems of core operation; even some problems that many aren't aware are common. What you'll read here is practical and all of the material works. We do it every day. Most of our success comes from a lot of hard work and a very dedicated staff. Dedicated staffs will always be needed, but I'm convinced that the work doesn't have to be that hard.

INTRODUCTION

The development of microarray technology has revolutionized the field of gene expression analysis. A growing number of papers use microarrays as part of their experimental protocol and a substantial literature exists regarding the proper methods for microarray data preprocessing and analysis. Sometimes lost in the technical discussions and the data presentations is the simple fact that someone, somewhere, has to print the arrays, scan them, and analyze the resultant image. For spotted cDNA microarray users, this means acquisition of arrays from commercial vendors or the development of in-house printing facilities. To date, the commercial market has not been kind to preprinted cDNA microarrays. This last year has seen a withdrawal from the market of several spotted cDNA or oligonucleotide arrays (Operon [geneomeweb.com, 6/1/2001], Corning [genomeweb.com, 10/18/2001], and InCyte [genomeweb.com, 10/25,2001], though there are currently rumors that Corning may re-enter the market [genomeweb.com, 4/26/2002]). At the same time, Motorola is promoting its arrays, though it has been reported that Motorola is also shopping around for a buyer for this business [genomeweb.com, 5/14/2002]. One article reported that Incyte technology had been licensed to Agilent [genomeweb.com, 11/5/2001]; this was followed two weeks later by a report on poor earnings at Agilent, though the earnings for the bioscience division were strong [genomeweb.com, 11/16/2001]. Although Incye had closed down its commercial microarray business in November, it had retained the operations until July 2002, when it was reported that the unit had been sold to Quark Biotech [genomeweb.com, July 8, 2002]. With this type of instability in the commercial market, and given the need for both reliable large-scale arrays and affordable focused arrays, the development of local spotted microarray facilities will be a recurring issue over the next few years.

In addition to the changes in the commercial preprinted microarray market, the market for hardware has been in, if you will, disarray. The current recession (this is being written in Spring, 2002) affected all industries, including biotechnology. Genomic Solutions acquired Cartesian Technologies, with some changes in personnel, but no noticeable change in operation [genomeweb.com, 12/20/2001]. Subsequently it was reported on June 14, 2002, that Genomic Solutions was at risk of delisting from Nasdaq as its share price remained below the $1.00 minimum for over 30 days [genomeweb.com, 6/14/2002]. The story continued into July, 2002, when Harvard Bioscience acquired Genomic Solutions (genomeweb.com, 7/18/02). GeneMachines laid off 25 percent of its workforce, primarily in response to low demand for major equipment [genomeweb.com, 6/17/2002]. A quick examination of the Affymetrix website finds no reference to its former pin-and-ring spotting technology, acquired from Genetic Microsystems. GenPak has changed its name to Genetix and appears to have discontinued the nicely built, but poorly accepted, Array21 microarray printer, although it continues to develop and sell its Qbot line of printers.

On the flip side, PerkinElmer (PE) continues to become a full-service powerhouse in cDNA microarray technology, having acquired Packard and its line of ScanArray microarray scanners (the ScanArray series began life as a part of the life sciences unit of GSI Lumonics). PE Life Sciences also markets the Tyramine Signal Amplification (TSA) kit for Cy5/Cy3 visualization, has a rapidly growing bioinformatics business, and a distribution agreement for Genomic Solutions microarray printers as well as its own Packard printers. Hitachi has entered the market with its unitized MiraiBio printing system. Enclosed systems of this sort, comparable to the Genomic Solutions line of printers (and others), provide a nice, relatively compact solution to preparing a printing facility.

Since the market is changing rapidly, this chapter is not intended to be a comprehensive discussion of microarray equipment. All of the above will be old news at the time this chapter makes publication, but it demonstrates the one outstanding fact about the cDNA spotted array field – it is, and will be for some time, a dynamic field. However, with all the change in hardware over time, the same problems seem to recur in each core and core personnel can anticipate spending substantial time troubleshooting before they begin accepting business. If this chapter can help groups past some of these perennial issues and reduces the average time to opening, then it will have served a purpose.

HEDCO/OREGON CANCER INSTITUTE SPOTTED MICROARRAY CORE AT OHSU

Spotted cDNA and oligonucleotide core development is a long, difficult, and expensive process. Anecdotal evidence from meetings and from messages on the UCSF Microarray listserver indicate that many cores have been forced to reduce services simply because the obstacles to operation are so high. To assist those just planning a new core, for those in the process of setting one up, and those just looking to compare their core to others, this chapter discusses many of the issues that were addressed and overcome as we prepared the HEDCO / Oregon Cancer Institute Spotted Microarray Core (SMC). The SMC is closely affiliated with the Oregon Cancer Institute, but it also provides spotted cDNA and oligonucleotide arrays to the entire OHSU campus. As a benchmark, our average output is normalized data from sixty to eighty slides per week. We are a full-service core, which means probe amplification, printing, target labeling, hybridization, post-hybridization processing (we use the PerkinElmer TSA kit for increased sensitivity), scanning, gridding, and preprocessing of the data. We can accommodate as many as 160 slides per week if the need arises, but our small staff size (two to three staffers who do all the real work and myself) makes that level difficult to sustain.

The SMC is part of a larger unit, the OHSU Gene Microarray Shared Resource, which includes, in addition to the SMC, the Affymetrix Microarray Core (AMC), and the Biostatistics & Bioinformatics Core (BCC), each with its own manager and staff. Both the SMC and AMC strive to be full service – RNA comes in one door and data goes out the other. Once the investigator has the data, statistical and bioinformatics support is available through the BBC. Not all cores will use this model. We considered a number of alternatives – including outsourcing target labeling to the individual laboratories and even having them do the hybridization, leaving us to do array printing and scanning. We decided that full service provided an additional, and quite valuable, level of quality control that might be lacking if each laboratory was responsible for labeling target and hybridizing the resultant cDNA to the arrays.

Because I work with our selection of equipment day-to-day, I'm biased in my understanding of some of the issues involved in making and using arrays. This will be the case for most people involved in using microarrays. Printers and scanners are expensive and it will be rare for one individual to become an expert at more than one or two models of each. For this reason, the discussion will be slanted toward our experiences with the equipment we use. I've spoken to enough other core managers to realize, though, that these problems are more general than might appear, so cores using other brands may find useful information here as well. Again as a benchmark, the following describes our core equipment, for better or worse. We have two Cartesian PixSys 5500XL array printers outfitted with TeleChem CMP-2 and CMP-3 pins, two Packard ScanArray 4000XL microarray scanners, a Qiagen BioRobot 8000 liquid handling system, and two MJ Research Tetrad thermal cyclers. As described above, we generated target cDNA using the PerkinElmer TSA kit and use the same kit for visualizing the bound material. We use BioDiscovery software throughout the arraying process – CloneTracker CE for designing our arrays and generating the scripts to operate the printers, ImaGene for gridding and quantification of the arrays, and GeneSight for data preprocessing, though many of our clients also use GeneSpring (Silicon Genetics) and OmniViz (OmniViz). The OHSU BBC uses its own selection of statistical software for in-depth analysis of microarray data.

The irony of establishing a microarray core is that, because the technique is so new and so expensive to import, few people are experienced at making microarrays. This means that a lot of equipment and supplies are evaluated and purchased based on the vendor's "propaganda" or on recommendations from other users, not on personal experience. Faculty advisors to the cores are usually end users with extensive knowledge of array analysis, but with little experience in the production of arrays. This may help with some of the issues a core faces, but making good arrays and generating good data often revolve around small issues, some of which will be discussed in this chapter. It is this information that makes or breaks a core, and the cadre of people with knowledge about these issues is small. As the field matures, the lack of general expertise will be less of a problem, though I suspect that experts at array production will always be uncommon. For the present, the best thing someone running a core or working in a core can do is tap into the collective pool of knowledge that exists among the various existing facilities and companies. The easy way to accomplish this is to subscribe to the Gene Arrays Listserv based at UCSF. One website describing it and describing how to get on the mailing list is http://ep.ebi.ac.uk/Links/Gene-Arrays.faq.html. Many of the authors in this book are frequent contributors and the information that can be gleaned from this source is generally quite valuable.

CORE SET-UP

Let us assume that your institution has decided to develop a cDNA and oligonucleotide spotted microarray core – one which will serve anywhere from 10 to 200 researchers (i.e., let's consider a core similar to the SMC). Funding will need to be secured, since it will cost anywhere from $500,000 to $750,000 to put together a good core. Having invested this much money in the core, the users will have a serious interest in getting it operational as soon as possible. Although one researcher objected to the SMC with the rationale that "all it takes to make microarrays is to build a printer, do some PCR, and buy some slides," spotted microarray operations, particularly when they use PCR products as probes, are notoriously complicated and difficult to make function smoothly. Such complexity makes it easy for the core to become delayed or, in the worse case, closed.

We all think we can create the most efficient facility around. The temptation to try this is powerful, but reinventing the wheel is not the best way to get a campus core up and running. The most efficient means of getting a new core operational is to model the set-up of an existing, well-functioning array core. Core operation is complex and if you start from scratch, you will be troubleshooting PCR protocols, getting the printers on line, getting the scanners on line, testing and selecting labeling protocols, training staff in RNA handling protocols, developing work-flow models, and, ultimately, even rebuilding the printers to some extent. All this will be in the context of a set of operational goals (e.g., 400 slides per week, 0.5 ug of total RNA per slide, less than 5% failure rate) that can prejudice the staff in its selection of equipment and protocols. It's a daunting task that requires good bench skills, good computer skills, and good mechanical skills. For much of the set-up, it helps to be familiar with equipment and to be handy with a screwdriver. However, save yourself the trouble of starting from scratch; find a core that that is successfully fulfilling a set of demands that is similar to the ones set forth for your own core, and then do one of two things.

First, ask the manager of the core to share all their secrets with you. If they agree, simply create a mirror of their core. This is not the time to be creative. Buy what they buy, implement as they implement. Don't purchase a Cartesian 5500XL if your partners have a Genetix Qarray. The Cartesian machine may work fine for you, but you won't get any advice from the other lab. If they have scripts for the printers or homebuilt software for data preprocessing, acquire it. Purchase arrays from the facility, if possible, to use in troubleshooting labeling protocols, and to compare to arrays made in-house – relying on your own arrays at this point is foolish. If things aren't working quite right as you get each process on line and if progress slows down, buy your mentor a round-trip ticket to your town, put them in the nicest motel in the area, and pay them a respectable consulting fee to spend a week watching your operation and telling you what to change. Once you get a few hundred arrays processed and people have stopped asking for proof that you can do what you say, you can start testing some of your own ideas.

Second, find a core that is doing what you need to do and join it for three months. Agree to come in and work at the bench, processing samples, scanning, gridding, whatever constitutes the normal workday. Bring money. Not just for yourself, but funds to defray the costs of training someone new. This might be as much as $10,000. It will be well worth it in the end. Note, though, that some cores will be reluctant to bring in a temporary worker that they haven't interviewed and selected and you need to be acceptant of this. Once a core's work flows smoothly, there is ample reason to resist anything that might interfere with it. Ideally, you would combine the two approaches. Since it will take almost three months to get all of the equipment, there is plenty of time to intern in a good core, then to maintain contact with it and use its manager as a consultant.

If the psychological need to create a core of one's own, as opposed to a mirror of someone else's, is a driving force, then at least make certain that a successful core operator has agreed to serve as an official advisor to the core. Formalize the relationship. It doesn't need to include money, but it should include a letter from the advisor stating that they agree to help. When you do this, make certain the advisor is involved in the day-to-day operations of their own facility. If they aren't, use their senior assistant as your consultant. Experience and understanding mean one thing to the manager who controls only the financing and scheduling and something completely different to the manager who has rebuilt the printers and designed most, if not all, of the procedures used in the core.

ARRAY PRINTING

Printing quality arrays is the single most difficult aspect of the core operation. The variables are many and often the problems are cryptic, requiring substantial time to understand and overcome. Once these problems are overcome, printing will become routine and the core is up and running. An excellent paper that is still relevant to many of the issues involved in array printing is that of Quackenbush and colleagues [Hegde, et al., 2000]. This paper is still quite current and it is an invaluable aid for core operations.

Basically, array printing involves four processes. The first issue is selecting and setting up the printer. The second item is selecting the type of slides to be used. The third issue is either purchasing or amplifying probes for printing on the slides. The fourth issue is putting all these pieces together and printing a quality array. The order of discussing these steps is arbitrary.

THE PRINTER

It is possible to build good microarray printers. Although not the route taken by the SMC, this method does ensure that the operators know the equipment very well and can maintain it without having to call in a service staff. There are classes in array printer construction, notably the summer course run at Cold Springs Harbor (www.cshl.org).

Good commercial array printers with core capacity will range from the mid-$70,000 to well over $100,000. Expense does not necessarily relate to the functionality of the machine. Consider also the value of purchasing two smaller capacity printers, which may cost only a bit more than one of the more expensive models. If you buy only one and it breaks (and it will break), then work stops until the printer is repaired. Buy two, and the chances that both will break at once is small (but not zero, as we discovered!). For a university core, this can be a significant issue, particularly if repair times are on the order to two to three weeks. If the only printer is down for two weeks, then all microarray work comes to a standstill; with two printers, throughput may be less, but not zero. The downside of this is that smaller printers are less efficient in their use of spotting material; the reasons for this are explained below. While examining printers, make certain you know what technology is being used in the printing steps. Most array printers in use are contact printers. Newer technology, such as non-contact printing, is worth exploring and if it fits your budget, then give it serious consideration. However, remember that first adopters of any technology pay the price with a steep learning curve and with no one ahead of them to find all the pitfalls.

Microarray printers have recently been reviewed (Fitzgerald, 2002), and readers are referred to this article (as well as Chapter 3 of this book) for information on available equipment and manufacturers. Key features that need to be considered when buying a printer are: slide capacity, microtiter plate capacity, storage options for plates and slides, printing technology (quill pin vs. solid pin vs. non-contact), printing speed, ease of programming, and servicing issues. One might also include the long-term prospects for the company, but, as demonstrated by the GSI Lumonics ScanArray scanners, a good product will move from vendor to vendor (GSI Lumonics to Packard to PerkinElmer Life Sciences), but will likely be around in some form or another for quite some time.

Purchasing a machine means you will usually have no choice regarding print head or pins. Manufacturers will preferentially use one type or another. Two primary technologies involved in spotting array pins are the solid pin, such as that used by MiriaBio, and the quill pin, such as those made and sold by TeleChem (www.arrayit.com), among others. Pin and ring technology, used in the Affymetrix cDNA printers, had its adherents, but appears to have disappeared from the marketplace. The quill pins are very common and, under normal operation, they are almost trouble-free. Quill pins do have the disadvantage of having a tip that is very fragile to any lateral impact. By contrast, they are amazingly resistant to impact along the Z axis. Our experience is that quill pins are most at risk during staff training and machine reprogramming, since it only takes an unguarded second to ruin an entire set of pins. Cartesian provides break-away pins for setting up protocols. These are good, but tend to be uneven in length, so they don't truly mimic the real pins. I prefer to test protocols with slightly damaged pins. It's no loss if they get further damaged and they provide a better representation of the way the real pins will operate.

Check each operation of the printer prior to starting production runs. Don't let the set-up crew escape before the machine in properly tested. In particular, if the printer uses quill pins and a wash station with a vacuum manifold for drying them, then it is absolutely vital to verify that the pins are dry at the end of the wash cycle. Most printers will wash and "dry" the pins multiple times before the final drying step. The initial "drying" steps don't need to dry the pin – indeed, they should primarily remove exterior solution and leave some water in the gap at the tip of the pin; this allows for more efficient washing by avoiding air bubbles that may block fluid entry. The final vacuum step should leave no trace of fluid in the quill. While the installers are on-site, set up a mock run and stop it immediately after the pins emerge from the drying station after their final drying step. Pull several pins and examine them under a microscope; the difference between a dry tip and one with water in it is obvious. If the split at the tip still has water, the length of the drying period can be extended until the pin is dry, but this is only adequate to a point. Ultimately, it may be impossible to find a reasonable period for drying the tips. This is a flaw in the machine, often a serious flaw in the basic design, and I have encountered two models (not individual machines, but entire printer lines), one each from two different vendors, that suffer from this problem. No machine should be accepted until this problem is eliminated. Failure to do this can result in poor quality arrays, cross-contamination of probes, and expensive or time-consuming work solving the problem.

The solution to this problem may not be simple. Both of our Cartesian printers suffered from poor drying of the pins. We managed to improve conditions by removing the small-bore mufflers from the vacuum pumps and replacing them with long runs of tygon tubing. Small-bore mufflers cause substantial backpressure that dramatically reduces pump efficiency at low vacuum levels (i.e., under standard operating conditions). The tygon tubing reduced the pump noise and significantly reduced backpressure.

Modifying the pumps improved the problem, but we still had problems getting the pins dried consistently. About eight months after we had the machines installed, Cartesian sold us a new product – a "riser" (Figure 1). This was a replacement

Figure 1. **Vacuum manifold riser for the Cartesian PixSys 5500XL.** The photograph shows a Cartesian riser in place on the vacuum manifold of one of the PixSys 5500XLs in the OHSU SMC. The TeleChem print head (sans pins) is to the right. The four pin slots on the right of the riser are covered with tape, as is the entire platform. The tape around the platform surface is necessary because the Cartesian manifold has no gasket to ensure that the edges are sealed. With the proper scripting, the print head steps the pins through the riser, cleaning each row of four pins (of a 4 x 4 matrix).

vacuum manifold with eight holes on a thin upright strip. Basically, the print head steps the four rows of pins through the riser, drying eight pins at a time. Having had so many other ideas not succeed, I wasn't particularly surprised that the riser didn't work as initially configured. Taping off four of the holes and stepping four pins through the riser at a time did work very well and solved the problem for us. Because we use a 4 x 4 pin matrix to print our arrays, the four-hole limit had no effect on the final layout of the arrays. I've spoken to another Cartesian user who added a second vacuum pump (there is a second vacuum outlet on the vacuum chamber). Either method works well, but unlike adding the second pump, the riser does increase the length of a print run. Fixing problems will usually create more; our software couldn't be programmed to recognize the presence of the riser. To operate the printers, we need to manually alter all printing scripts to step the pins through the riser. It's a good idea to test any such modifications with the pins removed to make certain that the pin always clears the riser before any x- or y-axis motion occurs. Testing printing protocols sans pins could be considered a general rule with any array printer when modifying a control script.

Before assuming that this is an inherent defect only in the Cartesian printers, consider that I've spoken with other core managers with other brands having the same problem. It is apparently surprisingly hard to get enough air through the quill tip to effectively dry it. At least the Cartesian machines can be corrected.

Closely associated with the problem of pins not drying is the problem of TeleChem pins sticking in the print head. My staff informed me that they were trained by the installers to simply tap these pins down when they stick. This is both impractical – it's impossible to sit next to the printer all night watching for stuck pins – and dangerous. Not only can it cause DNA to spray out of the tip of the pin, it can knock the head out of position and distort the array from that point on. The solution to pin sticking is very simple; keep the exterior of the pins dry. Moisture left on the pin shaft from sonicating prior to use will cause the pin to stick. This can be avoided by air drying the pin with a high quality blow dryer or, if the amount of water is very small, by repeatedly lifting the pin in the print head with a pair of forceps and releasing it until it drops cleanly. Water splashing from the wash chamber is the other cause of pin sticking. The flow into the chamber should be absolutely smooth and the level of the water should be above the top of the inlet port. If there is any turbulence or splashing in the wash water, it can spray up against the junction of the pin and the print head. When the pin is lifted by contact with the plate or a slide, this water is pulled into the slot with the pin and causes the pin to stick.

SLIDES

Some laboratories prefer to make their own poly-l-lysine coated slides. Our experience is that this is a cumbersome task with unreliable results, so it will be something each group has to try before going this route. Commercial slides have the advantage of consistency and, generally, availability. The most common types of slides used for printing are amino-silane coated slides, available from a number of vendors. A non-exhaustive list includes Corning, TeleChem, TurnerDesign, Erie Scientific, and a raft of others. Aldehyde slides are available for use with modified oligonucleotides. A third technology is epoxy-slides. These seem to retain more DNA, but the protocol for fixing the slides after printing was too involved to interest us. Slide coating chemistries are covered more fully in Chapter 1.

During printing a salt-bridge will form between the DNA and the amino groups that will retain the DNA in place until it is fixed. However, as described in the preface, under certain, apparently common, conditions, the typical amino-silane slide may be reactive for only 8 to 12 hours, depending on the manufacturer and the environment within the printer. Slides from most manufacturers suffered from this problem to some degree in our facility. The conditions are not well defined, but do not relate to humidity or exposure to light. One manufacturer has since suggested that the problem is oxidation and that the method of curing the coated slides prior to packaging may be the determining factor. If it is a problem in a given core, the symptoms are quite obvious by any method of general staining. Spots printed earlier form salt-bridges and protect the animo-silane from decline; DNA in these spots are retained after the baking / crosslinking step; spots printed later are unable to form salt bridges and are washed away. Some newer slides have overcome this problem. We particularly like the Corning UltraGAPS, though TurnerDesign also markets a slide that has long-term stability. Others will probably be available by the time this article is published. These problems are not universal. Following a listserv post regarding it, I received as much e-mail from groups saying they had no problem as I did from groups saying it was a serious issue for them.

The literature often suggests UV crosslinking to fix the DNA to the slides. It is incorrectly assumed that the crosslinking is between the DNA and the slide. According to the scientists at Corning, the crosslinking is between DNA strands. Baking results in a similar immobilized clump of DNA that is not covalently linked to the array. Consider this when using any technique that requires high-temperature liquid applied to the array - since it can strip the DNA free from the slide. Our experience is that UV crosslinking is very much dependent on the brand and age of the crosslinker. We've tried several and found that the technique is unreliable, particularly since age-related decline in crosslinking efficiency is difficult to monitor. Instead of using the crosslinker, we purchased a large oven and use it to bake the slides after printing. It's low tech and slow, but it has the advantage of being easily monitored for proper operation. Baking is mentioned in a number of protocols. The TeleChem protocol calls for 1.0 hour of baking, the Corning protocol for their UltraGAPS calls for 3.5 to 4.0 hours and does indicate that UV crosslinking is less efficient. We've found that all slides we've tried work best with 3.5 to 4.0 hours of baking at 80°C. One hour was not sufficient. One run was forgotten for eight hours, but hybridized no differently than arrays baked for 4 hours. We've successfully baked both PCR products and unmodified 70-mer oligonucleotides to the UltraGAPS, so this protocol can be used for any spotted DNA.

AMPLIFICATION

The price of 70-mer oligomer sets from Operon has become competitive with the price of amplifying and preparing PCR products from libraries. Serious consideration is d ue whether a core sh ould stay w ith established cDNA technology or begin a shift to well-defined oligonucleotide arrays. At this time, though, amplification is still the most common method of generating probes. Amplification is also probably the easiest process in the core, but it can be very labor i ntensive. PCR i s c ommon i n m ost laboratories, so f inding help troubleshooting problems should be relatively easy. Universal primers can be used for amplification of all but a handful of clones in most available libraries. Amplification requires no special treatment of the clone source – simply add one microliter of bacterial suspension to the reaction and let it go. However, keep in mind that one wants a good yield from the reaction, so the best option for the average core is still to do the amplification in 100 microliter volumes in 96-well PCR plates. This also provides multiple options for cleaning up the product in automated or semi-automated systems.

The most common high capacity thermocyclers, and arguably the most reliable, are the Tetrad systems from MJ Research. These cost about $25,000 each for the base and four alpha units. Again, if you can, buy two. Libraries may have from 15,000 to 40,000 clones or more. This is a lot of PCR. Two units will handle almost 800 clones per day with a single run; add in a night run and one can generate almost 6400 PCR products per week (just under 1600 per day for four days). It's possible to push the PCR production up to three runs per day, but the limiting factor then becomes the ability to process the samples and to run gels on them to verify the product.

We use the Millipore MultiScreen PCR plates to clean up our PCR products, as recommended elsewhere (Hegde et al., 2000), on a Qiagen BioRobot 8000. Most or all of the steps can be automated. We prefer a hybrid system. Staffers pipette the material into the Multiscreen plates. Each plate is then placed in the Qiagen robot, which moves them sequentially to the vacuum station (we had the manual wash screen modified to fit the vacuum rack of the robot) for a series of wash steps. The washed material is loosely bound to the membrane. Fifty microliters of solution is added and the plate is then manually sealed with standard sealing film and shaken on a microtiter plate shaker at top speed for 10 minutes. The material is pipetted into a fresh plate for storage. A small aliquot is removed and run on an agarose gel, one plate per gel. The ResGen human library has an accompanying file of gel images that can be used for quality control. Our gels are run in exactly the same format as the ResGen gels to facilitate comparison. Other libraries without accompanying gel images should be checked for the presence of product, the intensity of the product compared to a constant amount of standard, and for the presence of only one band. Flags can be entered into the data file to indicate products that don't pass muster.

We have tested a number of enzymes and found that the most successful one is DynaZyme EXT from MJ Research (manufactured by FinnZyme). This enzyme is highly processive, even with difficult templates. It is licensed for PCR, but, because only 0.5 units are needed per well, the cost per plate is competitive with unlicensed Taq and the product seems to be much cleaner. Although DyneZyme EXT is essentially template-independent in its efficiency and it has a very low failure rate, we have noticed that some plates seem to amplify less efficiently than others. This is likely a reflection of the growth conditions of the initial stock. These plates can be amplified twice and the resulting product can be combined for clean up in one Multiscreen PCR plate. Since the Millipore plate is a significant cost in doing large-scale PCR, combining multiple amplification runs for clean up can reduce the final costs. If one is efficient at all steps, the costs of amplification (excluding labor) end up being between $15 and $30 per slide, depending upon whether one is using one or two rounds of amplification per printing, and the number of genes printed per array. Each core will have its own mix of costs associated with array production. For the OHSU SMC, the major cost is the TSA kit, with amplification costs being less than half the cost of labeling and developing, so there is sufficient room in array pricing to allow some flexibility in the PCR protocols. Since the more DNA applied to the array, the more closely the measured ratio matches the actual ratio (Huibin et al, 2001), it can pay off in data quality to combine multiple PCR runs.

One issue only occasionally seen in the literature on printing of arrays is that the DNA from the PCR reactions should be concentrated before printing. We reduce our PCR products to dryness in a CentriVap from LabConCo, using their microtiter plate holder, which dries four plates at a time. We dry them at 60°C, which takes about four hours for a 50 ul recovered volume from the Millipore MultiScreen PCR plates, then resuspend in about 25 ul of DMSO / TE. The initial reaction volume was 100 ml, so this results in a four-fold increase in the concentration of the product, which seems to be adequate.

PRINTING THE ARRAY

After the PCR process, it is tempting to directly convert the volume of the final PCR product to a number of slides (what I, and probably others, call the array yield per amplification cycle). Overestimation is easy at this point. Note that array printing is inherently wasteful. The typical quill pin (TeleChem CMP-2 or 3) picks up about 250 nl of solution. We print two arrays on each of forty slides, with about 0.5 to 1 nl deposited per spot, or a maximum of 80 nl applied to the actual arrays. This leaves 170 nl, or 60% for preprinting and washing away at the end of the cycle. Large slide nests make the process more economical. If maximum yield per amplification is important, then it is a good idea to look for array printers that can hold as many as 100 slides. Printing two arrays per slide on a 100-slide machine will use 200 nl, resulting in only 50 nl of material being lost.

The printing plates are also very wasteful. We print from 384-well plates with round bottom wells (Whatman). We initially load our printing plates with 3 ul of PCR product. Multiple printings can be done from each 3 ul load of printing stock, but due to evaporation and well geometry, only 6 or 7 runs can be done before the ability to pick up material from the well becomes unreliable. This means that another 1/3 of the PCR product is discarded as unusable. The more tapered the well, the more material that is ultimately available for printing, but the best we can currently find have a problem as the volume declines into the 1 ul range. Combining the various limitations allows a quick calculation of about 1800 arrays per amplification, well below the theoretical maximum of 4000.

As implied, not all 384-well plates are equally useful for printing. The common plates have flat bottom wells and, in my opinion, these are essentially worthless for efficient printing. Many buffer systems will require 10 ul of solution to ensure that the bottom of a flat-bottom well is covered. This can result in both an increase in the number of necessary preprints, since the solution depth is greater, and a much larger waste of amplification product. Several vendors now sell 384-well plates with tapered or round bottom wells. Check with your general supplies vendor and see what they can provide. We like the Whatman plates, but others, such as those sold by Nunc, are probably just as good.

It is important to be realistic about printing throughput and machine reliability. It is certainly unwise to use manufacturer specifications for anything more than a very rough approximation of absolutely perfect operation in the real world (i.e., don't believe the numbers unless you can see one in operation). On the Cartesian printers, 16 pins printing 40 slides with two identical sixteen grid arrays per slide (one above the other) will take almost exactly one hour to print from a single 384-well plate. Printing 10000 genes per array will take almost 26 hours. If the typical printer is in operation for four days of the week – initial set-up on Monday and final run finishing on Friday – then the maximum number of slides that can be produced by the core is, obviously, four times the capacity of the slide nest. If the printer has a 40-slide nest, then the maximum number of slides that can be made in one week is 160. This becomes the highest average throughput per printer that the core is capable of, all other factors not considered. Not all print runs will be useful: pins will clog, the printer may mysteriously time-out, or the humidity control will stop functioning. The number of slides printed per week will also decline as the size of the array increases. Our 9400 gene arrays take about twenty-four hours to print. Combine this with set-up time and take down time and the printers can't print more than three times per week and then only if we are very efficient. That reduces the maximum production per printer to 120 arrays. We can currently process about 80 arrays per week, so this is not yet a limiting factor. If we become more efficient processing slides or acquire more staff, then the number of arrays we can print may become an issue. Realistically, given all the tasks involved in the core, and the reality of chronic understaffing, it is safe to assume that two 40-slide printers are likely to produce 160 slides combined per week on an ongoing basis, establishing the core's weekly maximum.

Given the long runs required for larger arrays, it is critical that all of the equipment is protected from power interruptions or surges. Invest in good uninteruptable power supplies (UPS) and connect these to an emergency power socket, particularly if the power is not of the highest quality. Your facilities department should be able to tell you how reliable the power in your building is. All but the most expensive power supplies are good for only a few minutes, being designed to permit proper shutdown, not long-term operation, during a power failure. Connecting into an emergency line can ensure that the printers keep running even during an extended outage.

A word of warning is due here. Many laboratories are wired like Christmas trees, with extension cables and power strips installed anywhere an additional line is needed. This is risky enough when working with the main power, but the potential for overloading emergency power is real and the results can be disastrous. The building or campus facilities management team should be involved in the decision to connect any equipment into emergency power. If too much load occurs as the power comes on, then the back-up system may shut down. Not only will a print run be lost, but every item on the line will be placed at risk.

The fact that a single print run can take over 20 hours also means that evaporation from the samples prior to and after the actual period of printing is a problem. Also, it is important to consider how the plates are going to be delivered to the printer. Many printers have a platform shared by plates and slides – the more plates used to provide DNA, the fewer slides that can be printed, or the maximum number of plates is limited to as few as five. It is possible to work with 1536-well plates with most current array printers. This means that printing 7680 PCR products will only require 5 plates, but working with 1536-well plates is not easy. Most groups still use 384-well plates for printing. The same print job using five 1536-well plates will require twenty 384-well plates. Regardless of how many plates are being used, if the plates share the printing chamber with the slides, then they will be exposed to relatively low humidity for the entire print run. This causes evaporation and decreases the number of printings one can get from a single round of PCR amplification. The external plate stackers of the Cartesian instrument, with closed chambers and humidity controllers, were a major factor in our decision to go with these machines. Because the stackers have independent humidity controllers, they can be maintained at high humidity during the print run, and each plate is exposed to the print chamber's low humidity only during the one hour it is being used for printing. Other machines have different, but very effective, systems for storing plates during the run, including some with temperature controls that also keep the plates cool.

The large capacity of the plate stack can cause additional problems. Array printers will generally perform a 10 to 20 spot preprinting to eliminate material on the outside of the pin. This occurs on designated slides on the nest. Unfortunately, our software is hardwired for one column (10 slides) of preprinting. This means that we have to change preprint slides in the middle of plate 17 (when the tenth preprint slide was filled). Since plate 17 usually printed at 2 am, this was not an attractive option. Although the machine could wait until the first staffers arrived around 7 am, that means that one plate sits at low humidity for 5 hours. We've learned to modify the printing scripts to change the preprinting protocol so that one column was adequate for any size of run. So, between the use of the riser and modification of the preprint cycle, we're doing a moderate amount of work of the machine scripts. This leads to the obvious: learn the printer control language as best you can; modifications to the protocols or to the hardware may require individual modification of each script generated by the layout program.

Once printed and baked, the arrays can be stored with desiccant in the dark for at least six months without loss of sensitivity. Before storing them, one or two slides from each batch should be examined for flaws in the printing. A number of dyes can be used for this. We use Syto 61, as described in Huibin, et al (2001). Alternatively we use a Cy5-labeled random 9-mer assay that tends to be more consistent across the slide.

HYBRIDIZATION

The most important part of the operation is getting labeled target onto the array and generating useful signal from it. In reality, the array source is incidental and a good core should be able to accommodate arrays generated from outside, especially the focused arrays available commercially. The only limitations are the obvious – the array must fit into the scanner and there has to be a data file that identifies the spots.

There are a number of different methods to get signal into the target. Each has its quirks. Most or all have been discussed in other venues, so I'll only touch on them here. The most commonly seen is direct incorporation of Cy3- or Cy5-modified nucleotides into the cDNA target. The advantage of this process is that it is simple. Incorporate, hybridize, wash, and scan are the only steps involved. The disadvantages are that the protocol can require large amounts of starting material – 50 to 100 ug of total RNA for each array. The problem of uneven incorporation of the dyes into the target has resulted in the standard dye swap assay.

A second protocol is the amino-allyl protocol. It improves on direct incorporation by only requiring 10 to 20 ug of total RNA per slide. By this method an amino-modified nucleotide is incorporated into both cDNA targets. After incorporation, the modified target is then "soaked" in a solution of activated dyes that bind to the amino group. After some washing the samples are applied to the slide. This protocol eliminates the problems of unequal incorporation of dyes. It is available in kit form from both Amersham and Stratagene (Invitrogen), and the dyes are available from Molecular BioProbes.

Small quantity RNA work can be done with one of two protocols, each of which is difficult to get working. The TSA protocol from PerkinElmer uses a modified immunohistochemical technique to generate the signal. The downside of this protocol is that it requires a lot of extra work to develop the slide. Again, this can be automated using, for example, the Dako autostainer (www.dako.com). Manual application of the TSA protocol can be a frustrating experience and at least one article evaluating methods for working with small amounts of RNA rejected it out of hand as too labor-intensive and irreproducible (Yu, et al, 2002).

Using the TSA protocol as described in the product manual IS a frustrating and time-consuming affair, with high background and variable signal. The trick is to not do it the way the protocol describes. Basically, the packaged protocol calls for pipetting solutions on the slide as it rests on the bench. If any part of the slide dries, even momentarily, background can almost ruin that part of the array. Residual wash buffer on the slide can result in concentration gradients that are very obvious when scanning. We modified the protocol so that the incubations are done in cheap, reusable 5-slide mailers and washes are done in TeleChem WashStations (Figure 2). Done this way, the slides never dry, background is virtually nonexistent, signal levels are high, uniform, and reproducible, and one staffer can process twenty slides in three hours.

Competing with the TSA protocol for the use of low amounts of target is the Genisphere protocol. By this method, a dendromer (highly branched molecule) containing up to 120 copies of Cy3 or Cy5 is hybridized to one end of the oligomer used to prime cDNA synthesis. Again this eliminates the dye bias of direct incorporation. As with the PerkinElmer system, this is a difficult protocol to get working, but those groups that use it seem very happy with their results. One study selected it as the most reliable method for working with small amounts of

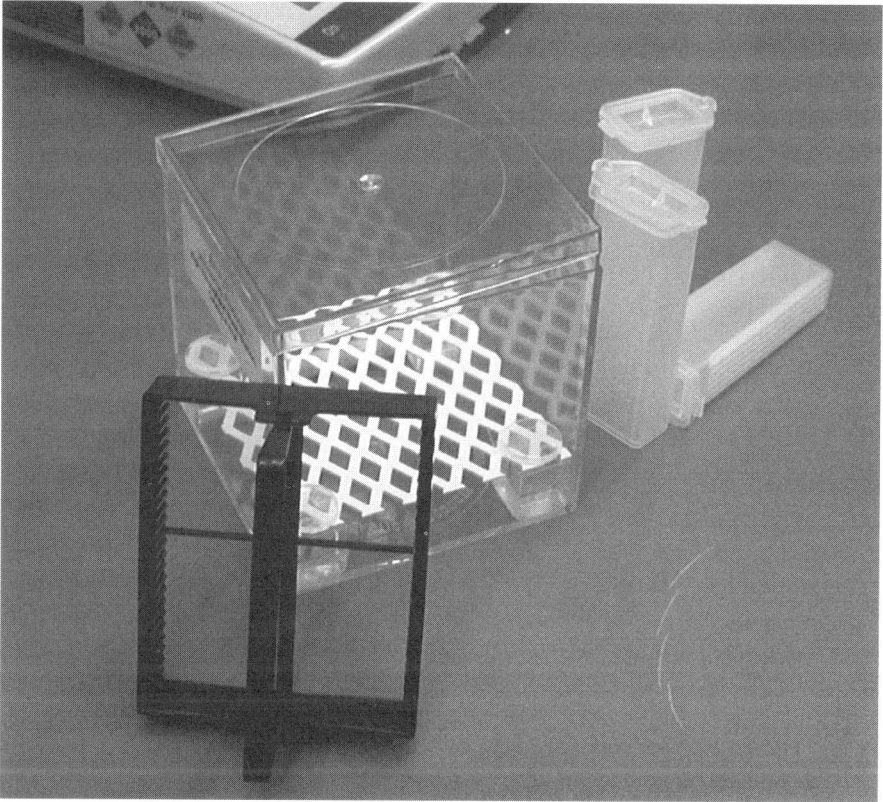

Figure 2. The TSA protocol from PerkinElmer is more consistent and of much higher throughput when performed with the correct equipment. Rather than purchase a $40,000 liquid handling robot that was recommended for the job, we spend for about $100 for a system that allows us to process 80 slides per week. The entire set up includes a TeleChem washstation (standard laboratory magnetic stirrer not shown) and a bag of 20 cent slide mailers, available from the Fisher Scientific clinical products catalog, or salvaged from the shipments of blank array slides.

total RNA (Yu et al, 2002). Recently a revised protocol has been developed that seems easier than the original one. The staff at Genisphere has been able to demonstrate very good results using arrays borrowed from the OHSU SMC, but we have never been able to replicate the results consistently. We have, however, had occasional success with the protocol and continue to work on improving our handling of it.

It should be noted here that the claims for the amount of total RNA required by these protocols might refer to small, focused arrays sold by that company. Being unaware of this leads to a lot of frustration and disappointment when "real-world" arrays are probed. The PerkinElmer claim of being able to use 0.5 ug of total RNA appears to apply to a 10 x 10 mm array used for their commercial focused arrays. If a standard array is used, one spanning 20 x 40 mm, then substantially more material must be used, but still less than other techniques. We can get nice results with 4.5 ug of total RNA on a 20 x 40 mm set-up, but this was prior to the recent improvement in the TSA protocol that we designed. We haven't examined how little RNA we can get by with when the background is so low. I have spoken to cores that can get by with as little as 2 ug of total RNA with this protocol.

There are a number of ways to increase hybridization efficiency. We use LifterSlips from Erie Scientific. These are cover slips with small Teflon strips along the outer edges that hold the slip away from the glass. We find more uniform hybridization when we incubate in a non-shaking water bath if these cover slips are used. It is important to check each batch to ensure that fluid easily migrates across the slip. Of our last four shipments, two were not usable for this reason.

Hybridization efficiency can also be improved by one of the commercial spotted array hybridization stations (e.g., Genomic Solutions GeneTAC stations). Unfortunately, they tend to be expensive and of limited capacity. As this article was being written, a new product from BioMicro Systems was announced that appeared to be a nice compromise between price and function. The company was selling a four-slide unit initially, which was too small for core use, but it indicated plans for a 20-slide unit in the near future. This may be a more efficient approach to cDNA array hybridization than the techniques now in use.

Another interesting approach that dramatically extends the use of one stock of RNA is to use two slides together, each serving as both cover slip and slide. This works best with bar-codes slides, which provide a small gap between the slides for diffusion of material. This technique entails printing with a larger top margin to accommodate the bar code strip (so one may prefer a slide brand with narrow bar-code labels). Presumably there are hybridization chambers on the market that accommodate two arrays in a sandwich. In their absence, a standard 50 ml conical centrifuge tube can be used in an oven.

SCANNER

Visualizing the hybridized array requires some form of scanner to generate a TIF image for image analysis. Scanners will range from about $40,000 for a basic two-laser unit to more than $100,000 for elaborate five laser units with autoloaders and batch-mode software. Non-laser scanners (i.e, Applied Precision) fall in the same price range, but provide additional flexibility in their use. Adding autoloaders to the more basic units can cost an extra $15,000 to $25,000 per machine. There are a lot of scanners on the market, so it will pay to shop around for features and price. A recent article reviews the current offerings (Cortese, 2001). The PerkinElmer ScanArray series and the Axon scanners are among the most popular at this time. Again, purchasing two scanners is a good idea in case one breaks down. In general, though, since scanning can be a significant bottleneck once the bench work becomes routine, two is a good idea, even when both are capable of autoloading. Be careful about assuming that batch loading will solve throughput problems. Some experiments may support batch mode, but it may not work well if samples from multiple laboratories are run at the same time – since RNA quality from group to group will have a significant effect on the scanning parameters. Autoranging can be used to adjust for this problem. However, autoranging works by keying on the most intense signals on the array. The zone for autoranging is generally user identified. Since the best region for autoranging can change from sample to sample, autoranging may not work as well as expected.

The basic issues regarding the scanners are: type of optics, resolution, scan rate, and number of lasers. The first two issues are the most important at the practical level. Most people are only working with Cy3 and Cy5 labeled target, so two lasers is adequate. Scan rate has some effect on image quality, but not as much as the optics and the resolution. The two major types of optics are confocal and charge-coupled device (CCD) cameras. Each has its advantages. On flat arrays the confocal will do well. The CCD camera works well on both flat surfaces and on textured surfaces, such as the FASTslides from Schleicher and Schuell, which have a nitrocellulose coat on them. Whereas the confocal imaging will only count photons from the plane of focus, the CCD will measure photons from the entire depth of the signal. The nitrocellulose layer on the FASTSlides is porous and binds PCR product in 3 dimensions. Adding labeled target, therefore, results in a much greater Z-axis range of fluorescence. Before assuming the CCD scanner is overall superior, consider that although the CCD scanners do allow access to a greater range of slides, the confocal scanners tend to have finer resolution. Also, the FASTSlides, and similarly designed slides, are not compatible with all labeling protocols. The point here is to match slides, labeling protocols, and scanners.

Most scanners will resolve down to 5 microns, and for many arrays 10 microns is adequate. As a rule of thumb, the scanner should be able to resolve areas that are, at the most, 10% of the spot size (e.g., under 15 microns for a 150 micron spot). Remember that scan resolution affects both scan time and image size.

For the ScanArrays, a 150-micron spot-size array can be scanned at 10-micron resolution with high fidelity in about 4 minutes. The resultant image is in the area of 3 Mb (assuming that the arrays are on the upper 40 millimeters of the slide). Increasing the scan resolution to 5 microns increases the scan time and the image size 4 fold. This may increase fidelity by a small percentage for spots of this size, but it will decrease throughput tremendously, an issue that a core facility must always balance against any concepts of perfection. However, if the arrays are packed as closely as possible, even if the spots are relatively large, the higher 5 um resolution may be necessary to get clean separation. At the risk of stating the obvious, 10-micron resolution means that each pixel is approximately 10 microns. If the spots are separated by 30 microns, this means, at best, only 3 pixels separate the spots on the image. If the spotting has been less than perfect, then separating spots may be difficult at this resolution.

Both types of scanners have the same general limitations, including four orders of magnitude dynamic range. This means that each pixel in the resulting image can range from 0 to 65,536 intensity levels (216), or the standard range for a 16-bit TIFF image. Given that invariably a small number of signals will be extremely high in intensity, autoranging functions of any scanner need to be monitored carefully. This is particularly important when working with brain tissue samples (Mirnics et al, 2001). As has been reported in the literature, most signal from brain-derived RNA is very weak, except for a small number of very bright genes. One has the option of restricting the dynamic range to include all spots, or to allow some spots to saturate while increasing the signal level of a greater number of weaker spots. Since the weakest signals are often the most interesting, this is usually a worthwhile trade-off, though it is important to ensure than normalization software excludes any saturated points from the process. An initial, low intensity scan can collect data on the highest signals while limiting the amount of photobleaching of the fluors. Increasing the amount of RNA applied to the array can reduce this problem as long as two-color analysis is being used. Large amounts of RNA are difficult to acquire from brain, but a single round of amplification can provide sufficient material, without adding significant bias to the process (Puskas et al, 2002).

Image quantification is a major issue for a core operation. The process of "gridding", or identifying the pattern of spots on the array can take a great deal of time. Good imaging software is essential. We use BioDiscovery's ImaGene, which can automatically adjust grids to match patterns, adjust grid boundaries to match the actual spot boundaries, allow for easy manual gridding, and give a reasonable number of quality control measures (which we have used in some of the in-house software we have written). Gridding can be vastly simplified if the layout of the plates places control plant genes at critical locations, such as the four corners of the first grid and at the upper left corner of each of the corner girds (Figure 3). Using a system like Stratagene's SpotReport allows one to include small amounts of plant gene RNA as template, ensuring that the necessary spots are always visible. This is also a good method for ensuring that the entire labeling process worked properly. D espite the c laims of a ny software manufacturer, we still find that human involvement in the gridding process gives a superior product when it comes to verifying the grid, flagging bad spots, and working with marginal arrays. The use of appropriate automation at other steps in the protocol can free staff time for manual gridding and imaging.

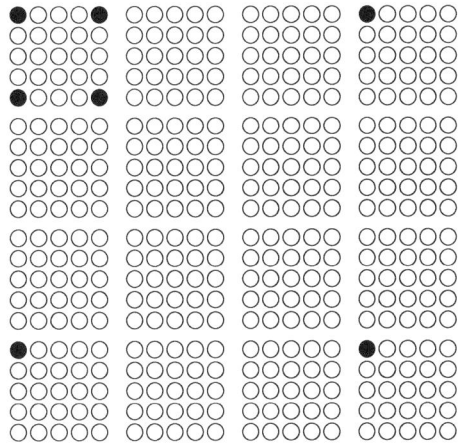

Figure 3. Placement of marker genes at the appropriate places on an array can m ake gridding w ith ImaGene remarkably easy. Arabidopsis genes from Stratagene can be spotted at the four corners of grid one, allowing an easy placement o f the b asic grid. Additional plant genes at the upper left hand corner of each corner grid allow the program to quickly snap a metagrid over the entire array.

OTHER EQUIPMENT

For those facilities that must prepare a "needs" list, the following describes the additional major equipment that we use. Of course we also have all of the standard laboratory equipment as well. The key pieces of core equipment are the printer, the scanner, and the thermocyclers (with whatever computers are needed to run them). However, a well-functioning core will have a lot of additional equipment necessary for proper operations.

I've alluded our use of a Qiagen BioRobot 8000 for clean up of PCR products. In many ways, this robot is overkill for the protocols we use it for, but some form of liquid handling robot is essential. Whether they are used for sample transfer from 96-well to 384-well plates, for cleaning PCR products, or for rearraying prior to regrowing library stocks, liquid-handling robots can free core staff for activities that need more active attention. Consider that a smaller robot can cost less than one staffer's annual salary plus benefits, yet may work for years doing work that would tie up the staffer almost full time, plus it won't complain about the work or change schedules to catch the World Cup games.

Libraries for amplification may be quite expensive. The basic IMAGE consortium human library from ResGen cost us, in fall of 2000, $30,000 for about 41,000 clones. A 21,000+ 70-base oligomer collection from Operon currently costs about $55,000. The mouse NIA 15k library was available for the cost of shipping, but only a handful of laboratories were able to obtain copies of it. Aside from the basic human IMAGE library, libraries cost about $6 / clone in 2000 and prices have reportedly increased substantially. Prices also increase for the level of sequence verification. All libraries must be viewed with some suspicion. Sequence verified libraries are often verified prior to growth. Sloppy growing procedures, misprogrammed robots, or a drop of rapidly growing stock into an adjacent well can nullify the value of the verification. Ideally only the final growth is sequence verified, but this is expensive since every growth batch must be independently verified. The core should at least make this information know to its users to avoid any liability for work performed based upon the annotations alone. Individual clones of interest identified by microarray analysis should always be verified by Northern blot or by real time PCR.

Each library used by a core will take up a lot of space. Our two main libraries came in almost 600 ninety-six well plates. The stocks are divided into two identical sets and all are frozen at –80°C. Amplification products and printing plates are stored independently at –20°C. This is a lot of freezer space. Standard frost-free freezers at –20°C are inexpensive, but each –80°C costs about $8000. It is necessary to have two to hold all of the plates that are used in the core. It is best to distribute duplicate stocks to different freezers so that if one freezer dies, the stocks are safe in the other.

A lot of small, dedicated equipment will be necessary. A tabletop centrifuge with microtiter plate carriers is essential. PCR products need to be dried prior to resuspension for printing. Some labs do this on the benchtop or in an oven, but we prefer to use a CentriVap (LabConCo) with microtiter plate carriers. This not only speeds up the process, but also prevents dust from getting into the samples.

Quality control is important for the PCR process. Ideally each product is examined by agarose gel to ensure it is the right size, present in the correct amount, and not contaminated with additional bands. Agarose gels are labor intensive. Having a stack of gel photos around is unnecessary, though, and the best approach is to use a digital gel camera. Remember to outfit it with the largest hood available, since gels that can run 96 samples at once tend to be quite large.

The amount of DNA placed on the array can have a significant effect on the ability to get useful data from a given spot (Huibin, 2001). I'm not convinced that the effect of intensity on Cy5 / Cy3 ratios described by Speed (Dutoit et al, in press) is not related in part to amount of DNA in each spot. The best way to adjust for the variable yield of PCR products is to measure and adjust the DNA concentrations. Measuring is best done using a spectrofluorometer that can read 96-well microtiter plates.

Software is another major issue. Database issues are dealt with elsewhere in the volume, so this discussion is about the software we use in the making and processing of arrays. We use software from BioDiscovery. We design our arrays using CloneTracker (purchased as an OEM product from Cartesian to generate scripts to run the printers), grid and quantify using ImaGene, and do data preprocessing with GeneSight. Overall software costs to start up will run from $25,000 to $50,000. Paperwork will be a serious drag on core operations, but MGED recommendations for information on array experiments require that a lot of information get stored. The best approach to this is a well-designed laboratory information management system (LIMS), which may run in the $75,000 to $100,000 category. Having commented on the commercial software we use, it is important to note that there are a number of free or inexpensive packages available that are quite good. We chose not to use them since we found some to be finicky about data formats and free software rarely has good customer support. Since we were responsible not only for our own data, but for the entire campus's data, we felt a need for well-supported software packages.

The big items are easy to identify, but it is often the small things that can slow a core down. Inventories need to be maintained. Advanced stocks of PCR products must be made, which means a significant upfront investment in enzymes and plastics. Microarray slides need to be stocked and additional pins must be available for the printers. Any kits that the core uses need to be stocked prior to use and maintained to avoid delays while awaiting shipping.

The best option for much of this is to centralize as much as possible. Find a good general vendor and use it. OHSU has a priority contract with Fisher and we are very happy with the service we've gotten from them. Shipping on all items is free for us. If we run low on a critical supply, the sales representative won't hesitate to ensure a drop-shipment (direct from the manufacturer) or overnight from the Fisher warehouse.

CONCLUSION

I hope that it's clear from this chapter that there is more to making microarrays than "building a printer, buying some slides, and doing some PCR." Any one of the problems described can tie up a core for months. The lack of clear answers to questions raised while solving the problems, both from the listserv and from the companies we work with, makes it obvious that even people experienced with microarrays still encounter problems that stump them. The basic nature of the problems encountered indicates as well that the field is still very young. The nature o f the e quipment problems s hows that t here is a lot o f room f or improvement.

Proteomics is the buzzword now and there is a feeling among end users that DNA microarrays are passé. This couldn't be farther from the truth. There are huge advances yet to be made in array production, in equipment, and in protocols. We've barely begun the work that DNA microarrays will do and part of the reason for that is the difficulty in getting arrays to everyone who needs them. The field will be mature when that difficulty is gone.

The most common comment I hear from vendors and product managers is "Good luck. If you're working with microarrays, you'll need it." Just as setting up a core doesn't have to be that hard, array work doesn't have to rely on luck. We now have a solid, if small, base of experts on printing and using microarrays – not the statisticians and informaticists, but core staffers and product specialists for various companies. If newcomers to the field draw on that growing base of expertise, very quickly spotted cDNA and oligonucleotide arrays will become routine for anyone wanting to use them.

REFERENCES

1.   Cortese, J. D. Microarray Readers: Pushing the Envelope. The Scientist, 15:36, 2001.

2.   Dutoit, S., Yang, Y.H., Callow, M.J., and Speed, T.P. Statistical methods for identifying genes with differential expression in replicated cDNA microarray experiments. Stat. Sin. in press.

3.   Fitzgerald, D. A. Microarrayers on the Spot. The Scientist, 16: 42, 2002.

4.   Genomeweb.com. June 1, 2001. Operon Discontinues Sale of Arrays in N. America Due to Concerns about Affy Patents.

5.   Genomeweb.com. October 18, 2001. Corning Axes Microarray Initiative as Part of Overall Restructuring.

6.   Genomeweb.com. October 25, 2001. InCyte to Slash 400 Jobs, Jettisons Custom Genomics Platform.

7.   Genomeweb.com. November 5, 2001. InCyte Licenses Key Microarray, Gene Expression Patents to Agilent.

8.   Genomeweb.com. November 16, 2001. Agilent Product Orders, Revenue Novedive in Q4; Firm to Cut 4000 Jobs.

9.   Genomeweb.com. December 20, 2001. Done Deal: Genomic Solutions Ties Up Cartesian Acquisition.

10. Genomeweb.com. April 26, 2002. Corning Develops Membrane Protein Chips, Ponders Reentry into Microarray Market.

11. Genomeweb.com. May 14, 2002. Motorola Looks to Sell BioChip Business.

12. Genomeweb.com. June 14, 2002. Nasdaq Gives Genomic Solutions 90 Days to Shape Up.

13. Genomeweb.com. June 17, 2002 GeneMachines Lays Off 25 Percent of Workforce..

14. Genomeweb.com, July 8, 2002. Incyte Sells Microarray Business to Drug-Developer Quark Biotech.

15. Genomeweb.com, July 18, 2002. Harvard Bioscience Buys Genomic Solutions for $26M.

16. Hegde, P, R. Qi, K. Abernathy, C. Gay, S. Dharap, R. Gaspard, J.E. Hughes, E. Snesrud, N. Lee, and J. Quackenbush. 2 000. A C oncise Guide t o cDNA M icroarray Analysis. BioTechniques 29: 548-562.

17. Huibin, Y., et al. An evaluation of the performance of cDNA microarrays for detecting changes in global mRNA expression. Nucleic Acid Res. 29: e41, 2001.

18. Mirnics, K. Microarrays in brain research: the good, the bad, and the ugly. Nat Rev Neurosci 2: 444. 2001.

19. Puskas, L. G, Zvara, A., Hackler, L., Van Hummelen, P. RNA Amplification Results in Reproducible Microarray Data with Slight Ratio Bias. BioTechniques 32: 1330. 2002.

20. Yu J, Othman MI, Farjo R, Zareparsi S, MacNee SP, Yoshida S, Swaroop A. Evaluation and optimization of procedures for target labeling and hybridization of cDNA microarrays. Mol Vis 8: 130. 2002.

Chapter 5

MICROARRAY DATA NORMALIZATION: THE ART AND SCIENCE OF OVERCOMING TECHNICAL VARIANCE TO MAXIMIZE THE DETECTION OF BIOLOGIC VARIANCE

Maureen A. Sartor, M.S. [a], Mario Medvedovic, Ph.D. [a],
Bruce J. Aronow, PhD [a,b]

[a] *Center for Environmental Genetics, Department of Environmental Health, College of Medicine, University of Cincinnati;* [b] *Developmental Biology and Pediatric Informatics, Children's Hospital Medical Center, University of Cincinnati*

NORMALIZATION: CORRECTING FOR TECHNICAL VARIANCE IN ORDER TO STUDY BIOLOGICAL VARIATION

A central goal in the analysis of microarray data is to identify and characterize genes and gene groups that exhibit differential and coordinate expression patterns as a function of biological state differences. This would be an easy task indeed if all gene expression measurements obtained from microarrays were perfectly accurate and consistent. Unfortunately, the reality of the situation is quite different. Microarray-based gene expression measurements are affected by a host of unwanted technical errors, linear and non-linear systematic biases, as well as additional or hidden biologic variances that pertain to individual or state variations that are not necessarily those that are to be evaluated in an experiment. Taking these effects all together, observed differences in gene expression levels between two samples for a given gene can be represented as the sum of two components: technical variance and biological variance. The goals of microarray data normalization are to identify and strip out as much technical variation as possible such that the most accurate expression levels associated with different biological states can be determined. Ideally, one would hope to understand the expression of each gene in relation to all other genes on a microarray, across a series of microarrays, as well as to those measured by other technologies, in other experiments, and in relation to the activity of the entire genome. In other words, exact absolute gene expression levels for all genes under any circumstance. Short of this, optimal normalization techniques seek to maximize the usability of all measurements per array and per array series. Correcting for any technical variation thus allows for the highest resolution view of biologically based variations in gene expression. Following optimal normalization procedures, the multiple origins of biological variation can be dissected in detail.

In any experiment, there may be one or more biologic variations of interest such as disease vs. healthy, good vs. bad disease outcome, good or bad medication side effects, all of which may be reflected in—or sometimes even be determined by—altered gene expression patterns. By identifying genes and expression patterns associated with these differences we can begin to trace the circuitry of biological systems that may permit us to more deeply understand life processes and to prevent harmful disease processes. Maximal biologic variance can only be obtained if technical variations affecting RNA preparations, labeling methods, arrays, and measurement methods are first removed. Technical variance can come from a host of sources: the individual array, fluorescent dye labeling, hybridization, array spot location, and gene-by-dye interaction, to name just a few. This chapter will discuss various methods for the removal of some of these technical variations, along with the effect that their minimization can have on the results of later analyses for the assessment of biological variation. Statistical software packages such as SAS or S-plus may be used to implement these procedures, or for some of the more common methods, gene expression microarray-specific software may be used, such as Genespring® or Spotfire.

First of all, it is important to understand the difference between random and systematic variation in an experiment. High random variation necessitates more technical or biological replicate measurements. Technical replication is the reanalysis of the same mRNA or mRNA pair as is measured on another array, whereas biological replicates are arrays that each use RNA samples from different individual organisms, pools of organisms or flasks of cells, but compare the same treatments or control/treatment combination. Depending on whether the mRNA is relabeled with the same dye, or the pair of RNAs are labeled with the opposite dyes for separate arrays (dye flipped), the variance is being measured with respect to either the array, the labeling reactions, or the variance caused by the specific dye being labeled. *Using the average of even a very large number of measurements can only give an accurate estimate of the true value of interest in the absence of a systematic bias.* Random error is easily detected, and confidence in your results can be calculated based on random error. Systematic variations (biases) are a cause for great concern because later analysis may not identify them. Correction for systematic bias can also greatly reduce the number of technical replicates necessary to achieve a reproducible measurement. With systematic biases, one may have consistent, but very inaccurate data, leading to false conclusions. Unless the cause for the bias is noticed and removed, it may cause a false sense of confidence. Figure 1a displays part of an unbiased array. More intense spots have higher signals, whereas smaller, duller spots have lower expression levels. Figure 1b illustrates a common bias due to fluorescent labeling. Besides this bias, you can also observe differences in spot size and shape in Figure 1; this is also a cause of variation in measurements. There are multiple ways to avoid biases and to design experiments that make it possible to quantify and remove them. Section 4 of this chapter will cover the role of the experimental design in the removal of technical variance.

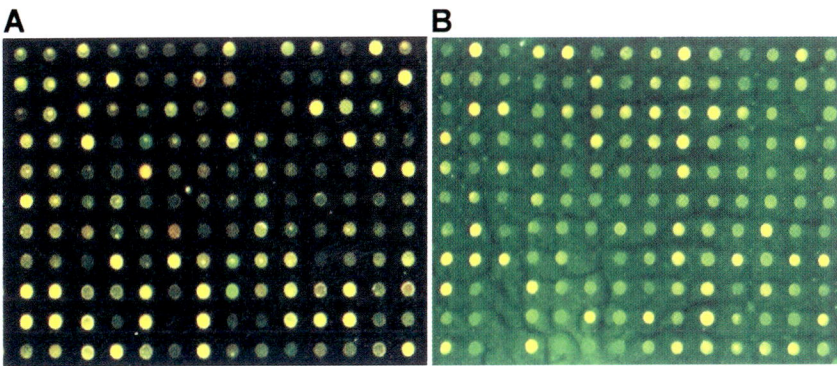

Figure 1. **Bias in spotted microarrays**. **A:** A portion of a two channel microarray. Greener spots indicate more mRNA in the Cy3 labeled sample, redder spots indicate more in the Cy5 sample, and yellow spots are close to balanced. **B**: An example of systematic bias and spatial bias on an array. Notice that Cy3 (green) is consistently more intense than Cy5 (red), and that the lower portion has a much higher Cy3 background.

Determination of candidate differentially expressed genes from microarray data can be broken up into two main steps, each step responsible for removing a different type of random error and bias. The first step is normalization; through this process, whole-array (or group-array) variations and biases in the data are removed, as well as overall dye effects if using a two-channel platform. The process of normalization removes systematic variations that can result in a systematic bias and an increase in the number of falsely implicated genes, and/ or in an increase in the experimental variability that would reduce statistical power of detecting true sample-to-sample differences in gene expression levels. The second step is the analysis, usually involving Analysis of Variance (ANOVA) or a similar type of model. This may be done on an experiment-scale, but is usually done per gene. Thus, this is the step that usually includes removal of gene-specific variations, some of which are gene specific array effects (individual spot effects) and gene specific dye effects. Dye-by-gene interactions occur due to differences in cDNA or oligo sequences interplaying with differences in dye incorporation or fluorescence. This may result in different genes incorporating or fluorescing Cy3 more efficiently than Cy5. At least one method has been proposed (Kerr *et al.*, 2000) for both normalizing and analyzing your data in one step.

For a detailed description of technical variance sources, see (Goryachev *et al.*, 2001). This paper takes a stepwise approach to the entire experiment, noting the technical variance that can be added at each step along the way. The result is a thorough overview of the relationship between concentrations of cellular mRNA (assumed ultimate goal), and the actual measurements that you receive from the phosphoimaging software at the end of the line. Between these endpoints lie several sources of error, broken into the categories of "chemical, optical, and computational factors." Examples of chemical sources of variance are variations stemming from "RNA preparation, labeling, hybridization, and washing". A source of optical error is background current in the photomultiplier tube (PMT) obscuring the actual fluorescence of lowly expressed genes. Computational errors may be caused by the inability to define a clear-cut line between spot fluorescence and background signal. This can be caused by irregular shapes and sizes of spots, which themselves are caused by inconsistent pressure in the spotting action of the pins. The importance of stripping out these random errors and biases can be seen by comparing the relative sizes of technical and biological variances in a typical experiment. Figures 2 and 3 show two different ways of visualizing the differences between technical and biological variance. Figure 2 displays plots of technical reproducibility, biological replication, and biological variation for an experiment in which four different RNAs are labeled and hybridized to five Affymetrix arrays (one of the RNAs was labeled twice) and compared using scatter plots of normalized expression data. The closer the spots are to the line y=x, the less variance there is. Similarly, the more spread out the data points are, the more variance there is attributable to the variance component that is being measured.

Figure 3 (below) shows box-plots for the variance of each experimental factor in a relatively simple experimental design in which there were two pairs of mice, each pair being compared in triplicate. Notice how much variation there is between mouse pairs compared to the variation between genotypes (the factor of interest), and that the dye effect and residual variance are on the same order of magnitude as genotype. It is easy to see that without measuring the effects of other variances, they would drown out the effect of interest, in this case, the effect of specific genotype.

Technical and Biological Reproducibility and Variance Measured on Affymetrix GeneChips
(shown are genes called "present" by Affymetrix MAS 5.0 algorithm)

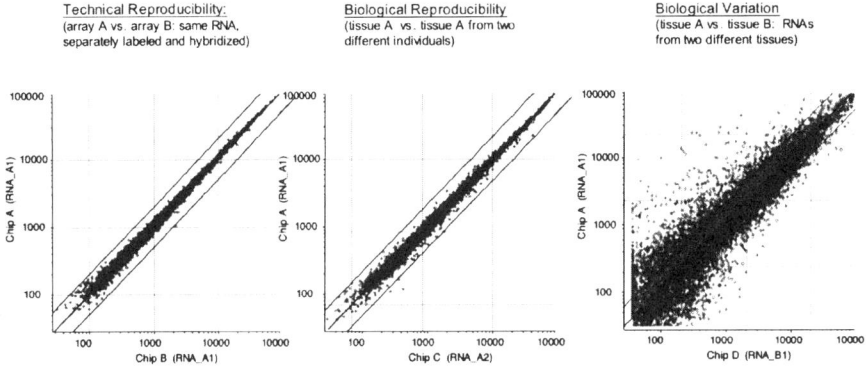

Figure 2. Scattergram depiction of the relative variance between signal obtained from four labeling reactions of three RNAs hybridized to four Affymetrix GeneChip ® microarrays.

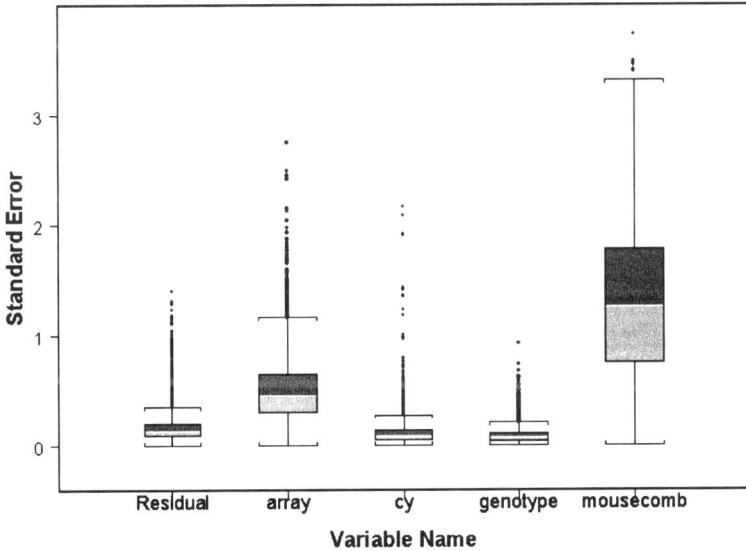

Figure 3. **Comparison of variance sizes.** Mousecomb is the biological variance between mouse pairs, genotype is the 'treatment' of interest, cy is the dye effect, array is whole array effect, and Residual is the residual variance.

The particular normalization method you choose for your experiment will first depend on the type of microarray technology with which you are working, the two main types being single channel and two channel arrays. Clontech membranes and Affymetrix GeneChips are examples of the one channel technology. Most other microarrays use two channels, with the dye Cy3 for one channel and Cy5 for the other. Whereas the single channel arrays allow the hybridization of only one biologic sample per array, two channel arrays incorporate the hybridization of two samples per array, allowing for more direct comparisons across experimental factors/treatments. Two channel arrays, however, do have an additional technical variance due to the dye effects. Because of these differences, the technical aspects of normalization of microarrays will be broken up into two sections of this chapter, with section 2 describing the detailed steps of one channel array normalization, and section 3 detailing the normalization techniques for two channel arrays.

SINGLE-CHANNEL DATA NORMALIZATIONS

Affymetrix GeneChip® microarrays are the most widely used single channel arrays. These arrays contain multiple spots for each gene, divided into perfect matches (PMs) and mismatches (MMs). Each PM is a short oligonucleotide probe corresponding to a specific gene transcript, and the MMs are the same except for one point mutation at the midpoint of each sequence. Each perfect match is paired with a mismatch, and the pairs are distributed randomly on the array in order to avoid spatial biases on the chip. There are usually some spots of an array that are bad, either due to specks of dust, smearing, or other technical errors. Before we begin the steps for normalization we assume that these spots have been removed from the analyses either by the imaging software or manually (for example by "masking" the image).

The first step in normalizing data from any single channel array is subtracting the background intensity. Background fluorescence is caused by nonspecific binding on the array and is a consequence of the chosen PMT level. This background noise can, and almost always does, vary across the surface of the array. Some software packages calculate an overall array background level, and suggest subtracting this from all measurements. More common are local (or block) background estimation and per gene background estimation. The Affymetrix Microarray Suite software divides its arrays into 16 blocks, and calculates the background as the average of the lowest 2% signaling spots within each block. Additional tweaking is done to create a smooth transition in background intensities between adjacent blocks. For per gene background estimation, it is decided whether or not each pixel is part of the gene spot. The median intensity of those pixels not part of the spot is then used as the background. The median is preferred over the mean, because it is less affected by outlying pixel intensities caused by specks of dust or other foreign objects. Once the background value has been determined for each spot, it is subtracted from each respective gene intensity. More background adjustment possibilities will be discussed in the next section.

The next step for Affymetrix arrays is obtaining pre-normalized estimates of each gene's overall signal, and there are various methods currently used for this. One is subtracting out estimates of non-specific binding. This can be accomplished by computing the "Average Difference" for each gene, which is found by subtracting the mismatch from the perfect match for each pair and then averaging these differences for each gene. Another possible formula to consider is the Log Average Ratio, which uses the ratio of PM to MM rather than the difference. While the Average Difference is most commonly used, other hybrid methods may be applied which use a combination of these metrics, and/or information on the variability among the PM/MM pairs. Since it is not known for certain whether PM/MM pairs have a multiplicative, additive, or other unknown relationship with the true amount of hybridized mRNA, neither metric mentioned above is necessarily wrong. Since most other array technologies use one spot per gene rather than PM/MM pairs, the above step is only relevant to Affymetrix array analysis.

Most often from this point on, the analysis is done with log-transformed data. That is, each signal intensity measurement on each array is transformed by the log-base 2 function. Other log bases may be used, but log-base 2 has the convenience of being easily interpretable when viewing fold changes. There are several reasons for performing this transformation prior to further analysis. One reason is that intensity values vary over several orders of magnitude and the dynamics are therefore difficult to view in any one range when graphing raw data. A second reason is that the distribution of raw data is skewed to the far left due to the majority of genes with low transcription levels, whereas the distribution of log-transformed data is approximately normally distributed (Figure 4). Furthermore, random error in microarray measurements increases with higher signal intensities, so that log transforming data helps to normalize the random errors. Data on the log scale still does not usually have equal random error across intensities, but rather variance in the lower range is usually slightly higher. It may be a slight stretch to say that log transformed microarray data is normally distributed, but it is usually close enough that many analysts make that assumption. The good thing about your data being approximately normally distributed is that it enables you to use many commonly used parametric tests often pre-programmed into software applications for further analyses. A final reason for log transforming the data, which is directly related to the other reasons above, is that it changes calculations from a multiplicative to an additive scale, which is easier to work with. Because many of the e rrors in m icroarray data a re assumed to be multiplicative with intensity, the log scale converts the errors to approximately linear.

Once data is log-transformed, you are ready to perform inter-array normalization in order to convert the data into a form that allows for sample-to-sample comparisons. Since the initial conception of microarrays, many normalization p rocedures have b een proposed. One of the first was the use of a predefined set of "housekeeping genes." T his method c ompared the average intensity of housekeeping genes among arrays, assuming that they should be equal. T he average intensity of housekeeping genes was calculated both per array and overall. E ach gene's intensity on each array was divided by

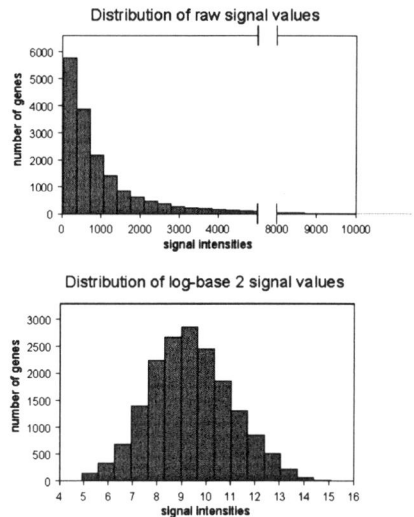

Figure 4. **Distributions of data from the same microarray on two different scales. (Top)** typical histogram shape for data on the original scale. Notice that the data spans several o rders of magnitude and i s skewed very far to the left.

that array's "housekeeping average." Finally, if desired, each gene's intensity could be multiplied by the overall average to return to the original scale. Since it was soon noticed that the so-called "housekeeping genes" did indeed change, this normalization technique was replaced by one with a more realistic assumption.

Mean normalization, also called the globalization method, rests on the assumption that each sample, or each array, should have the same total amount of mRNA. This normalization method is very similar to the housekeeping gene technique, except that all genes are used instead of a small predefined set. The formula for mean normalization is:

*Normalized log(G_{ij}) = [Raw log(G_{ij}) / Mean(A_i)] * OverallMean*

for the ith array, and the jth gene, and where Raw log(G_{ij}) is the log intensity of gene G_{ij} after background subtraction. The reason this method is not widely used is the mean can too easily be affected by outliers in the data. Instead, the median intensity, which is a more robust statistic, is more commonly used as the scaling factor in place of the mean. This procedure is appropriately called median normalization, and is probably currently the most common procedure used, mainly because it is easily implemented for all design structures and produces fairly accurate results.

Although the majority of journal articles dealing with normalization of microarrays use two channel arrays, there are a few to be found for single channel. (Hill *et al.*, 2001) offer a very different normalization procedure that involves spiking the mRNA mixture with known amounts of biotin before hybridization. These positive control spots can then be used to normalize the arrays against each other. This method has the benefit of not losing data whose average difference is negative, due to low expression and cross-hybridization with the mismatch. Since using only the spiked data would amplify error due to too few measurements, a method was developed to also use the array mean intensities in the computations.

There are various ways to visualize your data in order to view overall quality and/or the distribution of the data, but the preferred plot is the log-Ratio vs. log-Average plot (see Figure 5). The first step in producing this graph is to calculate the log-ratio, defined as $R = \log_2(X_1) - \log_2(X_2) = \log_2(X_1/X_2)$, where X_1 and X_2 are the signal measurements (background subtracted) of arrays 1 and 2 respectively, and the log-average intensity, calculated as $A = [\log(X_2) + \log(X_1)]/2$. Plotting R vs. A is essentially a rotation in axes and change in scale from plotting $\log_2(X_1)$ vs. $\log_2(X_2)$, or one array vs. another. A simple plot of one array vs. another, also on a log scale, is an alternative visualization of the data (Figure 2, for example). While it is essentially displaying the same information as log-Ratio vs. log-Average, expression level biases are more easily detected visually in the type of plot displayed in Figure 5. The advantage of the R vs. A plot is that biases attributable to labeling or hybridization variances between arrays can be detected visually as a group of data points somewhere along the x-axis that are not equally distributed above and below $R = 1$ (for example the circled areas in Figure 5). Note in Figure 5 that pairs of arrays were chosen such that differences in the factor of interest could be seen. Points above 1 or below -1 on this scale correspond to a greater than

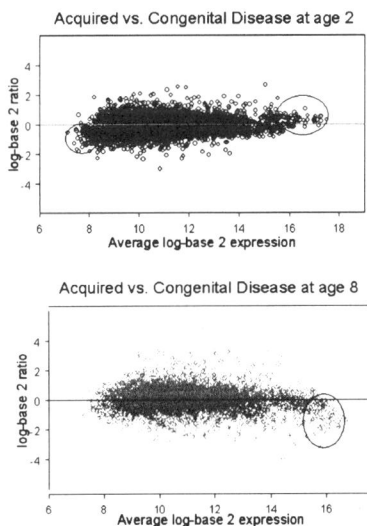

Figure 5. **Scatter plots of two pairs of Affymetrix arrays**; each circle represents one gene. Note the slight bias in the data at very low and high expression levels (**top**), and the strong bias in the opposite direction at high expression levels (**bottom**).

2-fold change, points above 2 or below –2 correspond to a greater than 4-fold change, and so on. Alternatively, replicate arrays could be plotted in this manner to obtain a visual estimate of the reproducibility of the data. In this situation, the ideal would be for all data points to fall on the line y = 0 (no difference), but since this of course will never happen you at least hope for less spread in this plot than is seen plotting treatment against treatment.

With single channel arrays, the method of normalization you use may or may not depend on the experimental design. For a simple treatment-vs-control or treatment x-vs-treatment y experiment with replicates, (Kepler *et al.*, 2002) offer a more rigorous method than those described above, which takes into account the non-linearity of the data, and corrects for the biases seen in Figure 5. They use a semi-parametric statistical method called local regression, which essentially smoothes the data across the average intensities so that for any intensity range, there will be approximately as many up-regulated as down-regulated genes. The method also includes a procedure that allows it to be robust with respect to outliers in the data. The only additional assumption made in this model is that the majority of genes will not change significantly between treatments. This will usually, but may not always, be a safe assumption to make. We will discuss local regression further in section 3, pertaining to the normalization of two channel arrays. Kepler's paper offers software and documentation free of charge at [**ftp://ftp.santafe.edu/pub/kepler/**] to implement their method. This type of non-linear procedure is highly recommended in the case when bias is seen in graphs similar to Figure 5.

A third alternative normalization method, introduced by (Zien *et al.*, 2001) and termed "centralization", involves normalizing ratios of treatment measurements, or the ratios of treatment means if there are replicates. Unlike the method described in the previous paragraph, this method may be used when there are no experimental replicates (although an experiment without replicates is strongly discouraged). Centralization has the weakness of having to discard data in any range where ratios are biased. For example, all the data to the far right in Figure 5 above would need to be discarded. On the other hand, it has the advantage of having very weak assumptions. The only assumption is that either most genes are not significantly differentially expressed, or that about as many genes are up-regulated as down-regulated.

(Yang *et al.*, 2000) give a thorough overview of various normalization methods available for both single and double channel arrays in "Normalization for cDNA Microarray Data." Yet another method for single channel normalization was described by (Schuchhardt *et al.*, 2000) where a dilution series of control spots are used in conjunction with pin-wise (block) normalization on a glass slide experiment, producing results found to be superior to array-wise normalization and two other normalization methods described in the article. Whether or not control spots are used, block normalization may still be beneficial especially if there is a spatial effect where average intensities in certain areas of the array are higher (lower) than others. If this is the case, a sort of global normalization may be done per array with blocks or pins replacing individual slides in the procedure.

NORMALIZATIONS OF TWO-CHANNEL DATA

Many of the same procedures and concepts from single channel normalization are also applicable for two channel arrays. For example, corrupted or otherwise bad spots must be flagged, background fluorescence must be corrected for, data is log-transformed, and similar plots are used for visualization of overall array quality and distribution of intensities. As expected, however, there are also many differences. Whereas single channel arrays are normalized in groups (between chip normalization), dual channel arrays are usually normalized on a per-array basis (within chip normalization). This is due to the fact that for dual channel arrays, the measurement of interest is the *ratio* between channels' signals, whereas in single channel array experiments, *absolute levels* of fluorescent signal were compared. For cDNA arrays (two-channel), the measurements of spot intensities are paired naturally by the design of the microarray (i.e. Cy3 paired with Cy5 measurements), rather than being paired by experimental factors as is the case with single channel arrays. Thus, plots of dual channel microarray data are done one array at a time. The visualization looks similar to the single channel arrays, except now either the two dye measurements are plotted against each other, as in Figure 6, or the average intensity between the two dyes are plotted against the Cy5/Cy3 ratio (log-Ratio vs. log-Average plot).

Similar to single channel normalization, we will assume that the bad spots have been flagged and removed prior to normalization. For further information of flagging weak spots, see (Yang *et al.*, 2001) who propose a statistical technique for flagging spots which they have found to reduce variance in ratios and improve the quality of later normalization steps.

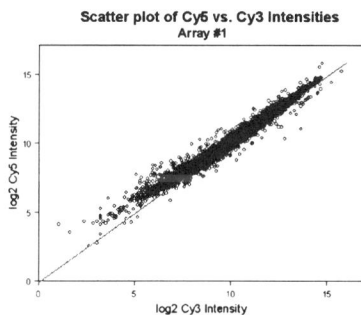

Figure 6. **Cy5 vs. Cy3 log2 measurements.** Each data point corresponds to one spot on the array.

The first step, as it was for single channel arrays, is background subtraction. The same choices are available as previously mentioned: overall, local, or per-spot. Background levels are subtracted for Cy3 and Cy5 separately. Using the median per-spot background subtraction, where the median of the pixels surrounding the spots were subtracted from the overall spot median of pixels, (Finkelstei *et al.*, 2002) found that background subtraction reduced the standard deviation of pixel intensities by 5.5% percent for Cy3 and 6.1% for Cy5 on average. Despite these positive results, subtracting background does increase the variance of Cy5/Cy3 ratios for low expressed genes. This may be due to the masking effect of the cDNA on the spots, causing a lower base signal within the spot than surrounding it, when a gene is low or absent (Applied Precision, scanner manual). This results in negative intensity levels after background subtraction for some, and unrealistically high (or low) ratios for others in the low range of expression. While background subtraction is by far the most common background correction method, alternatives have been proposed that are more complex, but may also be an improvement over the current standard. Noticing the higher variance in ratios after background subtraction, and the problem of having to discard genes with a negative intensity value, (Colantuoni *et al.*, 2002) suggest an alternative background correction method, which also adjusts for spatial biases on an array. The method will be discussed further below, but first we step through other normalization steps, which this method presumes are already done.

Once the intensities are corrected for background, you will want to log-transform the data, for the same reasons listed in the previous section, before proceeding with analysis. As mentioned before, log-base 2 is the most commonly used log base because it is easily mentally converted back to the original scale, but other bases may also be used. With your log-transformed data, you are ready for within-array normalization. Whereas with single channel array normalization, some of the more rigorous methods could only be used with certain experimental designs, two channel array normalizations do not depend on the design of the experiment. As in the previous section, one of the first normalization techniques involved the use of a predefined set of housekeeping genes. This method, however, is rarely used today. The simplest normalization method currently used (but not the best) is mean or median normalization, sometimes called the globalization method. In this procedure, the log ratios are calculated for each spot on each array as $\log(x_{i5}) - \log(x_{i3})$, where x_{i5} is the intensity for gene i on cy5, and x_{i3} with the intensity for gene i on cy3. For each array, the median (or mean) log ratio is then calculated, and this value is subtracted from each spot's log ratio. Using the median is usually preferred over the mean, because it is less affected by extreme outliers in the data. Figure 7 shows how median normalization centers the ratio around zero, so that as many genes are up-regulated as down-regulated.

The rationale for the median normalization method is that any offset in the median ratio from zero is caused by an overall array dye labeling effect in which Cy5 or Cy3 labeled more strongly. In practice, however, the dye effect is not constant for all genes. Variations in dye effects occur due to the differences in the relative efficiency of labeling of low and high abundance genes between Cy3 and Cy5 and to labeling or fluorescence d ifferences on s pecific genes between Cy3 and Cy5. Very often for two channel arrays, Cy3 and Cy5 do not have equal hybridization strengths throughout t he range o f expression levels. Therefore, intensity dependent biases from dye often occur, such as when one channel begins to saturate at a lower intensity than the other. This can be seen in the log-Ratio vs. log-Average plots where a whole r ange of genes appear to be up- (or down-) regulated. So, even though ratio biases dependent on average intensity are seen both in single and dual channel arrays, the cause is m ore easily c haracterized in d ual channel arrays, and the curvilinear bias is often stronger in this case. See Figure 8. Gene sp ecific dye e ffects will be further discussed below.

Probably the most common normalization procedure for dual-channel arrays is local regression, and has been described in (Dudoit *et al.*, 2000), . Local regression is a semi-parametric procedure that removes the curvilinear tendencies o f the d ata. It w orks by stepping through each point along the x-axis and taking a weighted average of ratios around that point. The size of the area aro und each point to use is designated by choosing a smoothing parameter (f). We have found that $f =$ 0.25 or 0.30 are usually suitable choices. Choosing too small of an f value can

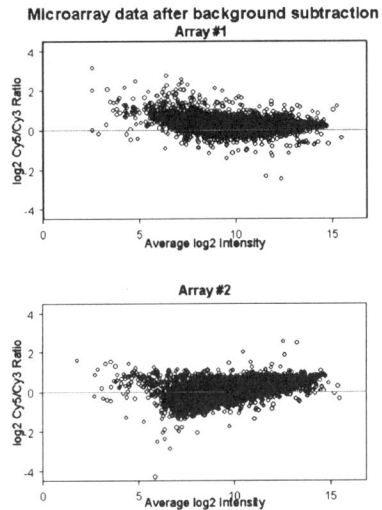

Microarray data after background subtraction

Figure 7. **Two examples of background subtracted data on the log scale.** High quality arrays will have appro ximately an equal number of g enes up a nd down-regulated at each point along the x-axis. Here we see examples of two types of bias in the ratios. Array #1 (**upper**) has a disproportionate number of up-regulated genes overall, especially in the low intensity range. Array #2 (**lower**) has an approximately equal number of up and down-regulated genes, however the l ow intensity g enes are more likely to be seen as down-regulated a nd vice v ersa for the high intensity.

cause a smoothing-out of real and significant variations in gene expression, while choosing too large of an f-value may underestimate the curves due to dye effects, resulting in a regression too close to linear. Local regression is performed for log-Ratios against log-Averages for each individual array, and the residuals of the fitted curve are captured as the normalized estimated ratios. This procedure can be implemented in statistical packages, and many genetic software packages. Figure 9 displays the post-normalized data for the same array as in Figure 8.

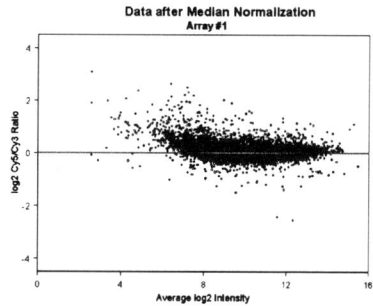

Figure 8. **Median normalization did not correct for bias in lower expression range,** but did shift all points down so that there are as many up as down-regulated genes. Same data as in Figure 7 **(top).**

Notice that the ratios are now centered about zero throughout the entire intensity range. Whereas median normalization centered the overall red-to-green log ratio, local regression centers the red-to-green log ratio throughout the entire intensity range. This is shown as a distribution plot in Figure 10, which shows (similar to Figure 7) that the overall Median Normalization procedure can leave an unequal distribution of intensities

for each dye on an array. The intensity distribution for each dye is different in the raw data, for median normalized data the distributions are shifted left and right so that they are centered, and for the data after local regression, the distributions are morphed so that they are not only centered but also have a similar shape. An alternative normalization procedure that is similar to local regression in that it also smoothes the data distributions was proposed by (Workman *et al.*, 2002). Their technique uses cubic splines with the quantiles of a reference array to normalize the data, where the data can be either from single or dual-channel arrays.

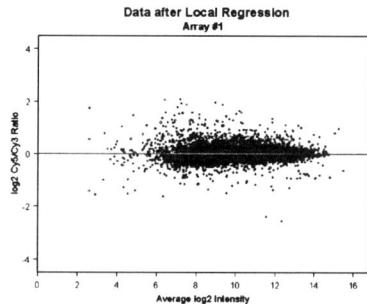

Figure 9. *Normalized ratios, after the bias at the lower expression range has been removed by local regression.*

Spatial bias is another type of bias seen in ratios; it is based on position on the array and may be caused by uneven washing and/or a differential in background levels across the array surface. Figure 11 illustrates an example of spatial bias on a cDNA array, where the raw average Cy5/Cy3 ratio is clearly dependent on the array region. As is the assumption made in (Colantuoni *et al.*, 2002), the spatial bias seen in Figure 11 was caused by a higher background for one dye than the other in certain regions of the array. In this particular case, the bias was so bad that the array was thrown out and redone, but it is a good illustration of the type of spatial bias seen in experiments.

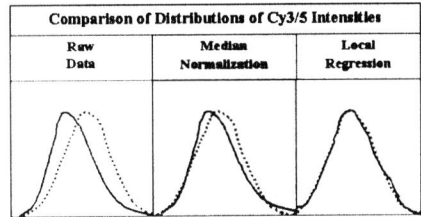

Figure 10. **Comparison of normalizations with respect to dye distributions.. A:** Distributions of cy3 (dotted line) and cy5 (solid line) are different. **B:** Median normalization shifts distributions left/right till centered. **C:** Shape of distributions are coordinated

When spatial bias is suspected in an array, or background subtraction is seen to greatly increase the variability in the lower expression range, the background correction described by (Colantuoni *et al.*, 2002) may be more appropriate. This method, assuming the globalization method, but not background subtraction, has already been used, performs a smoothing of average ratios across the array surface. This is similar to how local regression smoothes across average intensities, so that approximately the same numbers of spots are up-regulated as down-regulated in each region of the array. Local regression on average intensities can, and probably should, then also be done on the residuals of the above regression. This normalization technique has the advantage of not having to omit spots from the analysis that have estimated expression levels below background. This procedure can be done with SAS or R statistical language, but also may be implemented via a web-based interface at http://pevsnerlab.kennedykrieger.org/snomad.htm.

Planning a good experimental design can be more challenging for two-channel arrays, and because of this it is very important to keep in mind the statistical consequences of a design in the planning stage. Without replicates or arrays with samples labeled with opposite dyes, there is no way to estimate gene specific variability or reproducibility, or gene specific dye effects. Gene specific dye effects have been shown to be reproducible, and are important to remove if a comparison is to be made between the two samples of one array. Another thing to keep in mind when planning your experimental design is that pairs of samples for which comparisons will be made should not be separated by many steps through arrays. That is to say that if an important comparison in your experiment is A vs. D, it is much better to do an A vs. D array than the arrays A vs. B, B vs. C, and C vs. D.

Some other types of variance you may want to consider in your microarray

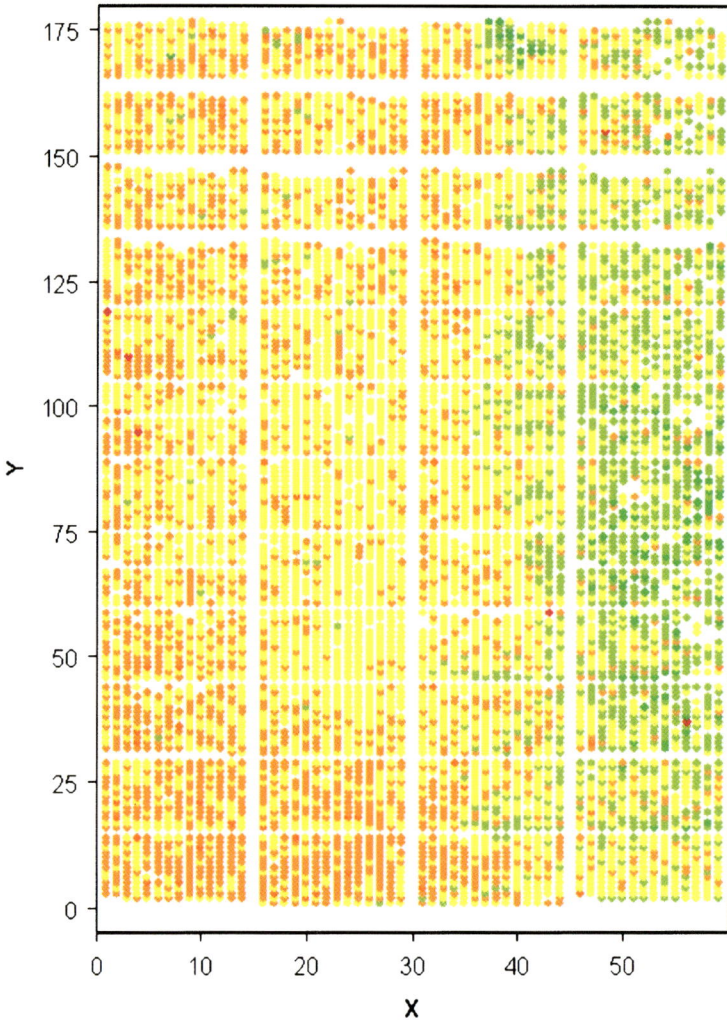

***Figure 11.* Spatial bias.** Red spots correspond to higher Cy5 expression; green spots correspond to higher Cy3 expression; yellow spots have approximately equal expression levels.

experiment are print-tip or block variation, and variation between replicated gene measurements on the same slide, if present. Print-tip variation is caused by slight differences or changes in the shape or pressure of the arrayer print head. This may cause a large area of an array to be biased more to the red or green dye than the rest of the array. A whole array normalization would not remove this type of bias. A spatial normalization method is an option for removing this type of bias, however, a print-tip normalization would be both easier and more effective. (Yang *et al.*, 2000) suggest simply using local regression separately for each print-tip group. Newer arrayer machines are able to do a whole array in a single print-tip, eliminating the need for this type of estimation of bias. For arrays with replicated spots for each gene, there are various ways to perform the normalization analysis. In their paper, (Tsodikov *et al.*, 2002) treat the replicated halves of arrays as two independent slides, but admit that it may be more beneficial to include an additional error structure to account for within slide replicate variance separately from between slide variance. With replicated spots on a single array, the array-by-gene interaction is no longer equal to spot effect, and so an additional variable must be added to the model to account for spot biases. For either method, local regression can be performed for each half-slide separately.

Over the past few years, microarray data has begun to accumulate from multiple experiments. More and more, researchers will want to perform meta-analyses, by combining the data from past experiments to obtain possible new results with more precision, and make further comparisons between treatments, tissue types, or other factors. In order to do this, the arrays from each experiment must be normalized against each other. One variance that is not often seen in a single experiment, but may very well be important across experiments, is the difference in average measured ratios. That is, some arrays may consistently measure ratios with a wider spread than others. In this case, an additional normalization method, one that transforms each array's average standard deviation to 1, may be applied in order to remove the bias of ratios that would occur if a per-gene analysis were to be done. In this case, each ratio from each array would be divided by its respective array's average standard deviation of ratios. Similar normalizations that attempt to standardize the per-array ratio variance will be discussed in the section 5, dealing with normalization techniques for gene clustering.

THE ROLE OF EXPERIMENTAL DESIGN IN THE REMOVAL OF TECHNICAL VARIANCE

Optimizing your design based on the experimental goal is an important part of a successful microarray experiment. In fact, the importance of pre-planning cannot be stressed enough. One question you may want to ask before designing your experiment is how much power you wish to have to detect differentially expressed genes with a ratio greater or equal to *x,* and how high of a false discovery rate (FDR) you are willing to allow (Hochberg and Benjamini, 1990).

This will determine the number of replicates you use. For example, (Wolfinger *et al.*, 2001) found that in order to have 85% power to detect a 2-fold change with a 1:20,000 false positive rate, seven replicates were needed. This number could change drastically depending on the type of array technology (single or dual channel), quality (reproducibility) of the arrays, the number of genes on each array, and the chosen false positive rate. Another question you want to ask is what are the most important samples, or comparisons you want to make, and how many experimental factors will be involved? For single-channel array experiments, it is obvious that more replicates should be done for samples of greater importance. For dual-channel array experiments, the many possible choices for designs make for a more complex problem. Depending on your answer to the above questions, you will choose from one of two main types of experimental designs- the universal reference design or the flipped dye design. The flipped dye design is more efficient for simple designs involving few factors, or for designs where one important factor of interest will be compared to many other factors, such as in a time series. Circular designs are complex versions of the flipped dye design and will be discussed only briefly at the end of this section. The universal reference design may be more appropriate for designs involving many factors of equal importance, such as comparing the expression profiles of a large number of tissue types, or for experiments that will likely be part of a larger meta-analysis in the future. In this section we will concentrate on the issues of experimental design as they relate to normalization and the removal of biases. For a more in depth look at issues of experimental design than is offered in this section, refer to (Yang and Speed, 2002a) and (Kerr and Churchill, 2001).

The rationale for the flipped dye design is that it allows for the estimation and removal of gene specific dye effects. These dye effects have been shown to be reproducible across independent arrays by the use of Control vs. Control arrays. Any deviation from a ratio of 1 in these arrays is due to either dye effect or residual or random error. Figure 12 shows ratio estimates for two genes on six arrays (3 sets of 2 replicates) using three different measurements: the raw data, the estimated ratio after dye effect has been removed by the use of control-vs-control arrays, and the ratio after dye effect has been removed by the flipped dye method (described below). As can be seen, the extra control array estimates of dye effect are very close to the effect estimated by the arrays themselves. The simplest flipped dye design consists of merely two arrays: one array with sample A labeled with cy3 and sample B with cy5, and the other array with sample B labeled with cy3 and sample A with cy5. This allows gene-specific dye effects t o be averaged out. The best estimate of the log transformed treatment effect for any gene x in this experiment would simply be the average of the ratios,

$$[log(A_1/B_1) + log(A_2/B_2)] / 2 = [log(cy3/cy5) - log(cy3/cy5)] / 2$$

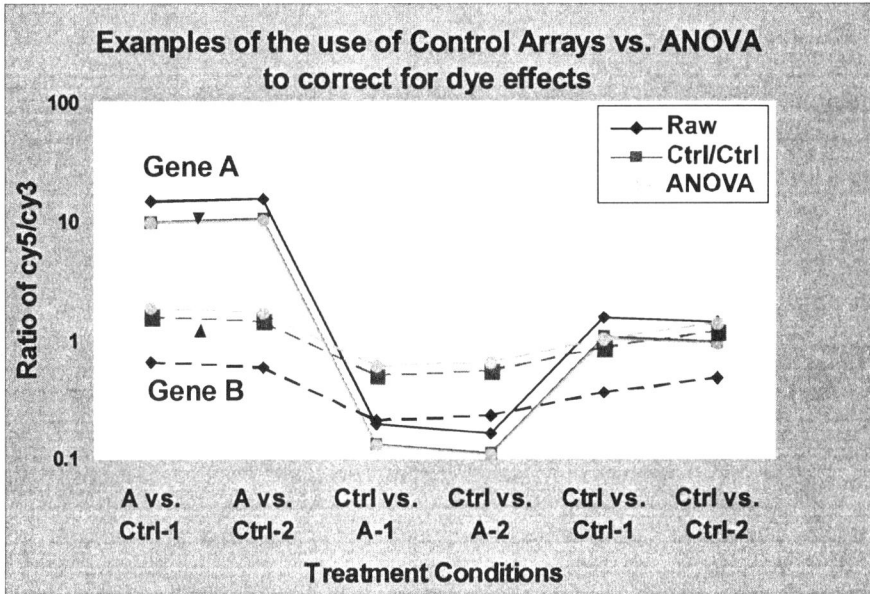

Figure 12. **Reproducibility of gene-specific dye effects for two genes.** For Gene A (solid lines) and Gene B (dashed lines), black lines are ratios of raw expression data between treated and control, dark gray lines indicate ratios after normalizing for dye-effects using control/control arrays, and light gray lines indicate ratios after normalizing for dye-effects using an ANOVA model without the control/control arrays. Note that both control/control and ANOVA methods give similar results.

and so the dye effects are averaged out. With more replicates, it is possible to obtain both an estimate of the dye effect, and a measurement of variance, or confidence, in your result. For any balanced design, where each sample is labeled with cy3 and cy5 an equal number of times, the ratio estimates may be calculated as averages, and the dye effects will be removed. For arrays done in triplicate, a simple average would weight one dye more than the other skewing the results; instead, the effects of dye could be removed using an ANOVA model with dye included as a factor, or as an alternative, Table 1 illustrates how dye effects could be normalized by hand. It is the same process as above with the added step of taking the average of the replicate ratios prior to averaging over the dye flipped estimates.

Figure 13 shows the steps taken in getting finalized ratio estimates of your microarray data by displaying ratios of genes from two flipped dye arrays. On

Table 1: Steps to remove gene specific dye effects in a triplicate experiment.

Array	Data (cy5:cy3)	Log transformed data	Average of replicates	log ratio estimate	Result
A:B	8:2	0.903 - 0.301 = 0.602	0.690	0.845	6.998
A:B	12:2	1.079 - 0.301 = 0.778	0.690		
B:A	2:20	0.301 - 1.301 = 1.000	1.000		

the left are the raw ratios after background subtraction, the middle shows the slight improvement provided by normalization by local regression, and on the right are the ratios after gene specific dye effects have been removed. As can be seen by comparison of these graphs, dye normalization also greatly improves the reproducibility of replicates per gene.

Including flipped dye arrays in an experiment is one requirement for having a statistically optimal design, which balances all the factors in the experiment-arrays, dyes, and treatments. (Here, treatments denote any other factor of interest, for example a toxin, mouse strain, tissue type, or age group.) Balanced designs do not confound any two experimental factors, meaning that the effect of each factor can be estimated and normalized out of the data. When factors are confounded with each other, their effects are indistinguishable. For example, an experiment done in triplicate without any flipped dye arrays will have

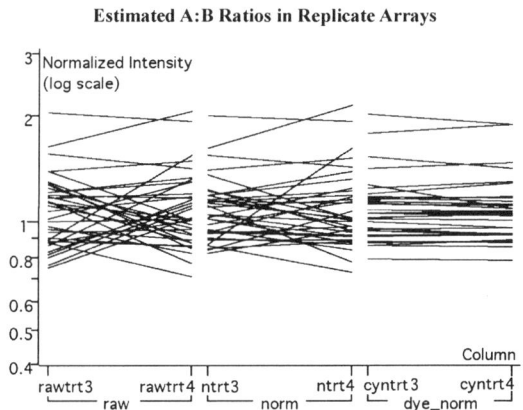

Figure 13. **Normalization Method Affects A:B Ratio Reproducibility.** Method: **A:** raw ratio, **B:** local regression, and **C:** local regression plus gene specific dye effect removal. (30 genes in a pair of dye flipped microarrays).

treatment completely confounded with dye effects, because each sample is only labeled with one dye each, and there is no way to separate their effects. Having confounding factors in your experiment is something to avoid, unless neither of the confounded factors are of interest. This leads us to the universal reference design, where dye is indeed confounded with treatment, but in this case all the treatments of interest are labeled with the same dye, so the dye effects with the reference can be divided out.

An example of a Universal Reference designed experiment is shown below (Figure 14) in which different regions of the mouse gastrointestinal tract are profiled with a series of microarrays described in Bates et al. (2002). This experimental design employed replicate arrays in which samples are labeled with Cy3, and universal reference which was mRNA obtained from a whole mouse was labeled with Cy5. Initial normalizations were median-based for each array, followed by Lowess local linear regression. Panel A shows the relative expression of each gene in relation to the universal reference, with many of the genes exhibiting relatively similar expression across the sample series, albeit higher or lower than the universal reference. Some of this effect could be gene specific dye effects and some of it could be due to consistently higher or lower expression in the GI samples relative to the whole mouse universal reference. To evaluate the relative gene expression across the sample series (i.e. to ask the question of what are the genes that are expressed in a GI segment-specific fashion) normalization is then performed versus the median ratio for each gene across the sample series. The result of this is shown in Panel B, which brings out the segment-specific differences in gene expression regardless of how strongly the gene is expressed in GI tract versus the whole mouse.

When choosing an experimental design, the goal is to minimize technical variance while increasing sensitivity to biological variance, but sometimes it may not be obvious which type of design will give the most desired results. *It is always a good idea to sketch the design, as in Figure 15, in order to assure yourself that all necessary factors will be able to be normalized, possible biases will be able to be detected, and gene specific dye effects can be removed especially for more complex designs.* In Figure 15, each box corresponds to a sample/factor combination, and arrows represent arrays, with the arrowheads pointing to the cy5 (red) channel. Ideas to keep in mind when designing an experiment are: keeping factors of interest balanced (not confounded), making sure there are replicates for each sample, and ensuring that the most important comparisons are either measured directly against each other on the same array, or close enough in the design that the multiplicative residual error from the connecting arrays does not end up overwhelming the effect of the factor of interest. The top portion of the experiment in Figure 15 uses a universal reference design, and normalization is accomplished with the per-array techniques described in section 2 (dual-channel array normalization) of this chapter. Gene specific dye effects are inherently removed by the structure of the design, since comparisons are made *between* ratios of treatment to universal reference. That is, the measured ratio for treatment A vs. universal reference is actually the true ratio plus dye effect (plus residual error), and the same is true for treatment B vs. universal reference. When these two values are subtracted, the dye effects cancel out.

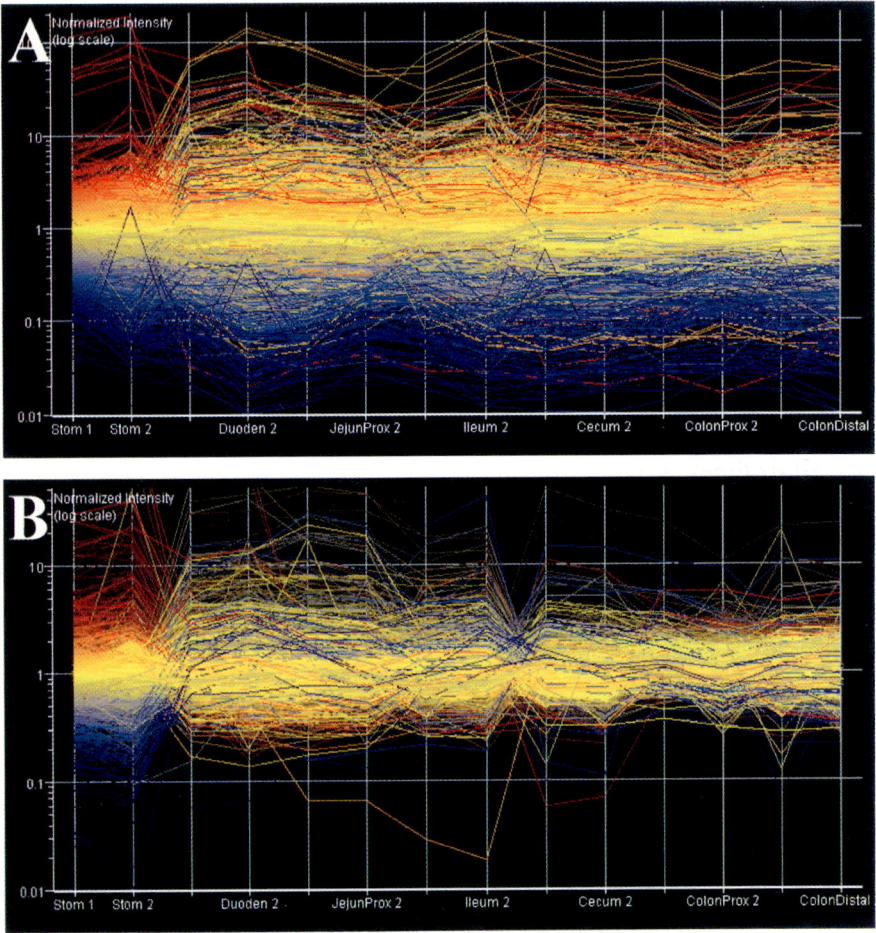

Figure 14. **The role of normalization in the characterization of GI tract segment-specific gene expression. A:** Segment-specific gene expression in GI tract samples relative to a universal reference of whole mouse mRNA. **B:** Same data as in panel **A**, but the ratio values for each gene in each tissue are normalized to the median of the gene's expression ratio across all of the samples. Note that the overall variance measured in the ratios is less than that of the whole-mouse normalized values because we are removing the variance attributable to the difference between the universal reference and the GI segments. Also note that the largest changes in relative gene expression patterns are those between the most prominent anatomic transitions of stomach to small bowel, and small bowel to large bowel.

This convenience is not true for the bottom portion of the design in Figure 15, which is a loop design with flipped dyes; in this case, there is an extra step in normalization to remove gene s pecific dye effects. The a dvantage of using the loop design is that it gives a lower variance for each comparison on average. Note that in the loop design each sample o f interest is measured six times, whereas in the top design each sample of intere st is measured three times. Yang and Speed (2 002b) re commend using the universal reference design in cases with 4 or more treatments to be compared. For 2 or 3 treatments, or for

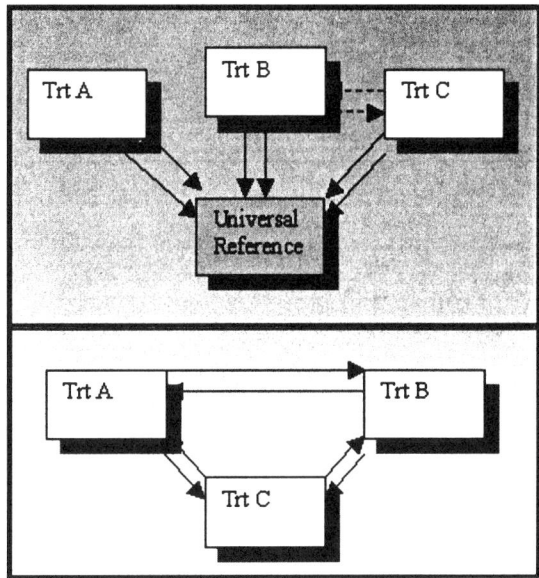

Figure 15: **(top)** Universal reference design with duplicate replicates.**(bottom)** Flipped-dye loop design also in duplicate. In this experiment, the loop design is likely to yield more precise results.

a small number of treatments in combination with other factors, a flipped dye design may be more appropriate. In cases for which there are a great number of treatments, including a few treatments whose comparison is more important than the others, additional flipped dye replicates may be added to obtain increased precision, as the dotted-lined arrows in Figure 15 indicate. Some of the more complex experimental designs do not fit into either of the above categories, but may instead be an amalgamation of the two. Whichever design you choose, the key idea to remember is to have more direct comparisons, and thus higher precision, for associations of the greatest interest.

GENE-SPECIFIC NORMALIZATIONS AND CLUSTERING

In the cluster analysis, the goal is to identify groups of genes with similar expression patterns across different experimental conditions or groups of biological states w ith similar o verall gene e xpressions. All p reviously described normalizations and transformation have obvious consequences on the cluster analysis. However, when identifying genes with similar patterns one can consider additional scaling of the normalized data in order to accentuate desired features of gene co-expressions. The most common goal of such additional transformations is to group genes based on similarities in relative changes in their expression instead of the similarities in absolute expression levels. This can be achieved either by transforming the data or by using the similarity measure that measure the feature of interest. Consider for example two clusters in Figure 16. The data consists of gene expression measurements across two cell cycles. Expression levels of genes in Cluster 1 peak in the S-phase of the cell cycle and expression levels of genes in Cluster 2 peak in the G1-phase of the cell cycle. Data was rescaled using the gene specific normalization by subtracting the gene-specific average expressions and dividing by gene-specific standard deviations. Rescaled data is shown in Figure 16A and 16B and non-rescaled data is shown in Figure 16C and 16D. From just a visual inspection of different clusters, it is obvious that rescaled data describes a very clear pattern of co-expression while the patterns in non-rescaled data are barely discernable.

This point if further demonstrated by performing the K-means clustering of the data in these two clusters using the transformed and non-transformed expression levels (Figure 17). Each line in the color-coded display represents expression levels of a single gene across 16 time points. Red color indicates high levels and green color indicates low expression levels. Label on the right side of the color display denote whether the gene originally belonged to Cluster 1 (labeled C1) or Cluster 2 (labeled C2). Clusters created from non-scaled data (Figure 17A) retain almost no original clustering structure. On the other hand, clustering based on the scaled data (Figure 17B) perfectly reconstructs the clustering shown in Figure 16.

When cluster analysis is performed using the hierarchical clustering approach, the similar effect can be achieved by using an appropriate similarity measure. For example, the classic microarray data clustering software described by (Eisen *et al.*, 1998) incorporates both "Centered" and "Uncentered" correlation coefficients as the similarity measures for the hierarchical clustering. Suppose that expressions of genes X and Y are observed at N experimental conditions and let x and y represent corresponding measurements under condition i=1,...,N. The similarity measure of expression profile can be then defined as

$$S(X, Y) = \frac{1}{N} \sum_{i=1}^{N} \left(\frac{x_i - x_{offset}}{S_x} \right) \left(\frac{y_i - y_{offset}}{S_y} \right)$$

where

$$S_G = \sqrt{\sum_{i=1}^{N} \left(\frac{g_i - g_{offset}}{N} \right)^2}$$

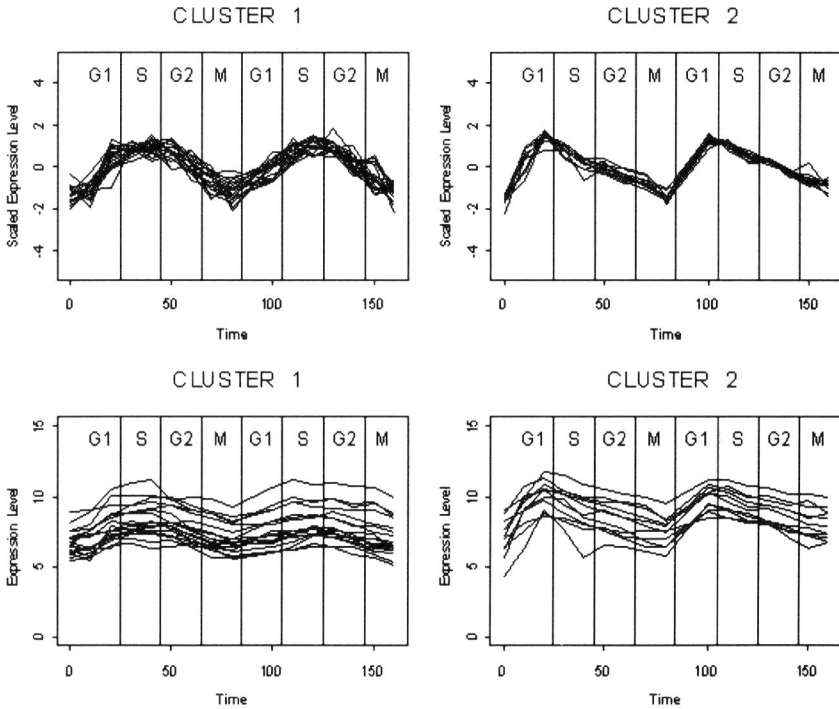

Figure 16. **Two Cell-Cycle Clusters** (Cho et al., 1998). **Cluster 1:** S-phase peak; **Cluster 2:** late G1-phase peak

The "mean centered" measure is obtained by setting g_{offset} to the mean expression level of the corresponding gene. In this case, $S(X,Y)$ is just the usual Pearson's correlation coefficient. "Uncentered" measure corresponds to setting g_{offset} to 0. Effects of centering on the results in a hierarchical clustering approach is demonstrated in Figure 18. When the data is scaled prior to clustering, results obtained by using either centered or uncentered correlation are identical (Figure 18A and 18B). In this case cluster analysis has successfully reconstructed original clusters. The more interesting results are obtained by using non-scaled data (Figure 18C and 18D). When centering was not performed (Figure 18C), the cluster analysis is not able to reconstruct original clusters. On the other hand, when "centering" is applied, original clustering is reconstructed perfectly and the hierarchical organization of expression profiles is similar to the one obtained using scaled data. To conclude, gene specific data transformations applied to normalized data prior to cluster analysis can have tremendous consequences on the results of the analysis. While choosing appropriate similarity measure in the hierarchical cluster analysis can sometimes substitute for the appropriate data transformation, revealed patterns of expression could still be more obvious when the actual transformation is performed (see Figure 16). Furthermore, if one wants to use some of the non-hierarchical clustering procedures (e.g. K-means), transforming the data becomes an imperative.

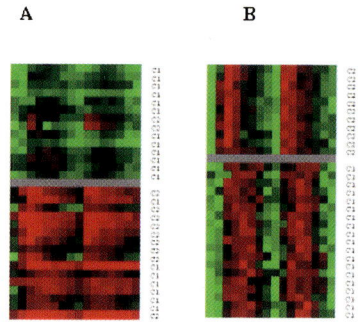

Figure   17.   **K-means clustering**. **A:** Non-Scaled Data; **B:** Scaled Data

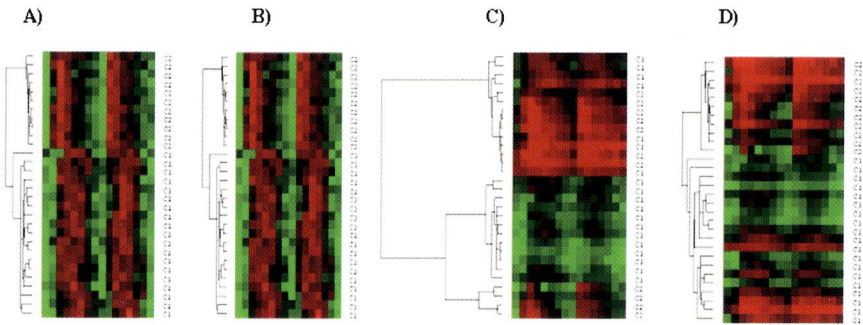

Figure 18. Two Cell-Cycle Clusters (Cho et al., 1998). A: Scaled Data Uncentered Correlation; B: Scaled Data Centered Correlation; C: Non-Scaled Data Uncentered Correlation; D: Non-Scaled Data Centered Correlation

REFERENCES

1   Bates MD, Erwin CR, Sanford LP, Wiginton D, Bezerra JA, Schatzman LC, Jegga AG, Ley-Ebert C, Williams SS, Steinbrecher KA, Warner BW, Cohen MB, Aronow BJ   (2002) Novel genes and functional relationships in the adult mouse gastrointestinal tract identified by microarray analysis. *Gastroenterology* 122(5):1467-82

2   Cho RJ, Campbell M J, Winzeler E A, Steinmetz L, Conway A, Wodicka L, Wolfsberg T G, Gabrielian A E, Landsman D, Lockhart D J and Davis R W (1998) A Genome-Wide Transcriptional Analysis of the Mitotic Cell Cycle. *Mol Cell* **2**: pp 65-73.

3   Colantuoni C, Henry G, Zeger S. and Pevsner J. (2002) Local Mean Normalization of
     Microarray Element Signal Intensities Across an Array Surface: Quality Control and Correc-
     tion of Spatially Systematic Artifacts. *Biotechniques* **32**: pp 1316-1320.

4   Dudoit S, Yang Y, M.J. and Speed T P (2000) Statistical Methods for Identifying Differen-
     tially Expressed Genes in Replicated CDNA Microarray Experiments. *Berkeley Tech Report
     #578 from Aug 2000.*

5   Eisen MB, Spellman P T, Brown P O and Botstein D (1998) Cluster Analysis and Display
     of Genome-Wide Expression Patterns. *Proc Natl Acad Sci U S A* **95**: pp 14863-14868.

6   Finkelstei D, Ewing R, Gollub J, Sterky F, Cherry J M and Somerville S (2002) Microarray
     Data Quality Analysis: Lessons From the AFGC Project. Arabidopsis Functional Genomics
     Consortium. *Plant Mol Biol* **48**: pp 119-131.

7   Goryachev AB, Macgregor P F and Edwards A M (2001) Unfolding of Microarray Data. *J
     Comput Biol* **8**: pp 443-461.

8   Hill AA, Brown E L, Whitley M Z, Tucker-Kellogg G, Hunter C P and Slonim D K (2001)
     Evaluation of Normalization Procedures for Oligonucleeuotide Array Data Based on Spiked
     CRNA Controls. *Genome Biol* **2**: pp 55.

9   Hochberg Y, Benjamini Y. More powerful procedures for multiple significance testing. Stat
     Med 1990 Jul;9(7):811-8

10   Kepler TB, Crosby L and Morgan K T (2002) Normalization and Analysis of DNA Microarray
     Data by Self-Consistency and Local Regression. *Genome Biol* **3**: pp 1-12.

11   Kerr KM and Churchill G A (2001) Experimental Design for Gene Expression Microarrays.
     Biostatistics **2**: pp 183-201.

12   Kerr KM, Martin M and Churchill G A (2000) Analysis of Variance for Gene Expression
     Microarray Data. *Journal of Computational Biology* **7**: pp 819-837.

13   Schuchhardt J, Beule D, Malik A, Wolski E, Eickhoff H, Lehrach H and Herzel H (2000)
     Normalization Strategies for CDNA Microarrays. *Nucleic Acids Res* **28**: pp E47.

14   Tsodikov A, Szabo A and Jones D (2002) Adjustments and Measures of Differential Expres-
     sion for Microarray Data. *Bioinformatics* **18**: pp 251-260.

15   Wolfinger RD, Gibson G, Wolfinger E.D., Bennett L, Hamadeh H, Bushel P, Afshari C and
     Paules R S (2001) Assessing Gene Significance From CDNA Microarray Expression Data
     Via Mixed Models. *Submitted.*

16   Workman C, Jensen L J, Jarmer H, Berka R, Gautier L, Nielsen H B, Saxild H, Nielsen C,
     Brunak S and Knudsen S (2002) A New Non-Linear Normalization Method for Reducing
     Variability in DNA Microarray Experiments. *Genome Biol* **3**: pp 48.1-48.16.

17   Yang MC, Ruan Q G, Yang J J, Eckenrode S, Wu S, McIndoe R A and She J X (2001) A
     Statistical Method for Flagging Weak Spots Improves Normalization and Ratio Estimates in

Microarrays. *Physiol Genomics* **7**: pp 45-53.

18 Yang Y, Dudoit S, Luu P and Speed T (2000) Normalization for CDNA Microarray Data. *SPIE BiOS 2001, San Jose, California.*

19 Yang Y and Speed T P (2002) Design Issues for CDNA Microarray Experiments. *Genetics* **3**: pp 579-588.

20 Zien A, Aigner T, Zimmer R and Lengauer T (2001) Centralization: a New Method for the Normalization of Gene Expression Data. *Bioinformatics* 17 Suppl 1: pp S323-S331.

Chapter 6

EXPERIMENTAL DESIGN AND DATA ANALYSIS

Eric Blalock
University of Kentucky, Department of Molecular and Biomedical Pharmacology

INTRODUCTION

Microarrays are extremely powerful tools for the analysis of genomic expression levels in physiologic and pathologic phenomena. Many early microarray experiments have been published in high-impact journals due to the sheer technical prowess of this methodology, as well as the future promise such technology h olds. S everal important f indings have b een r eported u sing microarrays, and the technology is becoming more and more accessible. There is no doubt that the technology itself is valid, and furthermore that important changes in gene expression levels are being detected with the various spotted and Affymetrix type arrays. Yet many researchers new to the field are quickly bogged down by the massive (and often seemingly inscrutable) results of their first studies.

In statistical terms, microarray studies are often woefully underpowered. That is, their ability to reliably discern a difference between expression levels in two treatments is inadequate considering the nature of the variance of that expression. In light microscopy, power refers to the ability to discern two points- the closer those points are to one another, the greater the microscope's power must be to resolve them. Likewise, the smaller the difference (or 'effect size'- see next chapter) between two groups, the greater the statistical power must be to detect differences. Increased power comes from two sources, decreased variability, and increased n (the number of samples measured in each group).

For underpowered studies, there may be too much within group variation within the subjects to reliably detect differences between groups, and/or there may be too few subjects in each group to provide an appropriate estimate of variance. Unlike cross-hybridization error, this does not mean that the signal intensities measured have somehow 'misled' the researcher into believing that a difference exists between two treatments when in fact there is no difference, but rather that the sample used to generate the signal was not an adequate representation of the population from which it was drawn. Many researchers argue, when this replication issue is brought up, that they are only using microarrays for gene discovery and that the microarrays themselves are too expensive for replication. However, there is a difference between expensive and wasteful. While a well-designed experiment with sufficient replication may be expensive, the money spent can lead to a library of data whose results can be mined for years to come by many researchers. On the other hand, misleading data generated from poorly designed experiments costs as much, if not more in terms of time spent analyzing data and planning and conducting experiments based on bogus data. Thus, despite microarray-based technology's many powerful attributes, the ability to overcome ill-conceived experimental design and poor replication is not among them.

One of the great challenges we are currently facing in microarray research is how to winnow the wheat from the chaff. There are such a large number of measurements on each chip that, in its native form, the data typically exceed human capacity to discern meaningful relationships. In mathematical, statistical, and bioinformatic fields, this exceedingly large array of potentially intertwined measurements is often referred to as "highly dimensional" (or the "dimensional curse"- see Chapter 8), meaning that once an experimenter has gone beyond examining the potential relationship among just three of the often tens of thousands of genes, they have saturated Euclidian space. Thus representations of these relationships are impossible without some sort of "dimensionality reduction". Actually, dimensionality reduction is used all the time, taking the average of a set of data or expressing the difference between two measurements as a percentage or fold-change difference are commonplace dimensionality reductions. With microarrays, however, such procedures fall well short of simplifying the data sufficiently.

In the following chapter I will discuss some of the strengths and weaknesses of microarray technology as it pertains to basic research goals. Because Experimental Design and Data Interpretation are fields unto themselves, it is not the purpose of this chapter to offer an all-inclusive overview these procedures. Instead, I intend to describe some of the more common experimental designs, the analysis techniques for which they are best suited, and their attendant problems.

MEASURING RNA

Presently there are four basic techniques that detect the amount of particular RNA species: Real-time polymerase chain reaction (RT-PCR), Northern blot,

RNAse protection assay (RPA), and *In Situ* hybridization histochemistry (ISHH). All of these technologies rely on the hybridization properties of complementary strands of RNA, and use "probe" strands of RNA designed to hybridize to areas that are representative of, or unique to, the mRNA of interest (target). The detection techniques can be divided into solution-solution and solution-solid phase hybridization events (Lemke, 1994).

Solution-solution hybridizations, assumed to follow simple bimolecular mechanics, are thought to provide the best measures of RNA abundance. Thus, RPA and RT-PCR are the "gold standard" approaches for the quantification of RNA. In solution-solid hybridizations, liquid phase probes invade a solid phase (a gel or membrane in the case of Northerns, or a tissue slice in the case of ISHH) that contains the target RNA. The dynamics for target-probe hybridizations in solution-solid interfaces are more complicated, because the solid phase in which the target RNA is imbedded can interfere with probe access.

Microarray technology takes advantage of the same hybridization properties, and is somewhat intermediate between full-fledged solution-solution and solution-solid interfaces. In early microarray experiments, researchers spotted RNA onto poly-L-lysine treated glass slides. Each molecule of RNA then non-specifically reacted with the treated glass surface (See Chapter 1). Thus one can visualize RNA adhered to slide surfaces in this manner like a plate of cooked spaghetti overturned on a counter top, allowed to dry, and those noodles that did not stick swept away. One surface of the longitudinal axis of each molecule spotted on the glass is unavailable for hybridization. If that surface is important to the hybridization process, then the signal will not be optimal. However, the rest of the molecule is exposed to the complementary-strand-containing solution, and in that sense this type of spotted array is superior to other solution-solid detection technologies because the target RNA is exposed, at least partially, to the solution phase.

In other spotted glass technologies, spacer molecules with predefined attachment moieties tailored to specific glass chemical coatings (See Chapters 1 and 2) are added to the 5' end of the RNA prior to spotting. Because the slide coating is designed to link to these moieties, only the modified region of the RNA adheres (covalently) to the glass surface. Thus, RNA molecules prepared in this fashion are often depicted as sticking up from the slide surface like kelp from the sea floor, and the ability for the slide-fixed and solution phase RNA to hybridize is less restricted (steric hindrance is reduced). It is difficult to determine whether or not this spatial distribution is true, and for all we know, the attached strands could wind together in complicated 3D structures (braids), that actually hinder interaction. This attachment technology probably suffers less 5' biased steric hindrance than do its direct attachment progenitors. The increased sensitivity reported by many authors indicates that these protocols offer definite improvement over non-covalent attachment chemistries. Affymetrix' photolithographic process confers similar advantages to their oligonucleotide GeneChip® arrays.

FOLD CHANGE SIGNIFICANCE

A casual search of published primary research based on microarray technology reveals that, from its very inception, microarray results have been reported much as the results of the technologies from which microarrays sprung. The ability to detect specific RNA has been with us since the mid 60's, and in the mid to late 70's pioneering work (Schechter, 1975; Arnemann et al., 1979), on mRNA quantification had begun. It was also at this time that "fold change"-based nomenclature came into vogue (Ramirez et al., 1975; Arnemann et al., 1979).

This was due in large part to the magnitude of the increases observed in these early studies. For instance, Ramirez et al reported a ~250 fold induction (albeit from starting concentrations that were barely detectable) in the expression of globin mRNA during the first 10 hours of erythropoetin exposure. Following this time course out, they subsequently found a 2 and 2-3 fold increase at later time points (in comparison to the next latest time point). Similarly, Arnemann et al reported a 50-100 fold increase in the expression of pre-uteroglobin following hormonal treatment.

Fold-change was preferred because comparative measures (e.g., control and treatment) were run side by side, thus a direct fold change between the two could be compared with another fold change metric calculated from another run of replicate or duplicate experiments, while the raw values were subject to technical variation between runs. In effect, the fold change calculation served as a control for the potential technical variation from one run to another. Furthermore, while attempts at counting actual copies of mRNA were made, researchers felt that, at the very least, the relative difference in expression level, as reflected by fold change, would hold true- while the technology available for counting actual copies of mRNA was not as reliable.

Later studies using more advanced technologies, such as two-color hybridization (Pott and Fuss, 1995; Schena et al., 1996), designed their experiments specifically to report differences on a fold change scale. In these studies, it became obvious to those involved that the primary source of variation was technological, rather than biological. That is, the largest differences in measurement occurred not because of inter-subject mRNA level variation, but because of the sensitivity of the detection techniques to subtle variations in procedure. For assessing quantitative differences between two treatments, then, this "technological sensitivity" could be disastrous, easily masking the often less robust biological differences.

Researchers attacked this problem with an essentially two-pronged approach. First, they would measure the same sample multiple times ("duplicates"). Because variations in duplicate measurement could be considered primarily technical in origin (as there is essentially no biological contribution to variability when the same sample is measured repeatedly) the variation across duplicates could be taken to represent technological variation, and procedures such as outlier removal and averaging could then be used to reduce the contribution of technological variation. Secondly, researchers limited their detection to only those genes that showed large changes across treatments.

Researchers attempting to detect miniscule amounts of genetic material, and further quantify that difference, must acknowledge that the number of mRNA copies they are interested in measuring is not the only thing affecting measurement. There are several steps in sample preparation, and potential technological error is introduced at each step. This principle is of course true of any measurement system, but due to the nature of microarray analysis and the tremendous importance placed on these fold-change based determinations, the subject deserves special treatment.

Basically then, fold change has been used as a proxy for statistical determinations of significance (Kerr and Churchill, 2001). The advantage of fold-change based analysis is the low replication requirement. The disadvantages are the absence of an estimate of variance, the assumption of a central tendency of the measurement, a bias towards selecting large changes that may or may not be as biologically important as those of lesser magnitude, and the instability of fold-change based measurements when at least one of the two measures to be compared is very small.

Calculating Fold Change

In a comparison of gene expression level in two conditions (A and B), fold-change is relatively simple to calculate according to the following logic statement:
$$If\ A > B,\ then\ A/B,\ else\ B/A*-1$$
Thus, every fold change is either positive if A is greater than B, or negative if B is greater than A, and there are no reported values between 1 and −1. However, calculating fold changes in this manner holds true only for designs in which there is a single measure for each condition, or in which the means are used to calculate the fold change. Consider a situation in which spotted arrays are run on 10 replicates for each condition (Table 1).

Note that there is a great deal of disparity among the various calculations for fold change (Figure 1; Table 1). Most scientists would probably prefer the simple fold change calculated based on the averages of the A and B categories (.85 /

Table 1: Fold change metrics.

Replicate	A	B	A/B	B/A	Sign	Log A/B	Log B/A
1	1.2	0.12	10.68	0.09	10.68	1.03	-1.03
2	0.6	0.88	0.63	1.58	-1.58	-0.20	0.20
3	1.5	0.69	2.18	0.46	2.18	0.34	-0.34
4	0.4	0.89	0.48	2.08	-2.08	-0.32	0.32
5	1	0.14	6.85	0.15	6.85	0.84	-0.84
6	0.4	0.05	8.90	0.11	8.90	0.95	-0.95
7	1.1	0.01	89.14	0.01	89.14	1.95	-1.95
8	1.1	0.46	2.43	0.41	2.43	0.39	-0.39
9	0.6	0.34	1.63	0.61	1.63	0.21	-0.21
10	0.6	0.74	0.78	1.28	-1.28	-0.11	0.11
Mean	0.85	0.43	12.37	0.68	11.69	0.51	-0.51
SD	0.39	0.35	27.22	0.72	27.57	0.69	0.69
Cvar			2.20	1.06	2.36	1.37	1.37
Geometric Mean						3.22	3.22

Note: A and B are the measures (in arbitrary units) of gene expression level in different treatments, grey boxes under ratio calculations (A/B and B/A) represent conditions in which the fold change value is < 1, a condition generally not reported in microarray analysis. This is compensated in the 'Sign' column using the logic statement above. Note that average log ratios (Log A/B and Log B/A) are identical (albeit with different signs). Cvar is the coefficient of variation (Standard Deviation/ Mean), and geometric mean for log ratio averages is provided by the antilog of the mean log scores (e.g., Mean Log A/B = .51, 10^{51} = 3.22).

.43 = ~2). Keep in mind though, that fold change is often viewed as descriptive, and of little value in the absence of some assessment of variance. Thus, this fold change calculated from the means, while descriptive of the process, cannot provide an estimate of variance. Furthermore, most spotted array software reports the fold change values for each spot, rather than the raw signal intensities, which could then be averaged to create a fold change. This 'per spot' ratio reporting helps control for variations in spot morphology, adherence, uneven

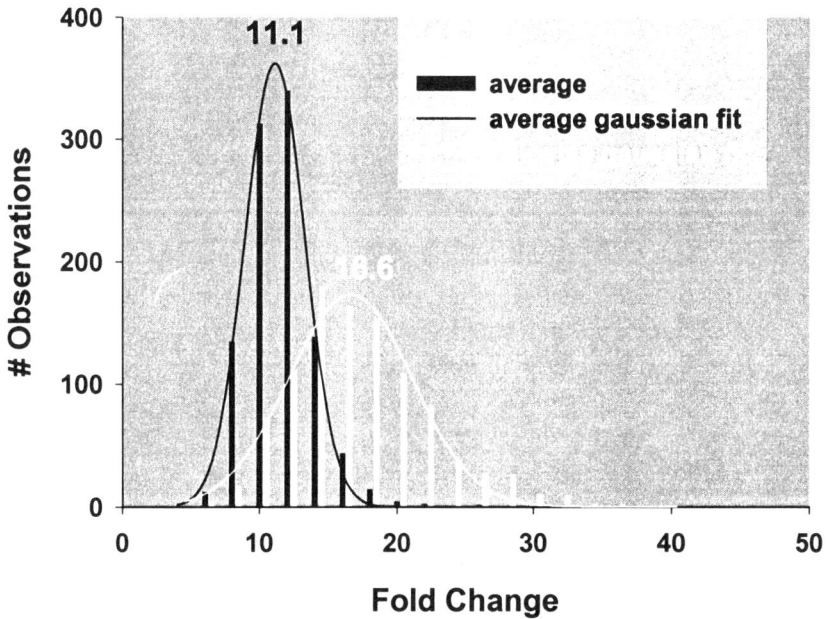

Figure 1. **Standard verses Pairwise fold change comparisons with model data**. On the x axis is the fold change level observed. On the y axis is the frequency of observations. The two groups compared were artificially set to have an overall mean 10-fold difference and data were randomly generated in Excel using the randbetween() function. The average fold change for each of a thousand iterations were calculated for the mean data (in black) and pairwise-based calculations (in white). The center of the resulting guassian fits is shown for each distribution. Multiple pairwise comparisons clearly showed a broader distribution and were biased to overestimations of the difference.

hybridization, etc., although a paired t-test would also control for such issues. These fold change values, if averaged strictly with one group as the numerator (A/B or B/A) show very poor agreement. In other words, deciding which treatment is to be the numerator has a dramatic effect on the mean fold change, standard deviation, and coefficient of variance (Table 1). Logging the ratios (Log A/B or Log B/A) results in log mean ratios that are equivalent except for the sign of the result. The antilog of the absolute mean of the log ratios results in the geometric mean, a reasonable estimate of the overall fold change (Nadon and Shoemaker, 2002). Note that in table 1 this estimate (3.22) is well above that of mean A / mean B (~2 fold), and well below that of the 'sign'- based estimate of fold change (11.69). The antilog of the mean log fold change is generally considered to be the superior method of calculation, except for very small values (weak signals) (Durbin et al., 2002; Nadon and Shoemaker, 2002).

There is considerable debate in the literature as to how large a fold change needs to be before it is considered significant. A fold-change cut off that generates a gene list of sufficient size for publication has become the pragmatic standard. Recently, the Limit Fold Change (LFC) model has been proposed as a way to objectively determine fold change levels that could be considered significant (Mutch et al., 2002). One of the problems with fold change is that weaker signals tend to generate larger fold changes and stronger signals tend to generate weaker signals. Mutch et al., fit this tendency with a linear regression, such that a fixed proportion at each expression level (in their examples either 1 or 10%) is determined to have a fold change sufficiently distant from the norm at that expression level. Thus, ratios calculated from lower intensity scores need higher values to achieve the same level of significance as ratios calculated from more intense signals.

Powerful Combinations of Information

There is no arguing with success. Molecular biology, low to no replications and fold change-based measurements notwithstanding, has contributed to a tremendous amount, if not the majority, of biological discoveries in the past forty years. So how can it be that such apparently flawed analysis techniques have resulted in such wildly successful work? One answer may lie in the way in which molecular biologists design their studies.

As an example, let's look at an experimental design that has met with great success in the hands of skilled molecular biologists. "Promoter bashing" has contributed a great deal to our present understanding of cellular physiology and gene regulation. Furthermore, it fulfils the above low rep/ fold change scenario [Michael K arin in S cienceWatch (ScienceWatch, 1999)]. In this type of experiment, a region of DNA known as a promoter (so named because its presence is thought to facilitate transcription) is attached to a reporter gene (one whose protein product is easily quantified- e.g., luciferase, ChAT, etc.). This new promoter-reporter construct is then transfected/infected into a biological system capable of expressing the reporter protein. Various manipulations are made to the genetic structure of the promoter itself (deletions, frame shifts, point mutations) and the difference in the level of reporter protein expression is taken as an indication of the degree of facilitation or inhibition invoked by these changes.

Thus, experimenters working with this protocol hold in their hands three important pieces of information. First, they know the direction of change invoked by the promoter's activation under ordinary circumstances. Second, they know which cells in their expression system are capable of responding to the treatment, and thus isolate those cells for subsequent measurement. Third, the researchers know which gene (the reporter gene) they should investigate. Additionally, the expression systems used to test these promoter modifications are often clones existing in a simplified (cultured) environment. Thus, many of the genetic and environmental contributions to variation among individuals may be greatly reduced. By combining expected change direction, low inter-individual variance, and *a priori* knowledge o f the g ene of i nterest, researchers c an make r eliable inferences about the effects of alterations in promoter sequence on gene expression. Furthermore, while additional replications of these experiments would undoubtedly statistically bear out these results, it is obvious from the success of this experimental design that such replication would indeed be wasteful.

However, this has lead some researchers to conclude that replication is itself wasteful, that statistical methods are largely superfluous for molecular research, and that granting institutions and journals requesting such statistical control are composed of molecular *illiterati* putting red tape in the way of scientific progress. While statistical replication has been empirically demonstrated to be unnecessary in the above promoter bashing scenario, this does not mean that a statistical analysis on replicates would fail to assign significance to such a result. If the researcher makes a true discovery (e.g., a treatment changes gene expression level relative to control) then the difference will be born out by statistical analysis at least 95% of the time (if the p value is set at .05 and the power- e.g., number of replicates- is sufficient for discovery). Certainly in an experiment in which one comparison was sufficient for discovery, it is unlikely that a statistical analysis done on more than one replicate would fail to have sufficient power. In fact, if a statistical analysis on three samples failed to agree with the single sample result, it would be more logical to conclude that there was more variance in the system than had at first been assumed. While statistical analysis in this narrow scope may be redundant, there should be no question as to whether statistical procedures are capable of detecting the same important differences found by empirical fold change approaches (Mirnics, 2001).

VARIANCE

Sources of variance can be broadly divided into two groups, technological and biological. Technological variance has multiple sources, but can generally be broken down into systematic and measurement error. *Systematic error* (or bias) refers to sources of difference in measurement that arise from procedural variation (sample preparation, RNA extraction, etc.), and is typically reduced by increasing familiarity with the steps involved. However, even the most practiced researcher's work will still suffer from some degree of systematic error, and it is strongly recommended that special attention in microarray experiments be given to the 'balancing' of these sources across treatments (Kerr et al., 2000; Wolfinger et al., 2001; Nadon and Shoemaker, 2002; Tiesman, 2002). Additionally, *errors of measurement* (one component of random error) include sources of variance based on the limitations of the tools used. Scanners, pipetters, amplifiers, and oscilloscopes- all measurement equipment suffers from some intrinsic variation.

Biological variation (the second component of random error), on the other hand, represents natural differences in the subjects being examined. While it may be desirable to increase the number of subjects in order to provide a more accurate picture of this variation, the variance itself is generally important information (thus the statistical term "standard error" may in some ways impugn valuable information by associating it with the term "error"). Additionally, there is the potential for interaction among sources of variance. For instance, a weight scale may be inaccurate at weights greater than 200 lbs. Thus, in a comparison of adult human weights, this scale may give a badly skewed representation of the natural variance in male weight and have much less impact on the natural weight variance of the female population (sources of variance are treated more rigorously in the following chapter).

So in any experiment, variation can be theoretically resolved into technological, biological, and interaction sources. Ordinarily in experimental design, technical sources are balanced across treatment groups ('counter b alancing'). For instance, in a two-color spotted microarray experiment, each treatment group should be given equal opportunity with each dye. If, on slide 1 treatment group A is labeled with Cy3 and group B is labeled with Cy5, then the dye labeling should be reversed for slide 2 to avoid potential dye-dependent measurement artifacts [so called "dye flip" or "loop" design- see the gold standard of this design (Kerr and Churchill, 2001)]. A more common example would be the practice of interleaving/alternating preparation of subjects from each treatment group, such that all the measurements from group A are not taken on one day, and those of group B on another.

Potential Sources of Variance (from presentations by Jay Tiesman and Grier Page at the 2002 CHI Microarray Data Analysis conference).

- Preparation (Major)
- Processing (Minor)
- Lot Number (Variable)
- Scan Order (Amplifies other differences)
- Technician
- Date
- Fluidics Station
- Location

Replicates vs. Duplicates

For the purposes of this chapter, I consider duplicates to be more than one measurement of the same biological material. In microarray experiments, the repeated measurement of gene expression levels from a single biological source (regardless of whether that source was contributed by a single or by multiple subjects) would be considered duplication. The variability inherent in duplicate measurements represents tool-based or technological variation (put simply- any variation that did not arise from the biological population). For microarrays, this information is hugely important. Before one can embark on scientific discovery using any new technology, it is important that the technological limitations be described. As with the earlier example of a weight scale that inaccurately measures weights greater than 200 lbs, it is important to define those limitations before meaningful interpretation of biological results can be made.

Replicates, on the other hand, are separate measurements from different biological samples that have in common the particular treatment being examined. Importantly, if multiple subjects have been combined to form a single biological source for a single array, and a different group from the same treatment group has been combined for another array, then these two arrays can also be considered replicates- this type of design is termed here "sub-pooling" and is discussed later in this chapter, as well as given a rigorous and formal treatment in the following chapter.

Thus, replication encompasses both kinds of variation, tool-based and biological, and is inherently more "noisy" than duplicate measurement if the two forms of variation could be considered additive. As the number of replicates in each group increases, the (hopefully) consistent contribution of tool-based variation, because it has been carefully balanced across all the treatments, will contribute proportionally less, as the biological variation becomes a more dominant source of variance among the measures. Biological variance represents the system being studied rather than that of the tools used to study it, and is the variation about which most people investigating biological phenomena are interested.

EXPERIMENTAL DESIGN

There are four factors to consider when constructing an experiment: control, randomization, replication, and balance. Most of these concepts are ingrained into us at the undergraduate and postgraduate level, yet a measure of insouciance has emerged in the microarray literature regarding the appropriate application of these principles. With the move in the past three years towards a more statistical investigation of differences in gene expression levels, statisticians are becoming more and more frustrated with us (basic science researchers) for our lack of adherence to these principles and the difficulty this lack of adherence engenders in statisticians' attempts to help us make sense of our microarray data (an often quoted statement from R.A. Fisher may apply *"To call in the statistician after the experiment is done m ay be no more than asking h im to perform a postmortem examination: he may be able to say what the experiment died of"*). In this section, I'll provide a very basic overview of these principles and their importance in microarray analysis.

Control: In its simplest terms, control refers to the arrangement of test subjects for an experiment- a baseline or control group and an experimental group are the bare minimum. The experimental group provides sufficiency for statistical testing. That is, is the treatment sufficient to bring about change in the measured variable? The control group provides a measure of necessity- was the treatment necessary for the change to occur?

Balance: Equally important but unfortunately often overlooked in microarray studies, balance refers to the equal representation of subjects within each treatment or condition. To use an elegant design like Kerr and Churchill's ANOVA design for microarray analysis (Kerr et al., 2000), each potential source of variation should be equally represented. If some of those groups are not equally represented, then i ndependent test s tatistics for e ach of t he factors a nd interactions are confounded. Without equal representation, the different groups under study will have estimates of variance that themselves vary- presenting a potential confound in analysis (Good, 1997). Most statistical procedures include in their formulae a component to control for different n's in different treatment groups, but this is generally in place to help control for inadvertent data loss in the course of experimentation, rather than the deliberate construction of an unbalanced design.

Randomization: The assignment of test subjects to treatment groups in a completely unbiased manner. Many obvious variables (e.g., gender, weight, age) are known before hand and experimental subjects can be parsed out into treatment groups in such a way that each of these factors has equal representation in each group ('blocking' the subjects). Randomizing test subjects' assignments to groups eliminates possible bias by those factors that are unknown or uncontrolled (such as the experimenter's subconscious bias towards selecting larger subjects first).

Replication: Replicates are the number of measured subjects in each treatment group. Importantly for microarray studies, it is the measuring event that leads to the assignment of replicate status (see below). Thus, if 10 subjects were assigned to one treatment group, and had their tissue pooled, their RNA extracted, and the pool of RNA tested on one or multiple chips, then that set of 10 animals becomes a single replicate. The act of pooling the data prior to measurement changes the statistical n of the experiment. In conjunction with randomisation, replication is essential to the assessment of experimental error, and is critical to the determination of the difference between background variation and that caused by the treatment (treatment effect).

Single Color Designs

With Affymetrix arrays, and other technologies that generate single intensity measures (that is, only measure one source on each array), discriminating between replicates and duplicates is straightforward because one array can hybridise to a single source. If this single source is measured on different arrays, then the results are duplicates. If different sources are measured on different arrays, then the measurements are r eplicates. For the purposes of illustration, a comparison between a control and a treatment group is considered.

A. Single subject and single chip per treatment. The most basic and least powerful design- a single subject from the control group is compared to a single subject from the experimental group. In this case no inference of variation can be made.

B. Single subject and multiple chips per treatment. An improvement because duplicate measures from each animal allow for estimation of tool-based variation. However, no biological variance within groups is included because only single subjects are measured, and this design is very vulnerable to the effects of outliers.

C. Pooling. Clearly, one major weakness of pooling approaches is their inability to detect outliers. The presence of such an outlier would be a disaster in the first "single subject, single chip, one subject per treatment" approach (and with thousands of measures for each subject, the presence of at least one outlier is not unlikely). The pooled design offers some relief, as outliers would be reduced by "dilution" with other observations. Unfortunately, while pooling offers some insurance against the influence of outliers, it is incapable of detecting their presence.

D. Pooling with duplication (Figure 2). More powerful than pooling alone, this design permits duplication-based estimation of tool-based variance. It is important to note that the derived estimates of variance do not contain any biological component. Thus, the variation detected between measures is probably inappropriate for statistical analyses designed to detect biological differences between groups, because no biological source is considered. In addition to the "pooled" design's increased confidence of fold-change-based discovery, "pooled with duplication" also helps to control for technologically derived outliers.

E. One subject per chip, multiple chips per treatment group (Figure 3).

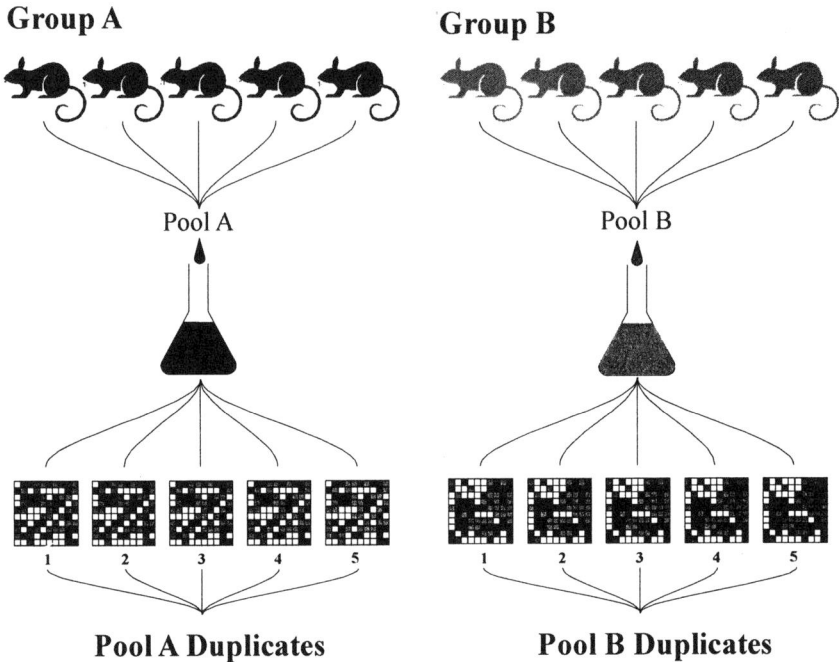

Figure 2. **Schematic of pooling with duplication.** The act of pooling the biological material from multiple sources reduces the potential n available among those multiple sources to that of the single pool. Microarrays prepared from this single source are duplicates.

This is probably the most conventional design, and is the first one discussed here that incorporates biological variance into the measure. On a per chip basis, variance is theoretically greater than that of the pooled designs. To increase the statistical power of the experiments, increase the number of subjects in each group. This method may provide data that is more universally applicable by other researchers, but the increased chip requirement for the detection of small magnitude changes may make this protocol economically inaccessible to many investigators.

F. Sub-pooling (Figure 4). This design also incorporates both biological and tool-based variance, and the pooling procedure reduces the amplitude of the variance while preserving its biological component. The disadvantage of this technique is that, owing to its highly individualized design, results may be less useful to other researchers.

Results generated using the design mentioned in "E" can be used to estimate the variance of particular gene expression levels, which can then be used by

Group A   **Group B**

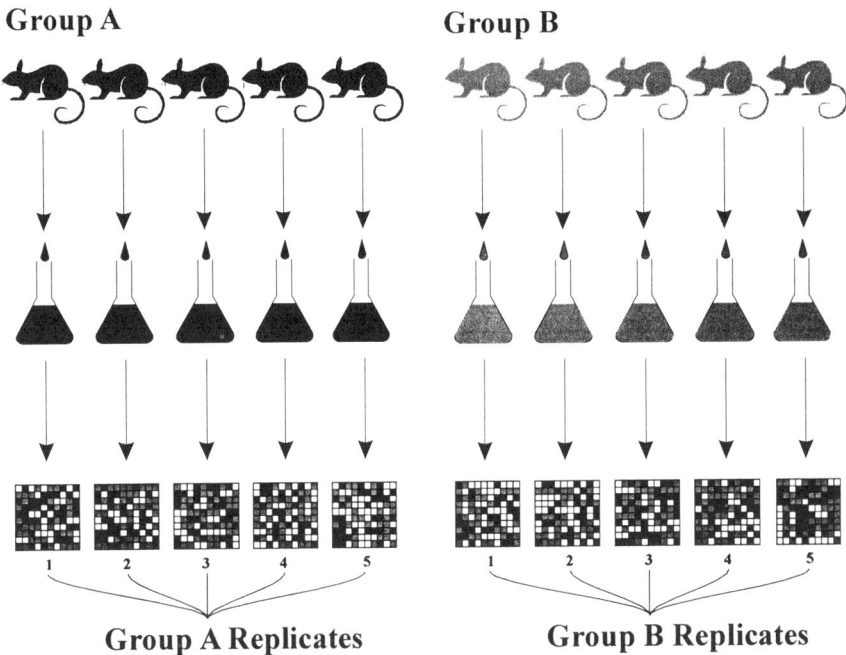

Group A Replicates   **Group B Replicates**

Figure 3. Conventional design, one biological subject per chip, multiple subjects per treatment group.

other researchers interested in calculating the number of chips necessary to detect a particular level o f change on a per gene basis (power analysis). Interestingly, with a few assumptions regarding the contribution of different subjects in pooling methodologies, data from pilot studies using the conventional approach can be used to help design sub-pooling strategies. The reverse however is not true. Results from sub-pooled experiments cannot be used to "back-calculate" gene variation on a 'per animal' basis. The following chapter discusses this in more detail.

   G. Combining replication and duplication. Duplication and replication can be combined in Affymetrix arrays, albeit at an increased cost in terms of arrays, by measuring each sample multiple times and measuring multiple samples (five duplicates for each of five different biological sources; 5 x 5 = 25 arrays). By averaging the duplicate measures for each biologically distinct source (and possibly by removing outliers), the contribution of tool-based variation is further reduced while comparisons across replicates preserves biological variation. However, this represents a profound increase in the "chip cost" (usually the most expensive and therefore limiting experimental component), which may outweigh the increased statistical gains.

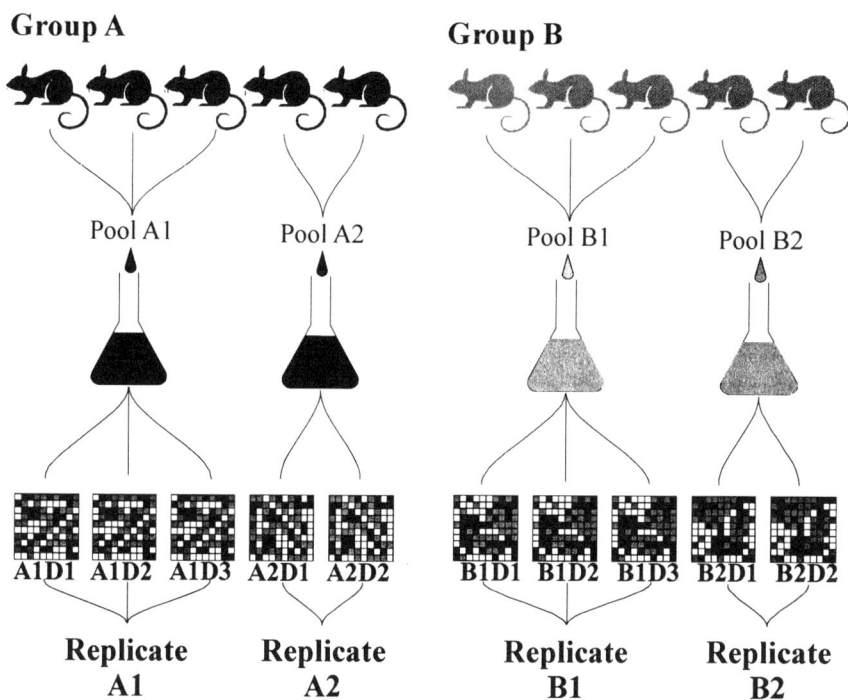

Figure 4. Sub-pooling offers a flexible alternative to conventional designs.

Two Color Arrays

In spotted arrays the situation can be a bit more complicated. Each array is hybridised to at least two different biological sources. These biological sources can vary in origin. For instance, it is not uncommon for researchers using spotted array technology to pool the two groups differently. All of the 'control' tissue might be incorporated using a pooled strategy, while the experimental subjects are measured individually. In addition, when spotting arrays, researchers often spot more than one copy of the array on each glass slide. These multiple copies can serve as duplicate measures. Admittedly they are not 'full' duplicates in the sense that hybridization is not physically separated as in a standard duplicate, and the different duplicate measurements are affected by the print run of the duplicate. Nevertheless, this technology can achieve mixed replication and duplication without the extra array expense required in Affymetrix-based technologies. For a fuller description of experimental designs, the reader is referred to advanced discourses on the subject (Kerr and Churchill, 2001; Yang and Speed, 2002).

A special note on measurements over time. The above descriptions reflect measurements in which only two groups, control and experimental, are compared, and are probably overly simplistic for the goals of most researchers. These descriptions are really intended to demonstrate the difference between designs that can and cannot provide estimates of biological variance. As the complexity of the experimental design increases, costly array requirements also increase. It is important for researchers to take a few things into consideration before designing a microarray experiment. The number of groups to be studied is determined by the researcher's discovery goals. The researcher's requirements for assessing biological and technical variance will influence design selection. And available technology will influence the decision to use spotted or Affymetrix-based arrays.

By far the largest increases in chip requirements occur with designs that are capable of estimating biological variance. For experimental designs in which treatments or phenomena are measured over time, each time point measured could also be considered a treatment or group. Therefore to assess biological variance in this type of experiment, multiple arrays in each group at each time point are required. Because the number of time points is generally more flexible than the number of treatment groups, researchers may want to calculate their chip requirements for the 'ideal' experiment, and reduce the specified number of time points until a satisfactory 'experimental cost' is achieved. In the following chapter, authors Stromberg and Peng describe an effective sub-pooling strategy for reducing chip costs.

VARIANCE AND FOLD-CHANGE

In the following section, I hope to demonstrate that the combination of fold-change based measurements and microarray sized data collection can lead to a profound weakening of the interpretability of microarray data. The type of variance demonstrated by a particular set of data can often be characterized by whether or not it follows a "normal" distribution.

Parametric statistical tests (e.g., t-test, One and Two Way Analysis of Variance, Pearson correlation) assume that the data being tested follow a normal distribution (Figure 5). Data that are not normally distributed are more appropriately analyzed by non-parametric tests (e.g., ANOVA on ranks, Mann-Whitney rank sum, Spearman correlation). Each gene measured on a microarray has it's own variance, some of them normal, many of them not (Grant et al., 2002); personal observations). Thus, within a single microarray experiment, the appropriateness of a given statistical test varies from gene to gene (as discussed later in this chapter, parametric ANOVA tests are fairly robust to violations of the normal distribution). That is, genes that are normally distributed are well suited to parametric statistical tests while genes that are not normally distributed would benefit from non-statistical tests.

On the other hand, non-parametric tests are generally less powerful than parametric tests in assigning statistical discovery to normally distributed data, while parametric tests used inappropriately on non-normally distributed data may, because of their underlying assumptions regarding a normal distribution, imply a significant result when in fact none exist. That is not to say that either type of test is always correct in its determination of a rejection of the null hypothesis (that the two groups are from the same population). Rather, these tests generally test for rejection of the null hypothesis with a user assigned probability (alpha level or p-value). Often set at .05, this value indicates that there is a 5% chance that the test used assigned significance when in fact none existed. Statistically this is termed a false positive [as opposed to a molecular false positive- generally

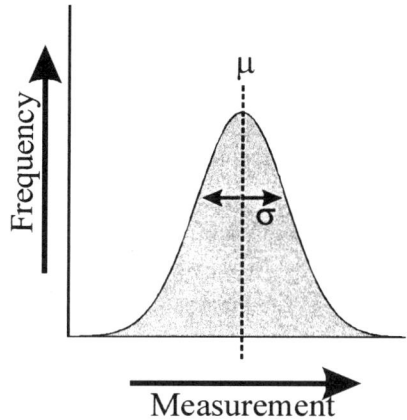

Figure 5. **Normal data frequency distribution**. On the X-axis are the different intensity measurements seen in a population, on the Y-axis are the number of times measurements in that range were seen (frequency). For normally distributed data, the center of the distribution (μ) is equally well represented by mean, median, or mode calculations. The degree of spread around that center is represented by variance (σ), and a gaussian equation accurately describes the frequency distribution.

the result of cross-hybridization and statistically quite replicable- see (Becker, 2002)]. It is important to note that a parametric statistical test bases this probability on the assumptions of underlying data distribution (Figure 6), and that these assumptions of probability are not entirely germane to parametric results applied to non-normally distributed data. It is difficult to appreciate the implications of data distribution when fold-change based analyses are used. Permutation tests (Tusher et al., 2001) (http://www-stat.stanford.edu/~tibs/SAM/index.html) provide a more robust, but computationally intensive approach to assigning significance. In addition, they can be used to calculate the false discovery rate (FDR- see chapter 7).

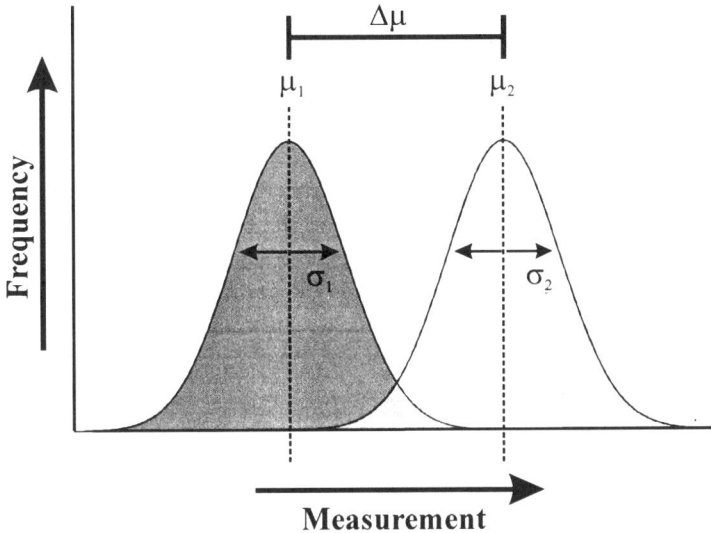

Figure 6. **Comparison of normally distributed gene expression levels in two treatments**. In the above, there is an obvious difference between the two groups, even though there is some overlap in the expression levels seen in the two populations (light gray area). Note that this represents only a single theoretical gene's population distribution across two treatments- each of the thousands of gene expression levels measured in a microarray experiment has its own normal or non-normal distribution, and these distributions may or may not be affected by treatment.

Single subject, single measure, one subject per treatment

First, let's agree that a single subject from one treatment group compared to a single subject from another treatment group (category A from the a bove discussion on experimental design) is probably not sufficient for gaining an understanding of the difference in the two groups, regardless of the comparison technique used. For instance, measuring the expression level of Gene A from one female individual and from one male individual will not give a valid comparison of differential gene expression between males and females in general because Gene A's expression level has not been measured in the general population; there are any number of differences between these two individuals, gender being only one. Additionally, even if thousands of genes from each individual were measured (as in a microarray), then what results is thousands of measures, none of which can be reliably attributed gender-based differences. In short, there is no way to take those thousands of measurements and make them any more securely "about" gender than the initial single gene observation- the fact of the matter is that you are not going to get a picture of treatment-based differences in a two subject experiment.

One important caveat to the above is the nature of inter-subject variation; as the relative homogeneity of a particular population is increased, then the amount of variation within that group is reduced, along with the number of subjects needed to achieve an accurate estimate of the group's gene expression levels. Thus, it is theoretically possible to have a system with cloned cells whose inter-subject variation is so small that the sample number requirements for successful discovery asymptotes towards a single subject in each group. Realistically, however, whether this criterion has been met is not statistically discoverable because variance cannot be assessed on single measures.

Even within a highly similar/ cloned population, individual differences may arise based on slight differences in the cyclic expression of certain genes or based on critical differences in the microenvironment of those genetically similar subjects. This theoretic single subject, single measure, one measure per treatment ideal is highly unlikely, and furthermore would have to be satisfied for each of the thousands of genes being tested. Although it is theoretically possible to achieve meaningful results with this lack of replication, it is highly improbable that such results would be practically applicable. Thus, while the two individuals do belong to the groups of interest, the fact that only one measurement has been taken from each population can lead to specious results.

RNA Extraction

The extracted genetic material from each subject is thought to comprise an equivalent representation of that subject's complement of mRNA, and the intensity values, taken from either pooled data or the average of replicates, theoretically represent the mean of the intensity values. It may be simpler to see this as an average concentration effect. If five different containers with different aqueous concentrations of NaCl are mixed together in equal parts, then the concentration of the resultant admixture will be the average of the concentrations of those five containers. By the same analogy, it is easy to see where this theoretical averaging of the individual subjects values can go awry. When extracting RNA, it is often assumed that the total amount of RNA in each of the tissues is the same- thus any change in total yield per tissue sample is considered a problem with the extraction procedure rather than a true difference in the biological subjects, and the final amount pooled from each extraction is then 'fixed' at a certain RNA weight.

In fact the situation can be a little more complex in actual microarray experiments. If some component of the difference in extraction is due to a REAL, treatment-based difference in the total amount of RNA present in the sample, then the assumption is violated and the representation is not of equal proportion from each group. The net result for our analogy is that unequal volumes are drawn from each sample, and the resulting final NaCl concentration/ gene expression level may not accurately reflect the mean of the subjects from which it was derived, it is biased to the sample in which the largest volume was drawn. However, for experiments performed at the pooled level, a single extraction performed on the pooled tissue from a treatment group may adhere more closely to assumptions of "average" representation than individual extractions. It is important to keep records of the tissue wet weights used in the extraction procedure and the amount of RNA extracted from each subject. Significant differences across treatment groups could point to important contributions from the extraction procedure, and appropriate weighting strategies applied to control for those contributions.

Estimating Error for "Pooling" and "Single Subject, Single Chip, One Subject Per Treatment" Experimental Approaches

For both "pooling" and "single subject, single chip, one subject per treatment" experimental approaches, fold-change is probably the *de facto* method of data expression because there are no replicates from which one could estimate error and generate statistical tests. However, replicates are not the only source from which one could estimate error. Silicon Genetics' GeneSpring program offers a "global error" model in which estimates of the error for a single measure are calculated based on the variance of other measures on the same chip that are of similar amplitude to the one in question. These estimates are a good shorthand method for determining which data on the chip may be unreliably detected, because data at or below detection level for the laser scanner, or near saturation for that detector (i.e., data outside the linear range of the scanner) will generally be more noisy than data within its linear detection range. Importantly, however, this kind of estimate can have no *biological* component as no biological variance has been assessed in this "single point" measurement. For sub-pooling approaches (see below) there is another type of estimate, the Bayesian prior, that can be used as part of an estimate of variance, and software such as Dchip (Schadt et al., 2001) and CyberT (Baldi and Long, 2001; Long et al., 2001) have been used with great success in the that regard (*e.g.*, Hamoen et al., 2002).

Modeling Fold Change

Using model data one can work backwards from a known result to determine the utility of different testing procedures. Unfortunately, a model that fully embraces the complexity of real microarray data has yet to be found, and the following model is grossly oversimplified. I only present it to illustrate a limited point. Instead of trying to detect legitimate differences between two groups, I will look at fold change another way- what are the chances that two individuals would be labeled as having arisen from different populations with fold-change-based analysis when in fact those two individuals really were drawn from the same population? To address this question, I used fold-change analysis to compare two lists of numbers drawn at random from a single normally distributed data set (model constructed in Excel 9.0.3821, SR-1; Microsoft).

With model data, the distribution is known *a priori*. This is a tremendous advantage in assessing the validity of certain analytical techniques. I chose the arbitrary and popular 2-fold cut off, and counted the number of genes achieving that criterion in model data transitioning from the "single subject, single chip per treatment" to the "pooling" layout (Figure 7). However, because these 2 fold changes represent falsely identified differences (remember that the numbers are being drawn from the exact same population) every comparison that crosses the 2-fold mark instead represents a mistake (or "false positive") of the fold change technique.

Importantly, increasing the number of subjects pooled per treatment group exponentially suppresses these false fold-change discoveries. Thus, going from a "single subject, single chip per treatment" comparison to a "10 vs. 10-pooling" comparison effectively reduces the false discoveries from 50% to 1% of the observations, and by the time "20 vs. 20" comparisons are performed, there are no 2-fold differences. Conversely, for actual experimental data where there are mixed "true" and "false" fold changes, as one increases the number of observations in each treatment group there will be not only an exponential decay in the number of false discoveries, but an exponential rise in the "true" discoveries if the treatment has an effect. Therefore, increasing the number of subjects pooled in each group in an actual experiment should serve to

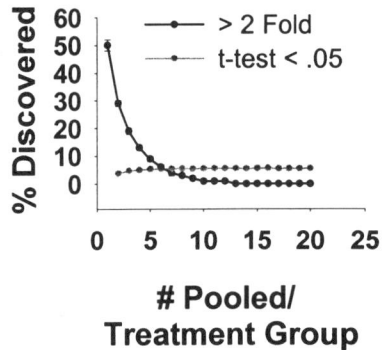

Figure 7: **A graphic relationship between the number of subjects pooled and the number of "2-fold" changes detected.** On the x axis is the number of biological subjects pooled within each treatment group, on the y axis is the percentage of genes (out of 10000) that were found to show a greater than 2 fold change.

thresh the true fold-change differences from the false ones. However, the number of chips needed to effectively exclude false fold-changes will depend on the degree of variance, and different genes show different kinds of variance as well as different variance amplitudes (Grant et al., 2002). In a real microarray experiment, determining how many observations are needed in each group is a much more involved process (see Chapter 7).

It is important to note that the constructed model used here has 1000 model genes per chip, and the expression level of each gene is modeled after normal variance. Thus the situation is greatly simplified over that in an actual microarray experiment, where multiple types of variance, fluctuating inter-gene signal intensity, background noise, etc. greatly enrich complexity. However, the model does demonstrate an important weakness of fold-change in that it cannot control for variance. Issues regarding the appropriate statistical procedures for determining genes whose expression levels have changed significantly are considered in the following chapter, and methods for clustering genes are considered in the final chapter.

AFFYMETRIX DATA

While several authors have discussed the procedures used in the manufacture, design, management, and analysis of spotted arrays, relatively little has been said regarding one of the most popular microarray platforms, the Affymetrix GeneChip. Although expensive, this product can offer some significant advantages over spotted arrays. Probably the most noticeable advantage is start-up time. A brief review of the postings on the University of California San Francisco Gene Arrays list server (LISTSERV@ITSSRV1.UCSF.EDU) reveals that the majority of posts requesting technical assistance with getting started or trouble-shooting involve spotted arrays (although thanks to the help of people like Todd Martinsky, those queries are becoming more and more routinely handled). While people working with Affy GeneChips generally start their questions with 'how do I analyze this?' Thus, Affy users often find themselves further along before they need help. To be sure, this increased ease of start up comes at a price as Affy systems and supplies are easily the most expensive. Whether that will translate into long-term increased costs is unclear. At our facility, the staffing for the Affymetrix core is much less than that necessary for the spotted array core. However, many researchers new to the field may not be aware of some of the issues with Affymetrix' chip architecture and some of the pitfalls in Affy-based analyses. Two of the more contentious issues are the handling of absence calls and the generation of a 'unified' metric for each probe set that reflects the expression level of the gene in question.

Probe Level Analysis

Affymetrix oligonucleotide arrays have undergone several evolutions in design, each seeking to increase the amount of 'total genome' coverage on a single chip. The oligonucleotides are built onto the chip using a proprietary photolithography technology
(http://www.affymetrix.com/technology/manufacturing/index.affx).
This results in a high degree of cross-chip compatibility and facilitates intra-experiment and intra-laboratory comparisons, although see Jay Tiesman's presentation (Tiesman, 2002) for interesting lot-based variations in data. However, this photolithographic process builds short oligonucleotides (25mers), which have relatively poor specificity and high cross-hybridization potential compared to the longer strands used for detection in many spotted arrays. To compensate for this, Affy arrays incorporate a multiple probing strategy. That is, between 11 and 20 unique 25mers are created for each gene and are designed to span unique regions of the mRNA of interest. Each of these probes is designed with Affymetrix optimized salinity/ temperature settings in mind, and is paired with a 'mis-match' oligo. The mis-match oligo is the same as the 'perfect match' oligo except for a mismatched nucleotide at the center position. Affymetrix has recently decreased the size of each feature (a square area on the chip that contains a unique oligo species) to 18-20 μm^2, further increasing the number of genes that can be analyzed on a single chip.

With Affymetrix GeneChips, data processing could be divided into three levels of analysis: *low*, *probe*, and *probe set*. *Low* level analysis involves the discrimination of two-dimensional space on the chip, the mapping of that space to the individual features printed on the chip, and the generation of single measures for each of those features. In other words, the overlaying of the grid that holds the coordinates for the different features on the chip (these are held in the Affymetrix library file- *.cdf), and the conversion of the mix of intensities measured at each feature into a single intensity for each feature. The scanned image (*.dat) of the RG-U34A chip contains 64 pixels per feature. The image analysis algorithm in MAS4 and 5 removes the outermost border of pixels and uses the remaining 36 pixels to construct a single intensity for that feature. This simplified, single-intensity-per-feature summarized version of the scanned image is referred to as the 'cel' (pronounced as 'cell') file (*.cel). The 'cel' file is then used in *probe* level analysis to generate *probe set* level data. There are many ways in which probe level data is converted to probe set level data, and these are discussed more fully in the following section 'Other Analysis Methods'. *Probe set* level data is analyzed by a variety of procedures, and holds the information with which most researchers are interested, an aggregate metric cobbled together from all of the information in the probe set about the expression level of each of the thousands of genes/ESTs on the chip.

One important and often overlooked exception to the above is pairwise analysis in Affymetrix' Microarray Analysis Suite (MAS). This procedure compares *probe* level data within each probe set across two chips rather than *probe set* level data. In its native form, this procedure offers little advantage to the researcher using a design with replication, because the comparison only runs across two chips. However, there are batch analysis modes in the MAS software that allow researchers to run multiple pairwise comparisons, and the results can be used to detect differences in expression level across different treatments. It has been shown using this approach that analyzing data at the probe level, rather than the probe set level, can enhance the power of the analysis (Welle et al., 2002).

The output of MAS 4 or 5 includes several different metrics. For the purposes of statistical analysis, the 'average difference score' (AD-MAS4) and 'signal intensity' (SI- MAS5) metrics provide an estimate of the relative expression level of a particular gene across chips. When using these measures, keep in mind that ' across gene' comparisons a re confounded b y differences in hybridization efficiency of the probe sets involved in the calculation of each gene's expression level. The AD for gene A may be much greater than that for gene B, but the disparity in AD (or SI) values from one probe set to another could be due to differences in the construction of their respective probe sets and the underlying hybridization efficiency arising from differences in the sequences selected for hybridization. Because this confound is not present across chips with regard to the same probe set, within gene comparisons of differences in SI or AD are valid.

However, there is quite a bit of debate regarding the analysis of GeneChip data. This is because different approaches to the analysis can yield very different lists of genes (Bakay et al., 2002; Hoffmann et al., 2002). To understand some of these issues, it is necessary to understand a little bit about the anatomy of GeneChips. More complete information can be found in the MAS5 manual, which is available at no cost from Affymetrix after (free) registration at their website (www.affymetrix.com). The composite expression score (either AD or SI) for a gene or EST is derived from a **probe set** (Figure 8). Depending on the chip design, these probe sets are comprised of between 11 and 20 **probe pairs**. Each of these probe pairs contains a **perfect match (PM)** and a **mis-match (MM) feature**. Each PM in the probe set is printed with multiple copies of a unique 25mer designed to hybridize to a specific region of the mRNA of interest. Affymetrix bases their decisions on what sequences are most suitable for detection of a particular mRNA by looking for unique regions of the gene sequence (as reported in GenBank) for which a single combination of salt, temperature, and time allow optimal hybridization. To control for the reduced specificity of 25mers

compared to longer oligos,
multiple features are used to
interrogate each gene or
EST, and each PM is paired
with an MM which contains
the identical 25mer except
for a single base mis-match
at the center of the oligo.

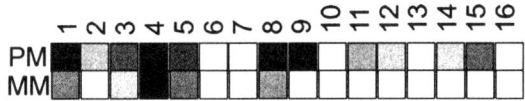

Figure 8. **Example probe set.** On older Affymetrix designs, probe sets were physically located in one area. On newer d esigns, perfect m atch (PM) a nd mismatch (MM) p robe pairs are located t ogether, but different probe pairs belonging to the same probe set are scattered about the chip surface.

Because of the stringency
of the hybridization reaction
and the presence of a
central mis-match, the MM features are expected to have less specific binding
than the PM. Importantly, this is not to say that MMs should show no signal, in
fact the signal generated by MMs may be considered to represent non-specific
binding, or cross- hybridizing events. The PMs, owing to their greater sequence
similarity, should bind to both the non-specific material to which the MMs bind
as well as the specific mRNA sequence for which they were designed. In both
MAS4 and MAS5, the difference between the PM and the MM is treated as if
it represented specific binding. This has lead to some debate in the research
community as to whether or not this design is performing as intended.

Negative Values

In the MAS4 analysis suite, negative values would result for probe sets in
which the MM values were greater than the PM values. One of the first concerns
among Affy users was the relatively high proportion of negative values in the
data set when using the MAS4 'empirical' algorithm. For instance, in our own
data set, among 29 different chips from three different treatment groups, 1566
probe sets showed a negative value across all chips. For many researchers, this
did not make biological sense- how could a gene have a negative expression
value? In fact, looking at the MAS4 algorithm it is easy to see how such an
event could occur if the MM intensities were higher than the PM intensities.
Consider a hypothetical Affy chip that could detect all of the mRNAs a multicellular
organism was capable of producing (a popular though unproven estimate for
mammals is ~ 30,000 genes). Any given tissue within that organism is assumed
to express only a subset of that total (for brain tissue, which is thought to express
a very high proportion of the potentially expressed genome, estimates run as
high as 50%). Thus there will be a population of probe sets on the chip, possibly
the majority of probe sets, whose intensity levels represent cross hybridization
and/ or noise because there should be no instance of their (unexpressed) target

within the biological sample. The discrimination between cross-hybridization and noise is important because noise is considered a random event, while cross hybridization is, by definition, specific.

Because both the PM and the MM for the 'noisy' probe sets are theoretically generating signal based on the same population (the PM-MM design of Affy is predicated on the assumption that the PM and MM are equally susceptible to random events in the absence of the relevant mRNA) then there should be a 50% chance for each probe pair that the MM will be greater than the PM. Thus, it should not be surprising t hat negative values c ome o ut of t he measurements. In fact, from a logical standpoint, for the fraction of probe sets with no target (the mRNA they were designed to detect is not present) and no cross hybridization, one would expect a 50:50 representation of negative and positive values. However, in the real world, it is very difficult to separate cross-hybridizing events from random events without rigorous investigation of the sequences involved. While negative values may not make sense biologically, they make perfect sense from the MAS4 analysis standpoint. The negative values for probe sets with no mRNA to detect may represent one half of a random distribution around noise.

Another potential source of negative calls is unintended specificity, or cross hybridization, in which either the PMs or the MMs specifically bind to another target. Using the MAS4 algorithm, it is easy to see that some negative values give significant results (348/4118 in our data). By chance at a = .05 level, one would expect ~ 206 positive results in a random data set (this expectation is biased in our set by prefiltering the data based on presence/absence -p/a- calls, but provides a rough estimate of the phenomenon- see section titled *handling absence calls* for further information).

We observed somewhat more significant results among the 'absent' data set than would be expected by chance alone. These could represent probe sets that are detecting important differences in gene expression levels for which the MM features are more sensitive than the PM features! Although we can be relatively certain that the PM signal is not as great as the MM signal, we have no idea which of these components is changing, and this is true for all MAS-derived expression level metrics. There is an *in silico* 'shell game' going on with the low level data. With either positive values (PM > MM), or negative values (PM < MM), there is no metric provided to help the researcher determine whether the majority of change occurred within the PM or the MM features across chips. Overall one can only be sure that there was some consensus of change between the two, and it is generally assumed to have been among the PMs.

To help mitigate the problems with the interpretation of absence calls, and also to provide a more refined estimate of expression level, Affymetrix developed a 'statistical' algorithm for MAS5 (Affymetrix, 2002a, b). Rather than taking the

simple average of the differences between the PM and the MM for all of the
probe pairs in a probe set, the median of the log(PM-MM) values is used to
construct a Tukey biweight estimate that reduces the contribution of PM-MM
differences as a function of their distance from the median value. Thus, the
median is used to construct an average whose contributors are weighted
according to their proximity to the median, and unlike MAS4, the probe level
data are logged to control for image intensity-based error. This weighting
procedure is probably a more appropriate procedure than that used in MAS4,
which was to ignore values that were rated as outliers, and the data were not
log transformed to control for image intensity issues. With regard to the negative
or positive value of the overall SI metric, the statistical algorithm employed in
MAS5 replaces MM values that are greater than their respective PM values
with an imputed Change Threshold (CT) value that is always less than the PM.
If the MM values for the entire probe set are greater than the PMs, then the
MMs are all replaced with numbers a little lower than those of their respective
PMs. The end result is that there are no negative values reported by the MAS5
statistical algorithm (Figure 9).

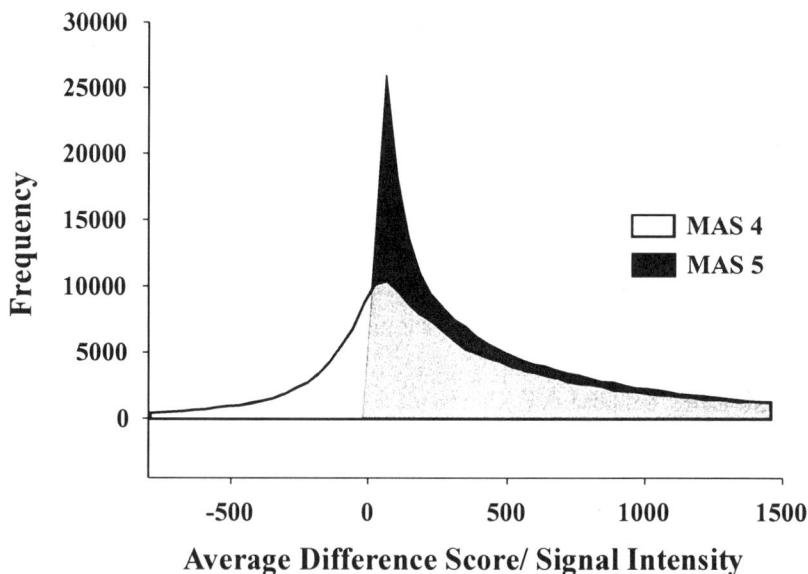

Figure 9. **MAS5 statistical algorithm eliminates negative values**. On the x-axis is
average difference score (MAS4)/ signal intensity (MAS5). On the y-axis is plotted the
frequency with which a particular measurement was found in all 30 chips in this data set.
MAS4 frequency distribution is shown in white, MAS5 in dark grey. Overlapping values
are shown in light grey. Note that MAS5 clips the negative values seen in MAS4.

Other Analysis Methods

Many researchers have developed their own analysis algorithms to deal with some of these issues. In most cases, the algorithms provide a 'single metric' representation of proportional expression level that is analogous to the Affymetrix AD and SI metrics with regard to interpretation, and, like the MAS5 algorithm, provide normalization procedures designed to reduce the contribution of systematic variances (Li and Wong, 2001; Naef et al., 2001; Bakay et al., 2002; Bolstad et al., 2002; Huber et al., 2002; Nadon and Shoemaker, 2002; Welle et al., 2002; Zhou and Abagyan, 2002). That is, some combination of mathematical and/ or statistical processes take information from some or all of the features on the probe set and summarize them with a single number. Several of these procedures are available as an Affymetrix exclusive package ('affy') from Bioconductor (www.bioconductor.org) and run inside the open source statistical language/ environment R (www.r-project.org).

One problem with the Affymetrix design is the subtraction of MM values from PM values. If one considers that both the PM and the MM values contain some aspect of the relevant signal, and some aspect of irrelevant noise, then whether subtracting MM removes relatively more signal or more noise from that measure is a source of some debate (Figure 10). It is likely that the 'noise removal' properties of MM subtraction vary among different probe pairs within a probe set, as well as among different probe sets. As mentioned in the last section, the Affymetrix MAS5 algorithm handles 'negative' results by replacing the MM measure with an imputed measure that is always < PM. Most of the alternative analysis procedures focus on 'PM only' data (i.e., without subtracting MM values). This successfully removes from consideration the sticky issue of negative value handling, and proves more accurate than MAS4 algorithms at extracting known differences from the data set (Bolstad et al., 2002; Goncalves, 2002). It should be noted that these procedures can dramatically alter the composition of the final gene lists (Figure 10).

Example of Probe Set Data

As an example of the kind of dynamics that exist within the probe level information, I selected a probe set from our data that showed a strong treatment effect using a one-way Analysis of Variance on MAS4 AD scores. There were 10 chips in each treatment group. For this individual probe set, then, there were 10 contributors to each feature in the probe set. Because there is presently no facility for looking at an 'averaged' image within each treatment group, I created a graphic representation of the averaged image intensity at each feature (a so-called 'heat map', Figure 11 A) by setting the maximum average feature intensity (PM intensity for Treatment group 3, probe pair 8) to pure white, and every lesser average feature intensity to a gray scale value, such that the lowest feature intensity (MM, Treatment group 3, probe pair 14) was nearly pure black. This was only done for graphic purposes, raw data (Figure 11 B) were used for statistical testing.

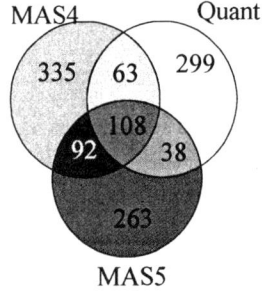

Figure 10. **Venn diagram reveals differences in gene lists resulting from different probe level calculations of composite intensity scores.** MAS4, MAS5, and 'quantile' normalization (bioconductor) procedures yield different gene lists. Ideally, one would hope that more powerful analysis procedures would yield ever-widening concentric circles of gene lists around those of less powerful techniques.

First note that there is a high degree of pattern similarity among feature intensities for this probe set within and across treatment groups- in general the same features are relatively brighter or darker in all three treatments. This is reflected quantitatively among the PM and MM data (Figure 11B; Table 2). PM features had the strongest correlations across treatment (p values typically $< 10^{-14}$). Correlations among MM values were only slightly less significant, and correlations between PM and MM features within treatment were the least significant ($p < .01$). This latter correlation points out some of the difficulties in dealing with MM features. To what component of the data does this correlation owe its significance? Is the difference in each feature's susceptibility to random hybridization, cross-hybridization, or target hybridization responsible for the correlation? Many researchers have concluded that MM feature intensity represents a complex mix of the above components, and that their use in the Affy algorithms as cross hybridization controls represents only one potential component of those relationships. Further, the relative proportion of this contribution may vary from probe pair to probe pair, representing a confound in the data set.

Another way to approach the analysis is to examine the raw feature intensities as a function of treatment. There is an intensity-dependent increase in variance among the features (Figure 11B), which is probably best treated by log transformation. This relationship is born out statistically (Pearson correlation coefficient = .842, $p < .001$, data not shown). Log transformation of the data reduces the strength of the correlation between variance and intensity across all chips, but the relationship is still significant (Pearson correlation coefficient = -.442, $p < .02$).

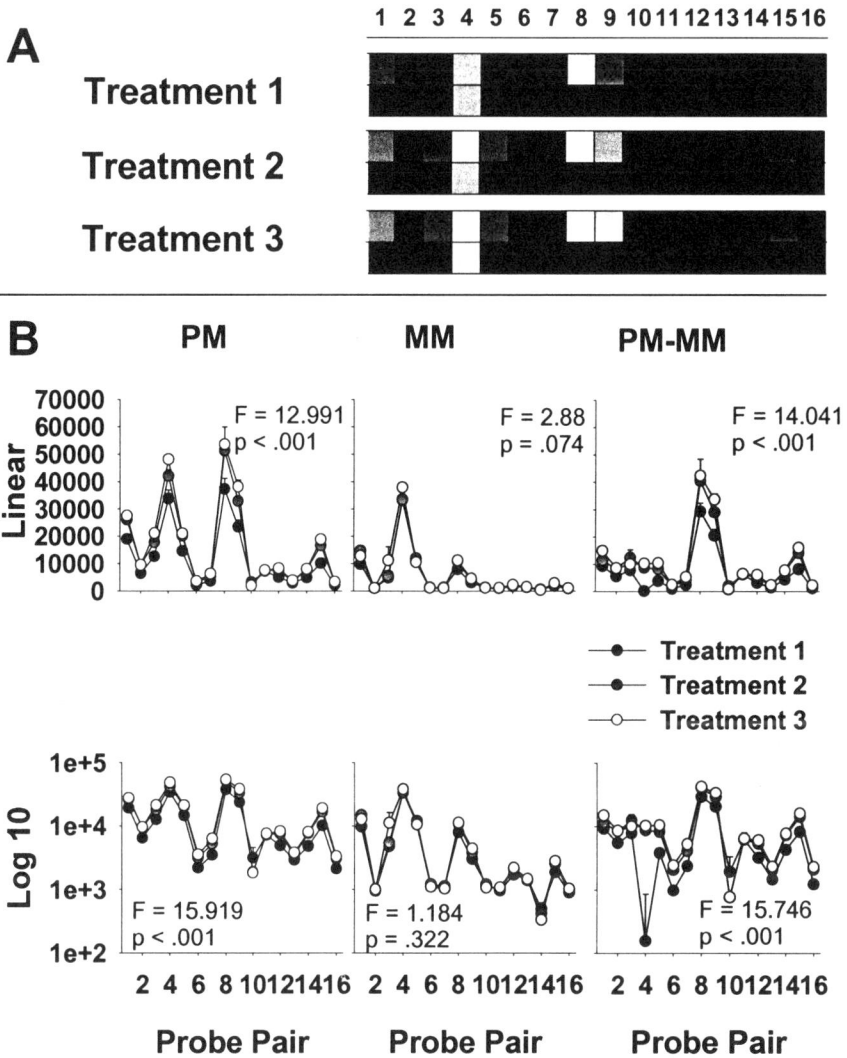

Figure 11. **Graphic probe set and plot summary. A**. Normalized representations of the average feature intensities for an example probe set from the three treatment groups included in this study. **B**. Graphs of the intensity values for PM, MM, and PM-MM values. Statistical results (2-way ANOVA on repeated measures p-values and F-statistics) for the main effect of treatment are shown for each group).

Table 2. Correlation among different treatment groups within probe set.

		Perfect Match			Mismatch		
		Treat 1	Treat 2	Treat 3	Treat 1	Treat 2	Treat 3
Perfect Match	Treat 1	1	.994	.993	.718		
	Treat 2		1	.997		.729	
	Treat 3			1			.738
Mis-Match	Treat 1	.718			1	.987	.985
	Treat 2		.729			1	.983
	Treat 3			.738			1

Note: Pearson's correlation coefficient reported for each of the performed tests at the intersection of the two correlated groups. Redundant test cells are grayed out to simplify the table. All of the reported correlations were significant.

For statistical testing of the raw data I chose the two-way Analysis of Variance on repeated measures because the same features were measured repeatedly in different animals and there were three treatment groups. This testing protocol seems intuitive when viewing the data as plotted in Figure 11. The F statistic and the p value are reported for the main effect of treatment in each panel of Figure 11B. It is clear that there is a main effect of treatment in the raw PM scores, and a nearly significant effect in the raw MM values. Interestingly, the raw PM-MM values show the greatest level of significance among raw scores. Thus for this particular probe set in this study, the MM features may represent a relatively greater component of noise. However, it should be noted that this probe set was chosen because of the high level of significance it demonstrated using MAS4 data, thus it is likely that this may have been one among a group of probe sets for which the Affy algorithms had preselected those with appropriate PM-MM performance.

Interestingly, log transformation of the data resulted in both increased PM data significance, and decreased MM significance, but had little effect on the PM-MM scores. For PM, MM, and PM-MM data sets, both raw and log-transformed measures failed to follow a normal distribution (Kolmogorov-Smirnoff test for normality, data not shown). Thus, although the log transformation alleviated to some degree the intensity dependent increase in variance, the transformation was not sufficient to generate a data set whose values could be appropriately tested parametrically (however, the ANOVA tests are fairly robust to violations of normality, so may still provide reasonable tests for significance). Thus, more advanced normalization techniques, applied at the probe level, may be necessary to convert the raw data into normally distributed data. Alternatively, non-parametric procedures could be used. Probably the most powerful alternative would be to perform permutation testing for significance. This is discussed further in the following chapter.

The two-way ANOVA on repeated measures tested for significant main effects of treatment and probe pair, as well as an interaction between the two main effects. Ordinarily, a significant interaction effect 'trumps' significant results in either of the main effect tests, because a significant interaction means that the assumptions under which the main effects were tested have been violated. For this probe set, every main effect of probe pair was highly significant, as was the interaction. Thus, *post hoc* tests may be appropriate to isolate which probe sets in which conditions are truly responsible for the significance call. Further, it may indicate that collapsing values of different probe sets in order to calculate a single metric for the analysis may reduce the power of the analysis. Procedures in which all of the probe pairs in a probe set are compared individually (Welle et al., 2002) may provide a more accurate picture of the relationship than methods in which a composite of the information contained within a probe set is constructed.

The purpose of this probe level example is not to claim that it is a representative example of the chip's overall behavior, but to show that there is a depth of information at the probe level of the analysis. Researchers new to Affymetrix GeneChips often take the probe set level data at face value for their statistical analysis, but differences in the interpretation of the raw probe level data by different algorithms can lead to drastically different lists of genes (Figure 10).

Handling Absence Calls

It's broadly assumed that a single tissue only expresses a fraction of the total capacity of its parent genome. General-purpose oligonucleotide arrays span as much of that genome as their designs allow on a single chip. Indeed, one goal of array designers is to produce a single chip capable of detecting all of the genes capable of being expressed in a single organism. Thus, for any given tissue that is hybridized to an array (if the assumption regarding fractional expression of the genome holds), only a proportion of the probe sets on that chip is reporting useful information.

The MAS4 and 5 algorithms generate a Presence/Absence (P/A) metric for each probe set on each chip to help researchers determine whether that particular probe set generated a reliable signal. In MAS4, probe sets are reported as Present (P), Absent (A), or Marginal (M). In MAS5, a p-value is generated for each probe set, such that $p < .04 = P$, between .04 and .06 = M, and > .06 an A. These parameters are tunable, in other words you can adjust the level of the p-value that is considered a P, an M, or an A in MAS5.

If you heed the advice in this chapter, and pursue an experimental design capable of assessing biological variability, then there will be replicate chips in each treatment group. Knowing that the more statistical tests performed, the greater the likely number of false positives (and the more severe the multiple testing correction), many researchers filter their data prior to statistical testing. It is clear from the data that the Affymetrix P/A call successfully discriminates between 'noisy' and 'quiet' data (Figure 12), but it is important to keep in mind that the P/A call is not completely based on the intensity of the features. Another component, the proportion of agreement among probe pairs in the probe set, is an important factor in the algorithm that is used to make the call (Microarray Suite 5 Manual, Affymetrix). Thus, there are really two things going on in the P/A call, intensity values that are not within the linear range of the scanner (too low or too high), and probe sets in which the multiple probe pairs do not agree as to the direction or amplitude of change. In simpler terms, the P/A metric 'calls' probe sets absent if the intensity of their features is too low, as well as if the probe pairs themselves do not appear to be performing in the manner expected. Thus, when using the Affymetrix algorithms, it is probably important to incorporate, at some level, the P/A calls, as they are labeling probe sets for which the signal intensity/ average difference metric is unreliable.

There has been some debate among investigators regarding the application of these calls. Among biologists it makes perfect sense to perform the statistical analysis after filtering P/A calls. But when researchers design experiments with multiple chips per treatment groups, there are P/A calls for each probe set on each chip. Thus, within a single treatment group, one probe set may show P calls on some of the chips, and A calls on others. How then does the researcher decide whether a particular probe set was worthy of inclusion in the statistical analysis? One could treat the probe sets with absence calls as 'missing values' and either use imputed values for those measures or treat them as blanks (in either case, the statistical test loses power as the number of replicates in a treatment decreases as a function of their removal by absence call criteria). The most conservative approach (with respect to the number of probe sets retained for further analysis) is to keep all probe sets for statistical analysis in which a presence call was made on at least one of the chips in the study. Less conservative approaches include the requirement that a greater proportion of the chips have a present call for that probe set.

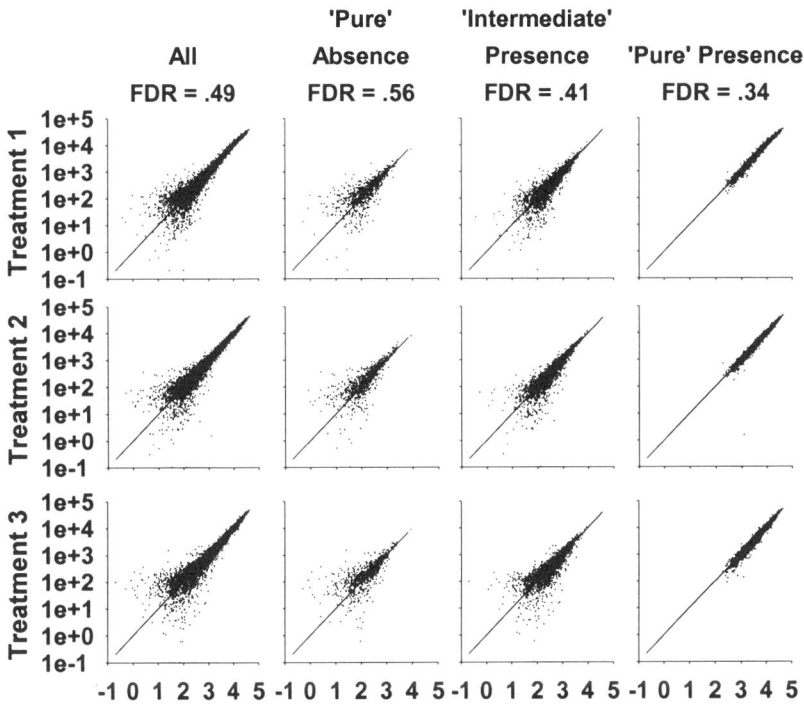

Figure 12. **Representative scatter graphs for each treatment group.** Average Difference scores from one chip in each treatment group (upper, middle, and lower rows) are plotted (x-axis) against the mean AD scores of all chips in the study (y-axis). From left to right; **All**- all probe set AD scores are plotted; **Pure Absence**- only the probe sets rated absent in all 30 chips; **Mixed Absence**- all probe sets rated absent in 1 to 29 chips; **Pure Presence**- all probe sets rated present in all 30 chips. The false discovery rate (FDR) is shown for each subset (note that this sort of 'selective' FDR calculation is inappropriate for usual studies- it is only shown here to illustrate the relative improvement in the FDR among Pure Presence probe sets).

One of the most controversial P/A based filtering approaches involves the use of the treatment group labels in the filtering procedure. For example, in our data set, if we required that at least 4 present calls be found in at least one of the three treatment groups, then we are biasing ourselves towards finding a subset of the data in which there are 4 presence calls in one treatment group, and no presence calls in the other treatment groups. This particular group is a subset of the group one would select by picking 'at least 4 presence calls in all of the chips'. It would seem that such a 'group label' based designation would bias the results towards significant findings, at least for the proportion of the data that fulfill those stringent P/A requirements.

By using the P/A calls prior to performing the statistical analysis, but especially when they are employed in combination with the group labels (e.g., "I must

have x presence calls in at least one of the treatment groups") the researcher is
'peeking' at the data before hand. This is a statistical violation, and its impact on
the results of the analysis can be quantified by doing permutation testing (Tusher
et al., 2001) (http://www-stat.stanford.edu/~tibs/SAM/index.html).

In our own data set, it was surprising to see that only six probe sets (out of
~8800) fulfilled the criterion of having at least four presence calls in one group
and all absence calls in the other treatment groups. Of these, only two probe
sets showed a significant result. Taking this reasoning out to other probe sets
whose inclusion was biased by the filtering procedure, probe sets that had at
least 4 presence calls in one treatment group, and between one and three
presence calls in each of the other treatment groups, overall there were 568
found, 53 of which were significant. Out of 568 tests done with alpha set at .05,
one would expect ~28 significant discoveries by chance alone. Indeed, the number
of significant results found in this group was greater than that expected by
chance: false discovery rate- 28/53 = .53 (Benjamini and Hochberg, 1995) but
much worse than the overall false discovery rate for the entire data set with no
filtering (.48), or that of the data set once filter was applied (.39). Thus, it is
possible that, although the filter should bias towards the finding of significant
results, it does not because it is primarily picking up 'threshold' data on the
border between presence and absence. Moreover, the use of 'group label' P/A
filtering may not be necessary, because omitting the group labels from the filtering
procedure (i.e., just taking all of the probe sets in which there were at least 4
presence calls overall) made little difference to the false discovery rate (.41).
Additionally, the incorporation of the P/A filter enhances the degree of agreement
among different analysis procedures (figure 13), and this agreement is further
enhanced by taking the data to the extreme of accepting for analysis only those
probe sets for which all chips showed a presence call.

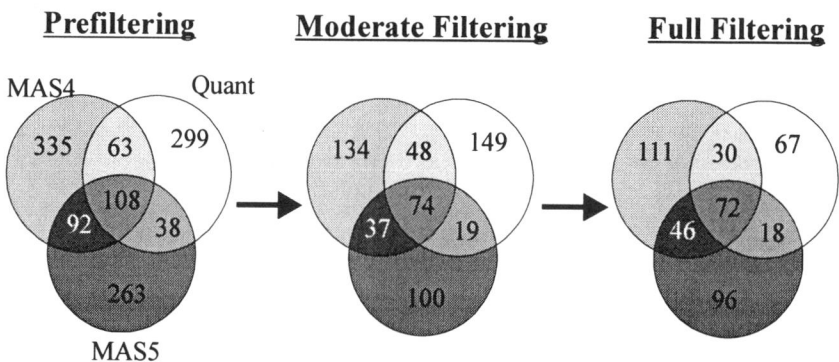

Figure 13: **Comparison of different analysis techniques with and without P/A
(presence/absence) filtering**. There is an overall 'constricting' of the values towards
overlap with P/A filtering in place (~3:1 ratio of unique to completely overlapped values in
'Prefiltering; ~2:1 ratio in Moderate Filtering; and nearly 1:1 ratio in the Full Filtering set).

It may be more appropriate to calculate an overall P/A score based on features from all of the chips in a treatment group, rather than individual P/A scores. For instance, if the P/A p-values for a particular probe set on all chips in a treatment group were centered around .1, they may still be called absent by the Affy algorithm, yet a unified approach would probably rate them as having a more meaningful representation than another probe set whose values were centered around .5 and also called absent. There is no commercially available software that uses such a procedure, although it would be a boon to researchers using Affymetrix chips in replicated experimental designs.

For the use of Affymetrix-derived metrics (SI or AD) the inclusion of P/A calls in the analysis may be very desirable. The safest way to employ them would be after the statistical analysis, however using them prior to statistical testing can help considerably with multiple testing issues. If used prior to statistical testing, then it would be best to use an approach that does not employ the group labels, as this is more obviously 'peeking' at the data. Additionally, the arbitrary nature of criterion (at least four presence calls, why not three, or six, or two?) is problematic. The most conservative (one presence call on any chip), the most extreme (present on all chips), or the imputation strategy would seem the least arbitrary.

WORKING WITH MORE THAN TWO GROUPS

There is a popular book from the late 80's entitled 'All I Really Need to Know I Learned in Kindergarten'. In it, Robert Fulghum demonstrates that the most basic rules of social behavior can provide a flexible framework for successful, moral living, even in environments far more complex than those in which the rules were originally taught. In a parallel fashion, all you really need to know about microarray experimental design you can learn from a two-group comparison. That is, incorporating randomization, balance, replication in any experimental design will yield valid results. Furthermore, difficulty in applying these principles should serve as an indication that the design could be improved.

However, including more than two treatment groups can dramatically complicate interpretation of results, even if those results are associated with an experimental design that has provided a good estimate of variance for each gene's expression level in each treatment group. In these situations, clustering tools and other forms of machine learning (as covered in Chapter 8) are indispensable tools for the further analysis of microarray data. While clustering methods are vital to microarray research, clustering results are sometimes difficult to interpret, and care must be taken with any of these techniques to control for the generation of 'ordered noise' (Mirnics, 2001). For instance, the resulting clusters are determined using non-linear equations in an iterative fitting process. Thus, gene membership within a cluster can change from one run of the clustering algorithm to another. Many algorithms, such as K-means and SOM, ask the experimenter to determine the number of clusters before hand, and this can be a divisive issue. In many instances the microarray experiment is designed as one of discovery, and *a priori* knowledge of cluster number can be difficult or impossible to estimate, leaving the researcher to guess at the number of clusters. Finally, for most of the common clustering procedures, estimates of variance are not included in cluster determination. Owing to the computational intensity of clustering procedures, this is not surprising. However, it does mean that these algorithms will treat genes with the same means but different variances similarly. For a better explanation of clustering and its considerable power and applications in microarray research, please refer to Chapter 8.

Increasing the Number of Treatment Groups Increases the Complexity of the Analysis

Consider that in a two-group comparison, some threshold for significance is chosen (parametric, nonparametric, or fold-change). Among this significant set of genes, researchers separate results into those that increased, and those that decreased with treatment. Thus, at a very basic level, two lists of genes that changed significantly with treatment are generated, those that went up, and those that went down. Effectively, these two directions of change constitute all of the possible patterns of change among the genes whose expression levels are found to change significantly with treatment. Importantly, the third 'pattern' of change, that is- no change, is excluded from consideration because the process of discovering significant genes has successfully filtered these results. The contrast between the two treatments can be considered to form a 'description' of each individual gene's behavior across treatment, and genes with like descriptions are categorized together. These descriptions, or patterns of change, are also present when there are more than two treatment groups.

Adding a third treatment group increases the number of possible patterns of change exponentially if we adhere to this 'treatment group vs. treatment group' (or 'all possible pairwise') comparison strategy. Suppose there are three treatment groups (A, B, and C), and we want to compare expression levels from each treatment to every other treatment (an all-pairwise comparison). There will be three pairwise comparisons to consider, **A vs. B, A vs. C**, and **B vs. C**. For each of these pairwise comparisons, there are three possible results: *no significant change, significant increase,* and *significant decrease*. The combination of all of the pairwise comparison results for each gene can then be used to form a statistical 'pattern' (or description) of that gene's behavior across treatment groups, and the number of elements in that description will be equal to the number of pairwise comparisons.

The number of pairwise comparisons is equal to $N(N-1)/2$, (which is equal to N choose 2 where N = the number of treatment groups (Triantaphyllou, 1993). Importantly, the number of patterns of pairwise results that can be generated by the data is finite, and can be given by P^r, where P is the number of pairwise comparisons, and r is the number of possible results. These are listed in Table 3. It is important to note that the number of possible patterns of pairwise results, P^r, is an upper limit (Ray and Triantaphyllou, 1998; Triantaphyllou and Shu, 2001). This calculation assumes that there is no interaction among the pairwise results. For instance, if we were comparing A v B and C v D and E v F, then the results of any one pairwise contrast would have no impact on the results of other pairwise contrasts, however, as shown in Table 3 (gray area), our comparisons are NOT independent, and some mathematically calculated patterns cannot logically exist. In this case, I define this inter-relationship among the pairwise comparison results as 'dependence', and further show in Figure 14 that as the number of treatment groups increases, the degree of dependence (or entanglement) among the pairwise comparisons increase, so that a relatively greater and greater fraction of the mathematically calculated results could not logically exist. However, by the time seven treatment groups are included, the number of possible combinations of results rivals the number of probe sets to be examined (~10,000), and any advantages of the pattern matching strategy are lost, as there is no effective consolidation of information with the strategy with larger numbers of treatment groups (e.g., in time course experiments).

Post-Hoc Pattern Matching

'*Post-hoc* pattern matching' (PPM) could be thought of as a 'bridge' technique, one that may be useful for researchers working with between three and six treatment groups. Another 'bridge' technique, template matching (Pavlidis and Noble, 2001; Reid et al., 2001; Cavallaro et al., 2002) is also useful, and, with appropriate *a priori* knowledge from the researcher, can be a very powerful way of associating genes with particular physiological/ pathological processes. Additionally, a combination of PPM and template matching procedures can help researchers establish groups of genes statistically identified as potentially important to a particular process.

PPM becomes hopelessly complicated as the number of treatment groups increases, because the number of pairwise comparisons increases as a quadratic

Table 3. **Possible combinations of pairwise results for a three treatment group comparison.**

A vs. B	A vs. C	B vs. C	A vs. B	A vs. C	B vs. C	A vs. B	A vs. C	B vs. C
↑	↔	↔	↓	↔	↔	↔	↔	↔
↑	↑	↔	↓	↑	↔	↔	↑	↔
↑	↑	↑	↓	↑	↑	↔	↑	↑
↑	↔	↑	↓	↔	↑	↔	↔↑	
↑	↓	↔	↓	↓	↔	↔	↓	↔
↑	↓	↓	↓	↓	↓	↔	↓	↓
↑	↔	↓	↓	↔	↓	↔	↔↓	
↑	↑	↓	↓	↑	↓	↔	↑	↓
↑	↓	↑	↓	↓	↑	↔	↓	↑

Note: Each combination of three pairwise comparisons represents a post-hoc pattern. Patterns highlighted in gray are those that have been calculated mathematically to occur, but because of the geometric/ triangular property of the data in question, are impossible in non-independent comparisons (Triantaphyllou, 1993). Additionally, the top post-hoc pattern in the right most section has been grayed out because the initial ANOVA-based significance screening should remove from consideration all genes in which there are no significant post-hoc comparisons.

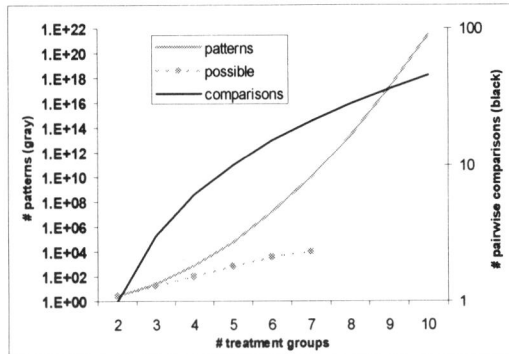

Figure 14: Quadratic increase between the number of treatment groups (X axis) and the number of pairwise comparisons (right Y axis) is shown in black. The exponential increase in the number of post-hoc patterns (left Y axis) is shown in gray. Note that by the time there are seven treatment groups, the data are capable of generating over a million different post-hoc comparison patterns. However, due to the lack of independence among the comparisons, the actual number of possible comparisons generated in 100 iterations of a 10,000 gene, 30 array model data sytem (dotted gray line) is considerably less.

function of the number of treatment groups involved in the study, and the number of patterns increases as an exponential of the number of pairwise comparisons (Figure 14) (this is true only when comparisons are independent). This effectively restricts PPM's use to smaller numbers of treatment groups. However, PPM includes statistical measures of significant difference in its assignment of different genes to different patterns of expression. Additionally, PPM has a predetermined maximum number of statistically defined expression patterns (crudely analogous to clusters) present in the data.

Example of the Use of PPM in a Three Treatment Group Comparison

In this example method, I provide detailed information on the PPM analysis procedures using Excel formulae. This information should be of interest to those wishing to use the PPM approach, but for other readers, I would recommend skipping to the end of this section.

The data set in the following example is taken from prior work (Blalock et al., In Press). **Design:** There are three treatment groups (n = 9-10/ arrays per group), and each array represents a single animal. Arrays are prefabricated Affymetrix RG-U34A GeneChips®. Tissue preparation, chip hybridization, and scanning were all performed according to standard Affymetrix procedures. For the present example, probe level analysis and background noise subtraction are performed using MAS5.

Arranging data for analysis in Excel: 'Presence p-value' and 'signal intensity' metrics are copied from the pivot tab form in MAS5 to two separate Excel worksheets within a single workbook (one worksheet for presence p-values, one for signal intensity measures). Within each sheet, I arranged data in a matrix such that each column represents the microarray results of a single animal, and each row represented a single gene/EST. There are 29 microarrays in the study, and each array holds 8,799 probe sets, thus the array on each worksheet was 8,799 rows by 29 columns, and the columns were grouped by treatment (columns 1-9 = treatment A; columns 10-19 = treatment B; columns 20-29 = treatment C).

In addition to these two worksheets, I added a third worksheet called 'Analysis'. I entered the filtering and statistical testing formulae in this worksheet (an example of this worksheet is included at the bottom of the methods section).

Filtering for present data: As mentioned earlier in this chapter, the handling of presence calls in Affymetrix data is a subject of some debate. In this instance, I counted the number of times each gene was found to be present (p-value for the presence p-value of < 0.05) in all arrays using the COUNTIF function. Genes were excluded from further analysis if they did not show at least 9 'presence' calls across all 29 GeneChips. The number 9 was chosen because the smallest treatment group (Group A) had 9 subjects/ columns. If some gene's expression level had been rated present in all 9 cases for treatment group A, and absent in all other treatment groups, then this information would be important to retain for further analysis. To guard against potential bias in gene selection, group labels were not used in the filtering procedure (in keeping with Rork Kuick's notions of 'agnostic' selection; http://bfx.kribb.re.kr/gene-array/ and personal communication). Roughly half of the original probe sets were retained for further analysis by this criterion (4029/8799).

Main statistical test: I calculated the p-value for 1-Way ANOVA on a per gene/row basis. Most statistical software generates tables for ANOVA tests, and in this case I need to generate a single p-value for each row (I do not want to generate 8799 tables of data). I made a formula in Excel that generates the appropriate p-value as follows:

=FDIST(((DEVSQ(ALL)-(DEVSQ(Group   A)+DEVSQ(Group B)+DEVSQ(Group   C)))/2)/((DEVSQ(Group   A)+DEVSQ(Group B)+DEVSQ(Group C))/26),2,26)

where the actual ranges set in Excel have been replaced with the names of the ranges (e.g., ALL = G2:AI2; Group A = G2:P2; Group B = Q2:Z2; and Group C = AA2:AI2). DEVSQ is the function in Excel that generates the sum of squares and FDIST is the function that returns the p-value of a given F statistic when provided with the degrees of freedom in the numerator (k-1) and the denominator (N-k), where N = number of subjects in the study and k = the number of treatment groups. P-value results using this formula agree with results generated using more typical statistical programs (S-Plus, SAS, SigmaStat, Excel's Analysis Tool-pak add–in).

Post-hoc statistical test: For results found to be significant across treatment by 1-Way ANOVA, *post-hoc* comparisons across all possible pairwise comparisons were used. As mentioned earlier, there are three pairwise comparisons when there are three treatment groups (A vs B, A vs C, and B vs C). There are a number of different *post-hoc* tests that might be appropriate here, including Scheffe's Test, Tukeys Honestly Significant Difference (HSD), and Fisher's Least Significant Difference (LSD). I chose Fisher's LSD as it is the least conservative of these tests, and I wanted to insure that, if the overall ANOVA showed a significant difference among the treatment groups, then at least one of the pairwise comparisons would also show a significant difference.

Therefore, I tested genes significant by 1-Way ANOVA for three different pairwise contrasts using Fisher's LSD. The LSD, which represents the smallest difference between two groups that must be achieved in order for a pairwise comparison to be considered significant, is given by the following general equation: $\sqrt{\dfrac{2*MSE*F_{\alpha,1,n-1}}{n}}$

Where *MSE* is the mean squared error as calculated by the ANOVA across all treatment groups, *F* is the value of the F statistic for some a level (typically 0.05) given *1* degree of freedom in the numerator and *n-1* (where *n* = number of subjects within a **single** treatment group) degrees of freedom in the denominator.

Because the *n*'s were not balanced across the three treatment groups (e.g., A = 10, B = 10, C = 9), I used the geometric mean of *n* across the three groups (cube root of 9*10*10 = 9.65) as an estimate for *n*. Thus, the F statistic for this example could be written as $F_{.05,1,8.65}$, and rounding the estimate of *n*, $F_{.05,1,9}$ = 5.12 and the equation for calculation of the LSD for any of the genes in this particular data set could be expressed as:

$$\sqrt{\dfrac{2*MSE*5.12}{9.65}}$$

In Excel, I used the following formula to calculate LSD:

=SQRT((2*((DEVSQ(Group A)+
DEVSQ(Group B)+DEVSQ(Group C))/26)*5.12)/9.65)

The basic test is then very simple, "is the difference between the average gene expression levels in two treatment groups greater than or equal to the LSD? If so, then the difference is significant". In terms of Excel formulae, this could be written as:

=ABS(AVERAGE(Group A)-AVERAGE(group B))>=LSD

where the ABS function will return the absolute value of the difference, and LSD is the calculated LSD value for that gene. If the difference between the means is greater than the LSD, then the formula result will be 'TRUE', if not, then the result will be 'FALSE'.

Arranging *post-hoc* tests: The *post hoc* LSD is useful for determining if a pairwise comparison yields a significant result, but does not reveal the direction of that result. In addition, the binary TRUE/FALSE result does not lend itself to the creation of a 'descriptor' that would encode the statistically defined pattern of expression across all possible pairwise comparisons. Thus, the results of the LSD test can be modified to do the following:

1) only perform the test on genes/probe sets previously established as significant by 1-Way ANOVA

2) reveal the direction of change for significant pairwise comparisons

3) report the results for each comparison in such a way that the results are easily combined into a unique descriptor of the statistically defined pattern of expression across all possible pairwise comparisons.

These modifications can be added to the pairwise contrast results in any number of ways. What follows is the formula I devised:

=((ANOVA<=0.05)*(ABS(AVERAGE(One Group)-AVERAGE(Other Group))>=LSD)*(AVERAGE(One Group)>AVERAGE(Other Group))*-10)+((ANOVA<=0.05)*(ABS(AVERAGE(One Group)-AVERAGE(Other Group))>=LSD)*(AVERAGE(One Group)<AVERAGE(Other Group))*10)

where ANOVA is the p-value of the 1-Way ANOVA, 'One Group' and 'Other Group' are being contrasted in a pairwise fashion, and LSD is as calculated above.

Conceptually, when Excel formulae are used to interrogate a cell value, they report 'TRUE' or 'FALSE'. For instance, the first part of the formula (ANOVA<=0.05) asks "is the p-value reported in the ANOVA column for this row less than or equal to .05?" If this were true, then Excel would report 'TRUE'. However, when these true/false reports are included in a mathematical calculation, they resolve to 1's (for TRUE responses) and 0's (for FALSE responses). Thus, ANOVA(<=.05)+10 would equal 10 if ANOVA <= .05 was FALSE, and 11 if it was TRUE.

The above modified pairwise contrast formula can be separated into two parts by the '+' sign. In the first part, I am asking (in order and separated by the asterisks) 1- is the overall ANOVA significant?; 2- is the difference between the treatment group means greater than or equal to the LSD?; and 3- is the average of the first treatment group less than the average of the second treatment group. This amounts to a series of true/false responses. Because they are multiplied together (prior to the '+' sign), any false response will collapse the entire result to 0. In the final calculation prior to the '+' sign, I multiply the result by 10. Thus, IF all three statements are 'TRUE', then the result will be 10.

After the '+' sign, the formula is basically repeated with two important differences. Instead of asking if the average of the first treatment group is LESS than the average of the second treatment group, I'm asking if the average of the first treatment group is GREATER than the average of the second treatment group. Furthermore, if all of the statements after the '+' sign are true, then a value of −10 is assigned.

Note that the entire statement before the '+' sign vs. that after are exclusive-both statements can never be true because the second group can never be BOTH greater than and less than the first group for any given gene's expression level. Thus, at least one of the two statements will always be 0, and a reported value of 10 would indicate a significant pairwise contrast with an increase in average expression level between the first and the second group, while a value of −10 would indicate a significant decrease in expression level between the first and second group.

For the first pairwise comparison, significant pairwise increases are set to 10, decreases to −10, and no change set to 0. For each subsequent pairwise comparison, results are set one order of magnitude higher, so that results from the first pairwise comparison are −10, 0, and 10; from the second pairwise comparison −100, 0, 100; and from the third pairwise comparison −1000, 0, and 1000. Alternatively, one could report the results of each test with a single character such as '+' for increases, '-' for decreases, and '0' for no change. The reports could then be concatenated (+-0). This method of reporting patterns is more visually intuitive, although it may be more difficult to arrange and sort in Excel, particularly when attempting to identify 'mirror' patterns.

Using Excel's 'conditional formatting feature', I set pairwise comparison cells to be white for a significant increase, black for a significant decrease, and gray for no significant difference (Table 4). This helps in visualizing and grouping the different patterns of gene expression.

Thus the sum of the results of these pairwise comparisons accurately encodes the statistically derived, *post hoc* patterns of gene expression in the data. Furthermore, the sums of equal but opposite sign represent exactly opposite patterns, and expression levels of those genes may be influencing or influenced by, similar phenomena. The genes can then be grouped according to these patterns. Grouping genes by similar post hoc patterns is the essence of post hoc pattern matching (PPM).

PPM can be a powerful technique for finding interesting patterns of expression. It is important to keep in mind that certain of these patterns are more likely to occur by chance than others (Figure 15). For instance, a pattern in which one mean is significantly different from both other means is more likely to occur than an instance in which all three pairwise comparisons are significantly different.

This is especially interesting given the results in Table 4. One of the most prevalent patterns of expression (-1110; analogous to Figure 15, right column, second pattern from the bottom) would be expected to have a relatively small contribution to the overall number of significant genes if the data had been generated randomly. Compare the results of Table 4 (from experimental data) with those of Table 4 (generated using random numbers). The fact that the - 1110 PPM pattern is prevalent in the data set is an important clue that the genes belonging to that pattern are not there by chance alone. Additionally PPM 1110, the opposite partner of −1110, is not nearly as prevalent. In random data, these opposite partners should have equivalent representation (for instance, -1100 and 1100 do show relatively equivalent numbers of genes). Thus, pattern −110 may also be of further interest, as its opposite number did not appear at this level of analysis at all- an occurrence that would be very unlikely by chance. Therefore, it is absolutely critical when using PPM analysis that some measure of the effects on random data sets be used. As with other analysis techniques used in microarray research, the fact that so many tests are performed at once can lead to some results that appear to be deliberate and biologically relevant but are in fact a product of the test used and reflect nothing more than random noise.

Table 4. PPM analysis applied to real data

order	P	ANOVA	LSD	AvM	AvY	MvY	PPM
3066	29	4.468E-07	3186				-110
2375	29	9.28287E-06	2690				-110
6302	21	1.13803E-05	84				-110
6369	29	4.77611E-05	1002				-110
5587	29	5.70736E-05	3372				-110
5821	26	6.32183E-05	383				-110
6437	11	0.000225219	231				-110
5475	29	0.000232867	395				-110
3403	29	0.000299049	689				-110
5854	29	0.000647326	443				-110
5005	29	0.000755099	765				-110
8089	29	0.000930322	76				-110
6368	29	0.000979465	2604				-110
2870	29	3.47843E-05	2717				-990
4323	29	0.000801234	650				-990
486	29	3.29621E-06	925				1100
6966	28	3.64062E-06	129				1100
487	29	8.88431E-06	1842				1100
799	29	1.17937E-05	619				1100
1978	29	3.73456E-05	361				1100
444	29	5.12402E-05	3970				1100
1976	13	0.000119257	169				1100
4079	28	0.00013796	188				1100
2432	29	0.000209977	364				1100
5547	29	0.000369785	284				1100
5762	29	0.000417066	120				1100
3333	29	0.000427247	626				1100
3628	26	0.000514952	94				1100
4318	29	5.86805E-06	648				1100
599	29	1.74616E-05	181				1100
681	29	3.15203E-05	1022				1100
6009	29	6.17802E-05	567				1100
7333	29	9.90389E-05	102				1100
6830	29	0.000141478	103				1100
2258	29	0.000242786	83				1100
2369	28	0.00024655	276				1100
5838	29	0.000259484	422				1100
7248	29	0.000339538	201				1100
3621	29	0.000489297	1245				1100
4024	26	0.000534964	148				1100
6560	17	0.00076699	181				1100
542	29	0.000799934	904				1100
8326	15	0.000824866	61				1100
6195	29	4.39524E-07	334				-1110
3454	29	8.92654E-07	1368				-1110
3419	29	1.88687E-06	394				-1110
2376	29	2.0035E-06	944				-1110
445	29	2.16695E-06	2514				-1110
6194	29	2.63183E-06	437				-1110
2953	29	3.54253E-06	844				-1110
5043	29	6.9653E-06	248				-1110
3402	22	8.73589E-06	435				-1110
6517	29	1.40073E-05	1002				-1110
6889	29	1.8719E-05	2295				-1110
5801	29	1.87977E-05	85				-1110
5634	29	0.00039783	4320				-1110
598	29	2.08949E-06	159				-1110

Note: order #- list of unique identifiers for each probe set on the chip; #P- number of chips (out of 29) for which the probe set was rated 'present'; ANOVA- 1-Way ANOVA p-value; LSD- Fisher's Least Significant Difference value; A v M/ A v Y/ M v Y- post hoc comparison results for each pairwise comparison in the study (white = increase, gray = no change, black = decrease); PPM- post hoc pattern matching sums for each gene. Results are organized by absolute PPM value, then by sign of PPM, then by ANOVA p-value. For the purposes of display, only genes significant at p < 0.001 by ANOVA are shown.

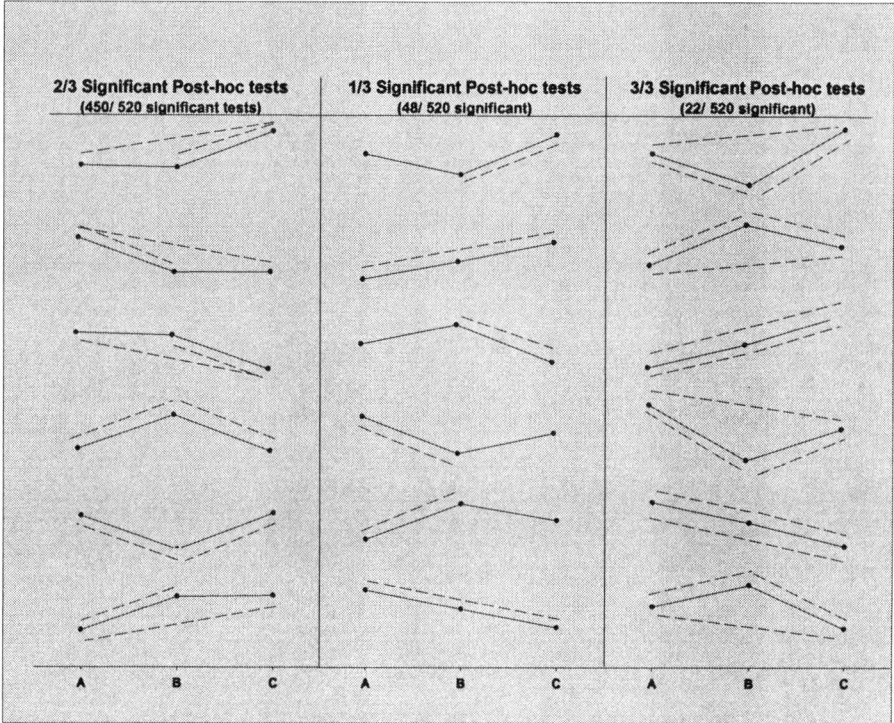

Figure 15. **PPM in random data sets; established patterns have different likelihoods of occurring.** Dotted black lines represent significant contrasts and solid gray lines represent non-significant contrasts. One hundred iterations of model data consisting of a 10,000 row by 30-column array of randomly generated numbers were used. On average, 520 of these fake 'genes' were found to be significant by 1-Way ANOVA. There were 18 possible patterns (see Table 3) to which these randomly significant genes could be assigned. There was a clear bias towards patterns in which 2/3 significant post-hoc tests (left column) occurred, while results in which 1/3 post-hoc tests (center column) were fairly infrequent and results in which all 3 post hoc tests were significant were rarest of all (right column).

As the number of pairwise comparisons increases, the complexity of PPM analysis increases. It is important to note that the simple exponential calculation of the number of possible PPM assignments ($3^{\text{# pairwise comparisons}}$) is an upper bound, and because of the dependence of the data being analyzed, not all of these patterns can occur. Additionally, as the number of pairwise comparisons increases, the likelihood of each PPM pattern occurring in random data changes. Thus, a random data array should be consulted for PPM analyses using different numbers of pairwise comparisons. (permutation and/or randomization tests may also be effective). Importantly, the p-value chosen for the *post hoc* tests can alter the prevalence of the detection of certain patterns of data. Thus changes in the alpha level of the *post hoc* tests should also be reflected in the assessment of random model data. As a shorthand way of looking at the data, one can always compare a prevalent pattern that has arisen in the real data with its opposite partner. If both are similarly represented, this does not necessarily mean that the genes in question are not 'really' changing, simply that the logic behind PPM analysis could not be used to persuasively argue for their importance.

In summary, PPM analysis may be a useful adjunct for microarray analysis. This may be particularly true in cases where there are more than two, but certainly no more than six, groups (at seven groups, there are > 1 million different potential PPM combinations). Once the number of possible groupings exceeds the number of genes being tested, it becomes less likely that the process will result in the desired dimensional reduction, and the procedure no longer makes sense.

However, for designs in which the number of groups is limited, this technique offers several advantages. First, PPM has a predetermined number of patterns, and the upper bound of that number is easily calculated. Second, PPM incorporates standard error into its analysis, allowing a measure of statistical validity to be included in an automated categorization system. Third, the likelihood that a particular pattern will arise by chance alone can be estimated using simple model systems, and the results of these model systems can be cross indexed to the actual results, allowing researchers to determine whether a particular pattern of expression is likely to occur by chance. Finally, the PPM number is a convenient way to summarize and encode patterns of expression, and the results of the various *post hoc* tests can be derived for each gene if the researcher keeps a record of the table design (e.g., the order of the pairwise comparisons).

FUNCTIONAL GROUPING

Functional grouping (categorizing genes based on their shared functions) is easily the most powerful component of microarray analysis. It adds to interpretation of microarray results because the researcher can make determinations regarding the effects of a particular treatment across entire functional groups of genes. The logic is self-evident; if multiple genes within a single functional category show the same pattern of change with a given treatment, then each of the genes in that category effectively 'borrows' significance from its neighbors, and the researcher can report that there was an important change for the entire group. A number of researchers have used this grouping notion to highlight important microarray-based findings (Lee et al., 2000; Lukiw and Bazan, 2000; Jiang et al., 2001; Mirnics, 2001; Prolla, 2002; Weindruch et al., 2002). Some researchers demonstrated that multiple genes involved in the same process (presynaptic machinery) agreed in direction of change, even if those changes failed to reach fold-change based criteria for significance (Mirnics et al., 2000; Mirnics et al., 2001). This last procedure, pulling additional information out of what would be considered statistical 'noise' based on a functional assignment, clearly demonstrates the effectiveness of this procedure. However, researchers new to the procedure should be aware that this procedure can be arduous, and there are some caveats to its use.

Difficulties with Functional Grouping

At a very basic level, functional grouping is a simplification of the data, collapsing the results of hundreds or thousands of significant genes into a handful of functional groups. Researchers are attracted to this strategy because of its power and utility. Functional grouping can reduce the complexity and size of the data set to the point that a table of the functional groups can be published in peer reviewed journals without exceeding their size limitations. Then the much more manageable list of groups can be discussed. However, functional grouping short falls have limited its utility. Three major issues must be considered when dealing with functional grouping strategies.

Manual Categorization

Assigning categories has in the past been largely the responsibility of the individual researcher (sometimes referred to as functional "groping"). The arduous task of looking up genbank accession numbers, doing medline searches and BLAST inferences, targeted web searches (e.g., SwissProt- http:// us.expasy.org/sprot/; Pfam- http://www.sanger.ac.uk/Software/Pfam/; GeneCards- http://bioinfo.weizmann.ac.il/cards/), and sometimes even web-wide searches (e.g., www.alltheweb.com, www.google.com) to get information on genes falls to the researcher. In many cases, the researcher's knowledge regarding a particular field of study will color their functional assignments. Furthermore, this subjective categorization will lead to disagreement among different labs as to which functional group a particular gene belongs. While some categories are easy to align from one research paper to another (e.g., immediate early genes or inflammatory factors) others are more subjective, and depend on the individual experience of each experimenter.

Representational Bias

Another criticism of functional grouping strategies in the past has been that, because the researcher is only categorizing those genes that were designated as significant, the procedure is unable to detect representational bias on the chip. For instance, a particular chip may be designed to detect a greater proportion of cytokine-related than intermediary metabolism-related genes (this may reflect the targeted design of the chip and/or the possibility that the genome itself does not represent each functional group equally). Thus, by chance alone, one would expect a stronger representation from cytokine than intermediary metabolism genes, and that chance is essentially uncontrolled in studies that do not assess the total functional group representation on the chip. Because manual categorization is such an arduous, time-consuming, and subjective process, researchers using manual categorization strategies generally do not attempt to assign categories to all of the genes, only the significant ones. Essentially, then, manual procedures that address only the significant genes are missing the denominator of the functional grouping fraction (Figure 16). Using Figure 16 as an example, a researcher who had no knowledge of this representational bias across the entire chip would be blinded to the fact that a finding of 8 significant white genes may be a more compelling result than 15 gray because 8/10 is a much greater proportion than 15/55.

A

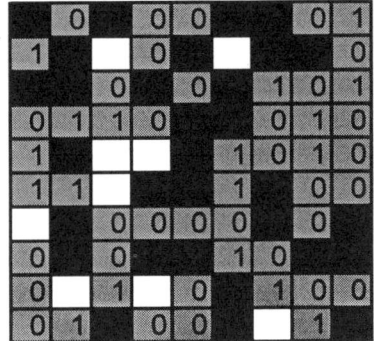

B

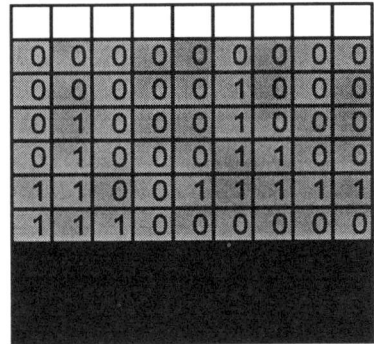

Figure 16. **Functional group representational bias**. Consider a simplified case with only three 'functional' groups (black, white, and gray). In **A**, each of the 100 genes (squares) has been assigned a functional group. In **B**, those genes have been rearranged to more clearly show the bias in representation among groups. If one were to randomly select a gene, there is a 10% chance it would be white, 30% chance it would be black, and 60% chance that it would be gray.

Genes to Functional Groups are Not 1:1 Relationships

A third consideration is the complicated relationship among genes and the functions of their products. The relationship among genes and the functional groups to which they are assigned could be referred to as a 'web', inferring interconnectedness among the genes. Obviously, if each gene were assigned its own unique functional category, there would be no use in performing functional grouping because there would be no resultant consolidation of data. However, this is not the case, and many genes can be assigned to a particular functional category, and the scope of that category widened or narrowed to suit the researcher ('enzyme' may be too general, and tyrosine hydroxylase too specific). However, appreciate that this multiple gene to a single category strategy works both ways- while there are fewer categories than genes, a single gene can be assigned to multiple categories. For instance, in the biochemical 'wiring' diagrams (available through KEGG- http://www.genome.ad.jp/kegg/kegg2.html#pathway), GenMapp-www.genmapp.org), a gene product such as aromatase can be shown to be involved in at least three metabolic pathways (fatty acid metabolism, gamma-hexachlorocyclohexane degradation, and tryptophan metabolism), and thus should have more than one functional categorization.

Unbiased Functional Grouping Procedures

To address these shortcomings (experimenter bias, representational bias, and complex functional categorization), researchers are developing unified databases that provide less biased coverage of the functional relationships among defined gene products (e.g., www.geneontology.org; www.genome.ad.jp/kegg; www.genmapp.org) (Rison et al., 2000; Stevens et al., 2000; Consortium, 2001; Pavlidis et al., 2002). These databases can be used to help address researcher bias by assigning gene categories without researcher input. Additionally, because the procedures can be automated, all of the genes on the chip can be categorized, allowing for proportional representation of each category to be assessed. Furthermore, these databases incorporate the complex inter-relationships among gene products and functional categories. Software such as Onto-Express (http://server1.openchannelsoftware.org/oe/onto-express_V2.php) (Khatri et al., 2002) can be used to not only assess proportional representation, but also to test for significant over- and under-representation of functional groups in the significant data set. In the following method, I will describe the nuts and bolts of a manual procedure using Excel, the Gene Ontology, and Affymetrix annotations to provide functional categorization and statistical testing for significant results. This procedure can be intense and laborious. In my opinion it is useful for researchers who are NOT working with 'popular' genomes (such as mouse, human, or yeast), yet still desire a less biased method of assigning functional groups and assigning significance to those groups.

Functional Grouping by Gene Ontology

The Gene Ontology is broken down into three separate hierarchies, Biological Process, Cellular Component, and Molecular Function (Figure 17). Each identified gene product may have one or more 'attributes' (with an associated, unique GO number) in each of these three ontologies, and each of these attributes may occur at one or more levels in the hierarchy of each ontology (Ashburner et al., 2000; Lewis et al., 2000; Consortium, 2001). Thus, each gene is conceptually fragmented into its attendant GO numbers for subsequent functional grouping analysis. It is important to remember that, after the genes have been used to generate the list of gene attributes, it is no longer the genes, but the attributes that are being tested, and it can be difficult to point to specific genes and claim that it was their contribution to the functional categorization that caused some level of the categorization to be significant.

Because each GO # occurs in a hierarchy of attributes, it is possible that, while two genes do not share the same GO #, they are grouped together at some higher level of the hierarchy. This may be more easily explained with an analogy. Suppose we were categorizing fruit, and under the category of 'apple' we had the following:
1) Apple
   a) Green
      i) Granny Smith
   b) Red
      i) Macintosh
      ii) Red Delicious

You may have an apple that is well defined, like a Red Delicious counted under '1bii'. Another, less well-defined apple could only be defined as 'Red' and so is counted under '1b'. An even more poorly categorized apple may be assigned to Apple '1'. In the final analysis, however, for assessing the number of all apples found, you would need to calculate a 'hierarchical sum' of all the apples found at or below each level of the outline. In this example we had 1 Red Delicious apple, 2 Red apples (one Red Delicious and one Red apple), and 3 apples in the entire category (2 Red apples and one with no assignment other than apple).

Building the Annotation for the Entire Chip: We used the Genome Ontology group's web-based gene ontology (GO) tree builder (Amigo; www.geneontology.org) to search for molecular function and biological process

HTML Tree View

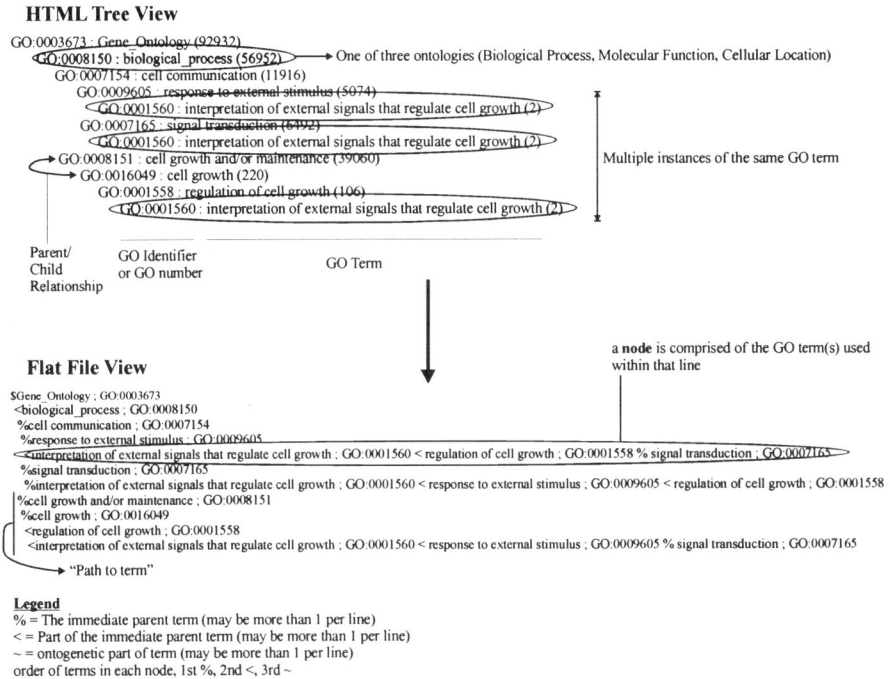

GO:0003673 : Gene_Ontology (92932)
 GO:0008150 : biological_process (56952) → One of three ontologies (Biological Process, Molecular Function, Cellular Location)
  GO:0007154 : cell communication (11916)
   GO:0009605 : response to external stimulus (5074)
    GO:0001560 : interpretation of external signals that regulate cell growth (2)
   GO:0007165 : signal transduction (6492)
    GO:0001560 : interpretation of external signals that regulate cell growth (2)
  GO:0008151 : cell growth and/or maintenance (39060)
  GO:0016049 : cell growth (220)
   GO:0001558 : regulation of cell growth (106)
    GO:0001560 : interpretation of external signals that regulate cell growth (2)

Multiple instances of the same GO term

Parent/Child Relationship   GO Identifier or GO number   GO Term

a **node** is comprised of the GO term(s) used within that line

Flat File View

$Gene_Ontology ; GO:0003673
<biological_process ; GO:0008150
%cell communication ; GO:0007154
%response to external stimulus ; GO:0009605
<interpretation of external signals that regulate cell growth ; GO:0001560 < regulation of cell growth ; GO:0001558 % signal transduction ; GO:0007165
%signal transduction ; GO:0007165
%interpretation of external signals that regulate cell growth ; GO:0001560 < response to external stimulus ; GO:0009605 < regulation of cell growth ; GO:0001558
%cell growth and/or maintenance ; GO:0008151
%cell growth ; GO:0016049
<regulation of cell growth ; GO:0001558
<interpretation of external signals that regulate cell growth ; GO:0001560 < response to external stimulus ; GO:0009605 % signal transduction ; GO:0007165

→ "Path to term"

Legend
% = The immediate parent term (may be more than 1 per line)
< = Part of the immediate parent term (may be more than 1 per line)
~ = ontogenetic part of term (may be more than 1 per line)
order of terms in each node, 1st %, 2nd <, 3rd ~

Figure 17. **Example of output from the Gene Ontology**. The view displayed on the web page (**HTML view**) is a summarized version of the '**Flat File View**' (this can also be downloaded as an Extensible Markup Language XML file- thanks to the Gene Ontology group for double checking the accuracy of this figure).

associations for the genes on the RG-U34A chip. To do this, I first searched for the most current information on these probe sets and the genes and ESTs whose expression levels they are designed to quantify. Current annotations for this chip design were downloaded on July 11, 2002 from Affymetrix' website (www.affymetrix.com). Of the 8,799 probe sets, 1,345 had associations within the GO: Biological Process category, and 1,211 had associations within the GO: Molecular Function category. A total of 6,799 probe sets on the entire chip had no GO associations, and of these, 3,292 had gene names/symbols representing 2,101 unique genes (several gene symbols/names were repeated).

This list of unique 'named but not GO associated' genes was then uploaded to the Affymetrix web site (500 genes at a time in tab delimited text files using the Netaffx 'batch analysis' tool) and other, more fully annotated chips (e.g., HG-U133- human, and MG-U74- mouse) were interrogated for supplementary annotation information. Using this process, we added GO: Biological Process associations for 976 additional unique genes, and GO: Molecular Function associations for 991 additional unique genes. Thus, the composite list contained GO: Biological Process associations for 3,232 probe sets representing 1,930 unique genes and GO: Molecular Function associations for 3,280 probe sets representing 1,948 unique genes. It should be noted that gene names are supposedly unique and represent a concerted effort by the scientific community to generate a list of gene names that are not species dependent (unlike GenBank IDs). However, there are several occurrences of overlap where different genes may have the same gene names and it is possible that some misinterpretation is included in the annotated list because of possible disagreements in the definition of gene name among different chip annotations.

This set of gene attributes represents the most up-to-date annotated information we have about the entire RG-U34A chip. There are several levels of evidence that have to do with the certainty of that association- 'experimental evidence' is probably the strongest support, while 'inferred from electronic annotation' is probably the weakest. For this study we chose the most liberal interpretation and included all annotations. This is somewhat mitigated at later stages of functional group analysis by only considering more basic associations.

It should be noted that the relationship between gene name/symbol and GO ID is not 1:1. For instance, agrin (Agrn) has two GO: Biological Process IDs with which it is associated (7009- plasma membrane organization and biogenesis, and 7268- synaptic transmission). Additionally, several genes may share an association (e.g., 7268 is shared by Agrin, Glutamate decarboxylase 2, Catecholamine O-methyl transferase, and 90 other probe sets on the chip). Thus each GO-associated gene had at least one GO ID with which it was associated, and each GO association had at least one gene with which it was associated, but the relationship between the two was not reciprocal.

Nearly 1/3 of the identified gene names/ symbols are repeated at least once on the chip. In the present work we counted these repeats as if they were separate genes, although more advanced analysis procedures use a combination of information from each of the repeats for a particular gene to construct a more reliable single metric (Pavlidis et al., 2002).

There were a total of 4,058 total GO: Biological Process associations on the chip, representing 504 unique GO #s. There were a total of 2,285 total GO: Molecular Function associations on the chip representing 727 unique GO #s. These unique GO #s were then uploaded into the Amigo browser (100 at a time into the 'advanced query' text box - http://www.godatabase.org/cgi-bin/go.cgi). With each GO ID, the browser built a representation of the entire gene ontology with only the levels for which there were GO matches expanded. In this way a tree or outline of functional relationships was built for each category. The resulting GO Tree was then saved as a 'flat file' (text file) and pasted into Excel for further analysis.

Keeping the Hierarchical relationship Among Attributes: For the rest of the example, I will focus on the *Biological Process* ontology. Similar procedures can be used on each of the ontologies. The flat file pastes as a single column in Excel. The first typewritten character on each line is either a '$' (first line only), a '<', or a '%' (Fig. 17), and preceding this typewritten character are a number of spaces. The number of spaces before the first character represents the outline level of that line/node.

Within the category of biological process, the RG-U34A had enough hits in the GO engine to generate a hierarchical list of 1033 lines. Remember that the ontology expands every line above and including the one on which the hit occurred (thus in our earlier apple example, a hit at the Macintosh level would expand three lines- Apple| Red| Macintosh, while a hit at the Red level would expand only two lines- Apple| Red). I pasted these into A1:A1033 in Excel. To determine the outline level of each line/ node, I entered the following formula into B1 and copied down to B1033;

=LEN(A1)-LEN(TRIM(A1))

where LEN reports the number of characters and spaces in a cell, and TRIM subtracts all spaces except those between words. Thus the length of the entire entry minus the length of the trimmed entry gives the number of leading spaces, which is equal to the outline level for that node.

Stripping GO #'s for further analysis: In order to use the GO #s in further analysis, they had to be stripped out of the text descriptions and isolated as individual numbers. To do this, I copied the column of downloaded GO associations from column A to a separate worksheet. I then used Excel's 'text to columns' feature (Data| Text to Columns| Delimited|) and chose the ':' symbol for delimiting. This split each row out into columns at each occurrence of ':' within each line/ node. Thus, a node with three GO #s would split into four columns, and a node with five GO #s would split into six columns. I then selected the entire area, searched for '_*' (underscore added to emphasize space, use a space instead of an underscore) and replaced it with nothing. Thus a node with two GO#s might look like:

GO:plasma membrane ; GO:0005886 % organelle organization and biogenesis ; GO:0006996

And split into:

GO| plasma membrane ; GO| 0005886 % organelle organization and biogenesis ; GO| 000696

After searching and replacing, it resolves to:

<blank>| 5886| 696

These stripped entries can then be added back to the original worksheet starting in column C.

Counting the number of GO# associations across the entire chip: For each GO# in each node, the number of genes assigned that number needs to be quantified. Remember that this number is different than the number of times that GO term occurs in the downloaded flat file. If two genes had the same GO#, then they were only counted once in the flat file. So I use the original annotated list that was uploaded to the Amigo website, and used the formula:

=COUNTIF(range of GO#s from uploaded list, GO # from flat file)

This will count the number of instances in which a particular GO# from the flat file was found on the entire chip. The sum of the COUNTIF results for all GO#s in a particular node will give the total number of occurrences for that node for the entire chip (the node sum).

Hierarchical sum: The node sum described above is not hierarchical, that is it does not include all of the hits at or below its level. To achieve this hierarchical sum, open a separate worksheet and paste the outline level results (as values) into column A, and the node total results into column B (as values). Label column A 'outline level', label column B 'node counts'. Label column C 'lower range' and enter the following formula in C2 and copy to the end

=IF(A2>=A3,ROW(),MATCH(A2,A3:A$1033,-1))

This formula finds the first instance in the list whose outline level is less than or equal to the level of the node beside which the formula has been pasted. It first checks the number immediately beneath the node being examined. If that number is less than or equal to the interrogated node, then it returns the row # of the node examined. Otherwise, it returns, via the MATCH function, the number of rows down from the examined node at which an outline value less than or equal to the interrogated node's outline level is found. This establishes a summation range. Thus, if the node being examined is in row 2, and has an outline level of three, then the formula 'looks' down the list in column A until it finds a value that is equal to or less than 3. For this example, say that it finds a '2' in row 12. The formula reports the value 12 (the number of rows after the interrogated node before it finds a suitable value). Now a range has been established for hierarchical summation- if the values in the B column from row 2 to row 11 are summed, that value will represent the hierarchical sum for the node in question. This can be generated by entering the following formula into Column D and copying down:

**=IF(C2=ROW(),B2,SUM(INDIRECT("b"&ROW()):
INDIRECT(("b"&C2+ROW()-1))))**

where the node count for that row is reported if the lower range is equal to the present row number, and Excel's INDIRECT function is used to assign a summation range in other cases (Table 5).

This procedure generates the denominator for a binomial statistical test (Khatri et al., 2002; Pletcher et al., 2002) that can be used to determine proportional representation. Performing the same procedure with only the **significant** gene

Table 5. Example of Hierarchical Summation

Outline Level	Node Counts	Lower Range	Hierarchical Sum
1	0	8	21
2	1	4	13
3	3	2	5
4	2	5	2
3	7	6	7
2	2	2	7
3	5	8	5
2	1	9	1
1	1	10	1

attributions generates the numerator. The total number of attributions found significant for the entire chip over the total number of attributions on the chip represent the fractional representation of significant attributions. Thus, at each expanded level of the Gene Ontology, there is a numerator for the number of significant genes attributions found, as well as the total number present at that level, and this can be compared to the overall significance score to determine whether a particular level of the GO is significantly over- or under-represented.

USING EXCEL

I lean heavily on Excel for microarray analysis (for which I am regularly harangued by my friends in the Statistics field). Initially this was because the only software out there that could handle replicated microarray experimental designs using conventional ANOVA tests with *post hoc* comparisons involved line coding (a skill I do not possess). This situation is changing- there are lots of well-written programs available that can handle many of these kinds of analyses. However, I still find that, at some point or another, I'll move my data into Excel for further analysis, storage, and/or sharing with colleagues (who often have neither the money to view data through expensive microarray analysis packages nor the patience to plow through analysis using software that is free to researchers).

You may also find that, at some point, you need to manipulate data in Excel. If so, this may be the first instance in which you find that you are taxing your desktop computer's r esources. Here a re a f ew tips t o help r educe the computational burden on Excel. Entered formulas are 'hot'- that is, they recalculate whenever Excel senses that information has changed (on saves, or any time data is entered). You can turn off this automatic recalculation by going to Tools| Options| Calculation and switching calculation from 'Automatic' to 'Manual'. Then you can initiate a calculation with the F9 key.

Some formulas add a lot more computational pressure than others- for instance; any formula that has to hold an array in memory (e.g., vlookup, rank, match) is going to take up more memory- the larger the array, the more computationally intensive. Generally, I keep these formulas to a minimum by calculating a column of formulas at a time, and then pasting the results as values except for the top row. This 'top row hot' method allows me to keep a record of the formulas I've used without burdening Excel with unnecessary recalculation.

Viewing significant data can be frustrating in Excel. Often researchers sort their data from most to least significant by some criterion, and copy the significant results to a new worksheet. This can get burdensome when there are multiple criteria by which the data could be filtered, or if that significance criterion keeps changing. If the data is entered correctly (the top row in each column has a description of what the column contains, there are no blank rows between the description and the beginning of the table) then put the cursor in any cell in the top row and try using Data| Filter| Autofilter. This causes a drop down menu to appear at the top of each column. The entire table can then be filtered based on criteria from as many columns as needed. If you have other questions regarding the use of Excel, the Mr. Excel message board (www.mrexcel.com) is an excellent resource.

ACKNOWLEDGEMENTS

I thank Arnold Stromberg and Xuejun Peng for valuable statistical (and philosophical) discussion; Chris Norris and Meredith Rauhut for critical reading of the manuscript; the people that work at and contribute to- the University of California, San Francisco Microarray List Server (UCSF List Serv; http://www.gene-chips.com/gene-arrays.html); the Mr. Excel website and message board (www.mrexcel.com); and Bioconductor.org.

REFERENCES

1 Affymetrix (2002a) Technical Note, Statistical Algorithms Reference Guide. In: Affymetrix (http://www.affymetrix.com/support/technical/technotes/statistical_reference_guide.pdf).

2 Affymetrix (2002b) White Paper, Statistical Algorithms Description Document. In: Affymetrix (http://www.affymetrix.com/support/technical/whitepapers/sadd_whitepaper.pdf).

3 Arnemann J, Heins B, Beato M (1979) Synthesis and characterization of a DNA complementary to pre-uteroglobin mRNA. Eur J Biochem 99:361-367.

4 Ashburner M, Ball CA, Blake JA, Botstein D, Butler H, Cherry JM, Davis AP, Dolinski K, Dwight SS, Eppig JT, Harris MA, Hill DP, Issel-Tarver L, Kasarskis A, Lewis S, Matese JC, Richardson JE, Ringwald M, Rubin GM, Sherlock G (2000) Gene ontology: tool for the unification of biology. The Gene Ontology Consortium. Nat Genet 25:25-29.

5 Bakay M, Chen YW, Borup R, Zhao P, Nagaraju K, Hoffman EP (2002) Sources of variability and effect of experimental approach on expression profiling data interpretation. BMC Bioinformatics 3:4.

6 Baldi P, Long AD (2001) A Bayesian framework for the analysis of microarray expression data: regularized t-test and statistical inferences of gene changes. Bioinformatics 17:509-519.

7 Becker K (2002) Deciphering the gene expression profile of long-lived Snell mice. In: http://sageke.sciencemag.org/.

8 Benjamini Y, Hochberg Y (1995) Controlling the false discovery rate: a practical and powerful approach to multiple testing. J Royal Statistical Soc Series B57:289-300.

9 Blalock EM, Chen KC, Sharrow K, Herman JP, Porter NM, Landfield PW (In Press) Gene Microarray Analyses of Hippocampal Aging:Statistical Profiling Identifies Novel ProcessesCorrelated with

10 Cognitive Impairment. J Neurosci.

11 Bolstad BM, Irizarry RA, Astrand M, Speed TP (2002) A comparison of normalization methods for high density oligonucleotide array data based on variance and bias. In.

12 Cavallaro S, D'Agata V, Manickam P, Dufour F, Alkon DL (2002) Memory-specific temporal profiles of gene expression in the hippocampus. Proc Natl Acad Sci U S A 99:16279-16284.

13 Consortium TGO (2001) Creating the gene ontology resource: design and implementation. Genome Res 11:1425-1433.

14   Durbin BP, Hardin JS, Hawkins DM, Rocke DM (2002) A variance-stabilizing transformation for gene-expression microarray data. Bioinformatics 18 Suppl 1:S105-110.

15   Goncalves J (2002) Advancements in Affymetrix probe level data analysis. In: Iobion (http://www.iobion.com/slides/CHI_2002/CHI_2002_Color.pdf).

16   Good P (1997) Chapter 8: Design of Experiments. In: Resampling Methods: A Practical Guide to Data Analysis, pp 133-156. Boston: Birkhauser.

17   Grant G, Manduchi E, Stoeckert Jr C (2002) Using non-parametric methods in the context of multiple testing to determine differentially expressed genes. In: Methods of Microarray Data Analysis (SM Lin KJ, ed), pp 37-56: Kluwer Academic.

18   Hamoen LW, Smits WK, Jong Ad A, Holsappel S, Kuipers OP (2002) Improving the predictive value of the competence transcription factor (ComK) binding site in Bacillus subtilis using a genomic approach. Nucleic Acids Res 30:5517-5528.

19   Hoffmann R, Seidl T, Dugas M (2002) Profound effect of normalization on detection of differentially expressed genes in oligonucleotide microarray data analysis. Genome Biol 3:RESEARCH0033.

20   Huber W, Von Heydebreck A, Sultmann H, Poustka A, Vingron M (2002) Variance stabilization applied to microarray data calibration and to the quantification of differential expression. Bioinformatics 18 Suppl 1:S96-S104.

21   Jiang CH, Tsien JZ, Schultz PG, Hu Y (2001) The effects of aging on gene expression in the hypothalamus and cortex of mice. Proc Natl Acad Sci U S A 98:1930-1934.

22   Kerr MK, Churchill GA (2001) Statistical design and the analysis of gene expression microarray data. Genet Res 77:123-128.

23   Kerr MK, Martin M, Churchill GA (2000) Analysis of variance for gene expression microarray data. J Comput Biol 7:819-837.

24   Khatri P, Draghici S, Ostermeier GC, Krawetz SA (2002) Profiling gene expression using onto-express. Genomics 79:266-270.

25   Lee CK, Weindruch R, Prolla TA (2000) Gene-expression profile of the ageing brain in mice. Nat Genet 25:294-297.

26   Lemke G (1994) Chapter 10: Gene regulation in the nervous system. In: An Introduction to Molecular Neurobiology (Hall Z, ed), pp 313-354. Sunderland, MA: Sinauer Associates.

27   Lewis S, Ashburner M, Reese MG (2000) Annotating eukaryote genomes. Curr Opin Struct Biol 10:349-354.

28   Li C, Wong WH (2001) Model-based analysis of oligonucleotide arrays: expression index computation and outlier detection. Proc Natl Acad Sci U S A 98:31-36.

29   Long AD, Mangalam HJ, Chan BYP, Tolleri L, Hatfield GW, Baldi P (2001) Improved statistical inference from DNA microarray data using analysis of variance and a Bayesian statistical framework - Analysis of global gene expression in Escherichia coli K12. Journal of Biological Chemistry 276:19937-19944.

30   Lukiw WJ, Bazan NG (2000) Neuroinflammatory signaling upregulation in Alzheimer's disease. Neurochem Res 25:1173-1184.

31   Mirnics K (2001) Microarrays in brain research: the good, the bad and the ugly. Nat Rev Neurosci 2:444-447.

32   Mirnics K, Middleton FA, Lewis DA, Levitt P (2001) Analysis of complex brain disorders with gene expression microarrays: schizophrenia as a disease of the synapse. Trends Neurosci 24:479-486.

33   Mirnics K, Middleton FA, Marquez A, Lewis DA, Levitt P (2000) Molecular characterization of schizophrenia viewed by microarray analysis of gene expression in prefrontal cortex. Neuron 28:53-67.

34   Mutch DM, Berger A, Mansourian R, Rytz A, Roberts MA (2002) The limit fold change model: A practical approach for selecting differentially expressed genes from microarray data. BMC Bioinformatics 3:17.

35   Nadon R, Shoemaker J (2002) Statistical issues with microarrays: processing and analysis. Trends Genet 18:265-271.

36   Naef F, Lim DA, Patil N, Magnasco MO (2001) From features to expression: high-density oligonucleotide array analysis revisited. In: Physics Abstract.

37   Pavlidis P, Noble WS (2001) Analysis of strain and regional variation in gene expression in mouse brain. Genome Biol 2:RESEARCH0042.

38   Pavlidis P, Lewis DP, Noble WS (2002) Exploring gene expression data with class scores. Pac Symp Biocomput:474-485.

39   Pletcher SD, Macdonald SJ, Marguerie R, Certa U, Stearns SC, Goldstein DB, Partridge L (2002) Genome-wide transcript profiles in aging and calorically restricted Drosophila melanogaster. Curr Biol 12:712-723.

40   Pott U, Fuss B (1995) Two-color double in situ hybridization using enzymatically hydrolyzed nonradioactive riboprobes. Anal Biochem 225:149-152.

41   Prolla TA (2002) DNA microarray analysis of the aging brain. Chem Senses 27:299-306.

42   Ramirez F, Gambino R, Maniatis GM, Rifkind RA, Marks PA, Bank A (1975) Changes in globin messenger RNA content during erythroid cell differentiation. J Biol Chem 250:6054-6058.

43   Ray TG, Triantaphyllou E (1998) Evaluation of rankings with regard to the possible number of agreements and conflicts. European Journal of Operational Research 106:129-136.

44   Reid R, Dix DJ, Miller D, Krawetz SA (2001) Recovering filter-based microarray data for pathways analysis using a multipoint alignment strategy. Biotechniques 30:762-766, 768.

45   Rison SC, Hodgman TC, Thornton JM (2000) Comparison of functional annotation schemes for genomes. Funct Integr Genomics 1:56-69.

46   Schadt EE, Li C, Ellis B, Wong WH (2001) Feature extraction and normalization algorithms for high-density oligonucleotide gene expression array data. J Cell Biochem Suppl Suppl 37:120-125.

47   Schechter I (1975) Region of immunoglobulin light-chain mRNA transcribed into complementary DNA by RNA-dependent DNA polymerase of avian myeloblastosis virus. Proc Natl Acad Sci U S A 72:2511-2514.

48   Schena M, Shalon D, Heller R, Chai A, Brown PO, Davis RW (1996) Parallel human genome analysis: microarray-based expression monitoring of 1000 genes. Proc Natl Acad Sci U S A 93:10614-10619.

49   ScienceWatch (1999) UCSD's Michael Karin Follows the Cellular Pathways. In: http://www.sciencewatch.com/march-april99/sw_march-april99_page3.htm.

50   Stevens R, Goble CA, Bechhofer S (2000) Ontology-based knowledge representation for bioinformatics. Brief Bioinform 1:398-414.

51   Tiesman JP (2002) Experimental Designs to Minimize Data Variation in GeneChip Experiments. In. Miami Valley Labs: Proctor & Gamble.

52   Triantaphyllou E (1993) A quadratic programming approach in estimating similarity relationships. IEEE Transactions on Fuzzy Systems 1:138-145.

53   Triantaphyllou E, Shu B (2001) On the maximum number of feasible ranking sequences in multi-criteria decision making problems. European Journal of Operational Research 130:665-678.

54   Tusher VG, Tibshirani R, Chu G (2001) Significance analysis of microarrays applied to the ionizing radiation response. Proc Natl Acad Sci U S A 98:5116-5121.

55   Weindruch R, Kayo T, Lee CK, Prolla TA (2002) Gene expression profiling of aging using DNA microarrays. Mech Ageing Dev 123:177-193.

56   Welle S, Brooks AI, Thornton CA (2002) Computational method for reducing variance with Affymetrix microarrays. BMC Bioinformatics 3:23.

57   Wolfinger RD, Gibson G, Wolfinger ED, Bennett L, Hamadeh H, Bushel P, Afshari C, Paules RS (2001) Assessing gene significance from cDNA microarray expression data via mixed models. J Comput Biol 8:625-637.

58   Yang YH, Speed T (2002) Design issues for cDNA microarray experiments. Nat Rev Genet 3:579-588.

59   Zhou Y, Abagyan R (2002) Match-Only Integral Distribution (MOID) Algorithm for high-density oligonucleotide array analysis. BMC Bioinformatics 3:3.

Chapter 7

MICROARRAY EXPERIMENT DESIGN AND STATISTICAL ANALYSIS

Xuejun Peng and Arnold J Stromberg
Department of Statistics, University of Kentucky

INTRODUCTION

Microarray technology is a new kid on the block for biological researchers, and it has shown a great potential in many fields. While biologists are thrilled by the power of this new technology, they are also over-whelmed by the enormous amount of data generated, and often feel uncomfortable about how to extract useful information from the data. On one hand, there is always some information to be mined in such big data sets; on the other hand, it is almost certain that spurious findings will result even if extreme care has been exercised in mining the data. This challenge mandates interdisciplinary cooperation among biologists, statisticians and computer scientists. Although research in this area is still in its infancy, it is our goal in this chapter to discuss some of the most prominent issues in experiment design and statistical analysis for microarray research and to present possible solutions to some of these problems.

Experiment Design Issues

When planning for a microarray experiment, it is important to schedule a meeting between biologists and statisticians to address the following issues: what microarrays can and cannot do; what treatment structure best suits the research goal; what type of biological samples will likely give best representation of gene expression levels while still satisfying technical and financial constraints; how many replicates of biological samples and/or arrays are needed to achieve statistical control. The goal of such a meeting is to find an optimal design that will possibly balance the above concerns. Three important principles of statistics need to be followed: randomization, replication and balanced design. These topics will be discussed in the section titled "Designing a Microarray Experiment."

General Statistical Analysis Procedures

Once the data have been generated from microarray experiments, roughly three stages of statistical analysis are necessary to extract the information from the data: data processing, statistical testing and exploratory analysis, and gene functional analysis. These procedures are covered in the "General Procedures" section.

Multiple Testing

One common goal of microarray experiments is to identify genes that are differentially expressed between or among different conditions. Due to the expense of such experiments, typically only a few replicates are affordable for each condition. This causes a problem with false negatives due to lack of statistical power. On the other hand, when multiple statistical tests with several thousands genes are conducted, we can expect to find some genes that are truly differentially expressed, but we are also prone to false positives just by chance alone. The statistical problems with multiple hypotheses testing and some solutions will be discussed in the section titled, "Multiple Hypothesis Testing for Microarray Experiments."

Analysis of Variation

This issue will be discussed separately for cDNA two color arrays and high density oligonucleotide arrays since their designs are quite different. The emphasis will be put on analysis of variance using a general linear model and/or mixed model approach. Whenever possible, variation due to biological sources should be differentiated from that due to technical sources. The "Analysis of Variance" section covers these approaches.

Other Issues

Realizing that this book serves as an introduction to researchers who may have little experience with microarray experiments, we illustrate our ideas by using both real examples and computer simulations. Also, some other statistical issues with microarray data are briefly discussed in the last section and interested readers are referred to sources both on- and off-line.

DESIGNING A MICROARRAY EXPERIMENT

Before a microarray experiment begins, a meeting between biologists and statisticians should be scheduled to discuss the following issues.

What Can Be Expected From a Microarray Experiment?

There are many ways that a microarray experiment can help a researcher to investigate biological phenomena, although there are many limitations as well. We can use microarrays to help find genes that are differentially expressed under different experimental conditions, discover genes with positively (and negatively) correlated expression patterns that hint at functional relationships, screen for target genes to generate new research hypotheses, and classify and predict subtypes of samples with gene profiling (so as to assist early diagnosis of cancer), etc.

However, we must understand that there are inherent limitations of this technology due to the principle and the implementation of the currently available versions of the microarrays. These limitations include cross hybridization due to short sequence of probes, indirect inference of the biological functions since mRNA rather than protein is measured, possible occurrence of alternative splicing, and uncertainties of measurement because of mRNA instability. Additionally, arrays provide a 'snapshot' of gene expression, not a 'movie.' Therefore microarrays can only provide a rough picture of the biological processes in the cells; it is unlikely that complex biological systems can be reverse engineered from the limited information gained in microarray experiments.

What Treatment Structure Can Be Applied to a Microarray Experiment?

Design of treatment structure for microarray experiments can be roughly classified as: one way layout, factorial design, and time course design. In a one way lay out, the treatments are parallel to each other and there is no interaction among them. The simplest case of this is a control vs. a treatment. It can be extended to three or more treatment groups, e.g. when several groups treated with different chemicals are compared in the same experiment. In a factorial design, two or more factors are included and there is possible interaction between factors. For example, in an experiment designed to study the effect of a drug on gene expression, one factor may be drug treatment vs. no drug treatment; the other factor may be genotype I vs. genotype II. An interaction between the two factors for some genes could exist if the effects of the drug are different for the two genotypes. A gene expression may be down regulated by the drug in genotype I but may show no difference or even be up regulated in genotype II. Thus factorial design may help identify gene expression patterns under different treatment combinations. A time course design is one with gene expression measured at different stages of the cell cycle or different time points of the biological process under investigation. The motivation of such a design is to gauge the dynamic expression patterns of genes chronologically.

What Biological Samples Should be Used?

There are several concerns when choosing biological subjects to which the treatments are applied. Firstly, the underlying biological problem is very important. You may have a research problem that can only be studied using human subjects; or you may have some treatments that cannot be applied to human subjects. These conditions determine your selection of experimental samples. Now suppose there is some flexibility, what other issues do you need to consider? You may need to consider financial constraint and statistical properties. Generally speaking, human subjects are more expensive than animal models, which in turn are more expensive than cell lines. The main statistical property that needs to be considered is the variability among subjects. Typically, the variability among cell lines is smaller than that among animal models, which is smaller than that among human subjects. This means that with the same sample size, cell lines will give the best precision and hence the best statistical power. However, experiments using cell lines are in vitro and hence the conclusions may not be readily extended to in vivo situations. Additionally, animal and cell line subjects are typically more amenable regarding issues of balance than are human subjects. All of the above concerns need to be addressed when choosing the optimal subjects for the experiment.

Sample Size and Power Estimation

How many samples are needed? This is usually one of the first questions asked when an experiment is planned. While molecular biologists traditionally paid little attention to this issue, they are more and more aware of its importance for microarray experiments now. The fact that thousands of genes are interrogated simultaneously does not obviate the planning of sample size. This is because statistical inferences are fundamentally made at the individual gene level rather than at the array level.

Traditionally, either a pilot study or information from prior work is needed to get an estimate of how many replicates are needed. For a specific experimental design, we need to assume a probability distribution of the variable being measured, get an estimate of the effect size (standardized difference), establish the maximal type I error rate that we are willing to tolerate and define the minimal power we want to achieve. With regard to microarray studies, prior work is often limited and pilot studies can themselves be expensive. Also there are several thousand genes to be considered. Thus it is a big challenge to estimate effect size. One possible solution is to specify a minimal effect size that we want to detect among all of the genes without actually estimating the effect size for specific genes. The distribution of the variable of interest (e.g. cy3/cy5 ratio for two color cDNA array, or probe set signal for oligonucleotide array) is not easy to guess either. Empirically, normal distribution after logarithm transformation seems to be a good approximation. This may not hold true for your specific experiment. Given all these problems, the estimated number of biological replicates needed per treatment group (which is a function of minimal effect size to be detected, maximal tolerable type I error rate, desired minimal power and the probability distribution of the variable of interest) is best treated as an approximation, and it is not necessary to adhere religiously to these estimates.

Due to the expense of gene chips, generally not very many replicates are affordable in microarray experiments, but we suggest at least 3 replicates per treatment group. Can we improve statistical power with so few replicates of gene chips? The answer is yes under certain conditions. Peng et al., (2003) investigated how sample pooling can help improve power by using statistical theories and computer simulations. We report that pooling several mRNA samples from the same treatment group together in equal amounts, before they are hybridized to the array, can reduce the within treatment variability and increase effect size and hence improve power (i.e. the probability of detecting the genes that are truly differentially expressed). Theoretically, in a two sample t test scenario, pooling 24 mRNA samples into 6 arrays is equivalent to using 20 mRNA samples and 20 arrays in terms of false positive rate and false negative rate provided the underlying distribution is normal. Hence, pooling samples in a technically correct way can reduce the total cost of the experiment and help with situations in which the amount of mRNA extracted from a single subject is insufficient for a single microarray. Empirically, we have seen smaller variability among subjects within the same treatment group from those experiments with correct pooling schemes compared to others who did not use any pooling when the number of arrays was approximately equal. This indirectly supports our claim.

GENERAL PROCEDURES FOR STATISTICAL ANALYSIS OF MICROARRAY DATA

After you have done the experiments and generated a colored or black-white image for each of the arrays, how do you decipher the biological information from them? Statistics play a central role in extracting such information. Roughly speaking, we can apply statistics in three stages: first we need to get the data into good shape, next we need to check differentially expressed genes by hypothesis testing or parameter estimation such as mean and standard deviation (which can be done with each individual gene being analyzed separately or with multiple genes analyzed in the same model) and we can also explore the quantitative relationship among genes or samples using multivariate methods, and finally, biological functions of groups of genes can be analyzed statistically with extra information and annotations from data bases currently available to the public (e.g. SwissProt, Pfam, GenMapp, Gene Ontology, etc.).

Stage I. Data Standardization and Normalization

The raw data obtained by the computer after scanning the images are typically messy and noisy. The simplest way to check data quality is by visual inspection. This should give first impressions on the background intensity, major artifacts such as scratches, and overall quality of the data. Commercial software provided by microarray venders typically generates metrics that can be used to help objectively determine data quality. Once image preprocessing has been done, normalization and transformation are usually needed. Please refer to Bruce Aronow's Chapter 5 for discussion on these issues.

Stage II. Statistical Testing and Exploratory Analysis

Once the data are in good shape (at least we hope so), we can have more "fun" with statistical testing and/or data mining. Roughly speaking, the following statistical tasks can be performed. Preliminary analysis, such as descriptive analysis, scatter plot of the data, etc., help us gain an abbreviated understanding of the data and detect obvious problems such as unusual genes/arrays or some heretofore undetected systematic variance. Hypothesis testing, which is more rigorous in statistical flavor, usually outputs results with a probability argument (p value). Exploratory analysis (such as clustering analysis) attempts to group genes or samples together with some similarity or dissimilarity measures (See Willy Valdivia Granda's Chapter 8 for a detailed discussion). Classification and prediction, which try to build models that classify samples by using expression information from subsets of genes, and then use the models to predict to which class a future sample may belong. This could be especially useful for early diagnosis of certain types of cancers. Last but not least, data visualization tools are very helpful because one good picture (graph) is worth a thousand words. Even mathematicians cannot comprehend pages of numbers without good graphs.

Stage III. Interpretation of the Biological Functions

After statistical analysis, you typically end up with lists of genes that satisfy certain criteria ('genes of interest'). What do you do with them? You need to interpret them biologically. The Gene Ontology (GO) Consortium, Onto-Express, GOminer, etc. may provide assistance to you in this regard. You may want to conduct more experiments to confirm the results or test new research hypothesis. For further information, please refer to Eric Blalock's chapter 6.

MULTIPLE HYPOTHESIS TESTING IN MICROARRAY EXPERIMENTS

Pitfalls in Multiple Testing

Multiple testing refers to the testing of more than one hypothesis at a time. It is a subfield of the broader field of multiple inference, or simultaneous inference, which includes multiple estimation and multiple testing. In general, when testing any single hypothesis, conclusions based on statistical evidence are uncertain.

We typically specify an acceptable maximum probability of rejecting the null hypothesis when it is true, thus committing a Type I error, and base the conclusion on the value of a statistic meeting this specification, preferably one with high power, i.e. low Type II error rate. Type II error is defined as failing to reject the null hypothesis when in fact it is not true. Power is defined as one minus the probability of committing a type II error. It is desirable for a test to have a low type I error rate and a low type II error rate simultaneously. Ideally, in microarray studies, this means that within our list of significant genes we have those that are truly differentially expressed while avoiding those that are not. In reality, however, we have to compromise between the two types of errors because reducing one generally means increasing the other. Table 1 illustrates the concepts of false positives and false negatives.

Table 1. Illustration of false positive and false negative in a two-sample test.

H0: $\mu 1 = \mu 2$	Truth: $\mu 1 = \mu 2$	Truth: $\mu 1 \neq \mu 2$
Reject H0	False Positive (Type I error)	Correctly rejected
Do not reject H0	Correctly not rejected	False Negative (Type II error)

When many hypotheses are tested, and each test has a specified Type I error probability, the probability that at least some Type I errors are committed increases, often sharply, with the number of hypotheses. For example, if 100 hypotheses were tested simultaneously and the type I error rate of each test were controlled at 0.05, then the probability that at least one type I error would occur would be $1 - (1-0.05)^{100} = 0.994$. For microarray experiments, we are typically testing several thousand genes simultaneously and this may have serious consequences if the set of conclusions are taken without adjustment. Numerous methods have been proposed for dealing with this problem, but no one solution will be acceptable for all situations.

Table 1b. Possible outcomes when m genes are tested simultaneously.

	# of H_0 Not rejected	# of H_0 rejected	Total
# of True H_0	U	V	m_0
# of False H_0	S	T	m_1
Total	A	R	m

Note: m is the total number of genes being tested; m_0 is the number of genes with null hypotheses being true and m_1 is the number of genes with null hypotheses being false.

Reviewing literature results in the following definitions and approaches of addressing the complex issues of false positives due to multiple testing (Table 1b):

1. Per-comparison wise (type I) error rate (PCER): Pr (V $\geq$ 1), which is the probability that at least one false positive occurs among all m tests. If we use the same α level for each individual tests without any multiple testing correction, then PCER $\leq \alpha$. For large m, this is usually too liberal since we end up too many false positives among the "significant" list.

2. Family wise (type I) error rate (FWER): E (V/m), which is the *expected* ratio of false positives over all m tests. This has been one of the most investigated properties among multiple testing literatures. It tends to control the false positive rate on the conservative end. Procedures such as Bonferroni, Holm, Sidak, etc. are aimed to control FWER at a certain level. The difference among these procedures is mainly the power (of each individual test).

3. False discovery rate (FDR): E (V/R|R>0) Pr(R>0), which is the *expected* rate of false positives among the list of tests claimed to be significant. This was first proposed by Benjamini and Hochberg in 1995 and a procedure was given in the same paper that works for independent tests. They later proposed some modifications of the procedure for dependent tests.

Two examples will be given in this section to illustrate multiple testing problems and to compare the performance of various multiplicity adjustment approaches. First, we conduct virtual microarray experiments by using computer simulations. Although it is not exact, the advantage of computer simulation is that we know the truth. Suppose we generate a data matrix with 10,000 rows and 2 * n columns, we treat the first n columns as replicates from one condition and the last n as from another condition. We can generate 8,000 rows (genes) with no differential expression between the 2 conditions. For example, each row can be a random sample from the same normal distribution with same mean and same variance. To better resemble microarray experiment, each row should have its own mean and variance. The other 2,000 rows can be used to simulate genes with differential expression levels. To make it simple, we can assume that for each row, the first n and the last n are both from normal distributions with same variance but with different means. Further, we assume that the effect size (i.e. the difference between the 2 means divided by the common standard deviations) is a constant for all of the 2,000 rows that do not share means. This is an over-simplified version of microarray experiment, but it helps us assess the false positive rates and the false negative rates. For the 8,000 'no change' genes (each of which was generated from the same normal distribution with same mean and same standard deviation), we know that we should detect no change in gene expression between the 2 groups, and furthermore, that any positive findings by our statistical test will be false positives. For the other 2,000 'yes change' genes, any negative findings will be false negatives. We can apply a two-sample t test on each row and record the corresponding p value. Next, we record the number of genes claimed as 'differentially' expressed for different a levels. We repeat the simulations 1000 times (that is, repeat the same experiment 1000 times with different randomly generated measures in each iteration) and record the average false positive and false negative rates. Tables 2 and 3 and Figure 1 show the results from the simulation described above.

Table 2. Simulation example of false positives (FP) and false negatives (FN) at a level = 0.05 for 10000 genes (n = 5 per group).

# of genes	$\mu1 = \mu2$	$\mu1 \neq \mu2$	Total
the truth	8000	2000	10000
statistical non-rejection of H0	7632 (TP)	760 (FN)	8392
statistical rejection of H0	368 (FP)	1240 (TN)	1608

Note: The simulated false negative rate of the above example = 760 / 2000 = 0.38; the simulated false positive rate = 368 / 8000 = 0.046; the simulated false discovery rate FDR is 368 / 1608 = 0.229.

From Tables 1-3 and Figure 1, we see some interesting points. 1. False positives and false negatives do occur and can cause serious problems in interpretation of the results; 2. False positive rates without adjustment are approximately the same as nominal α levels with relatively large sample size; but they are smaller with smaller sample size. However, this advantage of small sample size is at the cost of increased false negatives; 3. As sample size increases, false negative rate decreases but false positive rate tends to increase; 4. We can use simulation to compare different approaches of adjustment in multiple testing and assess their relative merits empirically (see Tables 2 and 3).

Many procedures have been proposed to adjust for the false positive problems associated with multiple testing and we review them in next section.

METHODS BASED ON P VALUE ADJUSTMENT

Bonferroni Correction

This is one of the classical approaches to control the *family-wise* type I error rate. The adjustment is very simple: if we do *m* tests simultaneously, we compare

Table 3. Simulated False Positive Rates (FPR) at different nominal   levels with different sample sizes.

n	False Positive Rate at α =			False Negative Rate at α =		
	0.001	0.01	0.05	0.001	0.01	0.05
3	0.0004	0.0054	0.0352	0.9756	0.7941	0.3737
6	0.0007	0.0084	0.0466	0.4562	0.089	0.0091
9	0.0009	0.0093	0.048	0.0586	0.0035	0.0001
12	0.0009	0.0096	0.0487	0.0037	0.0001	0
15	0.0009	0.0097	0.0494	0.0001	0	0

(Two sample t test: n = sample size per group) *FPR: False Positive Rate; FNR: False Negative Rate; nominal α = 0.05; effect size = 2 when alternative hypothesis is true; 100 iterations..*

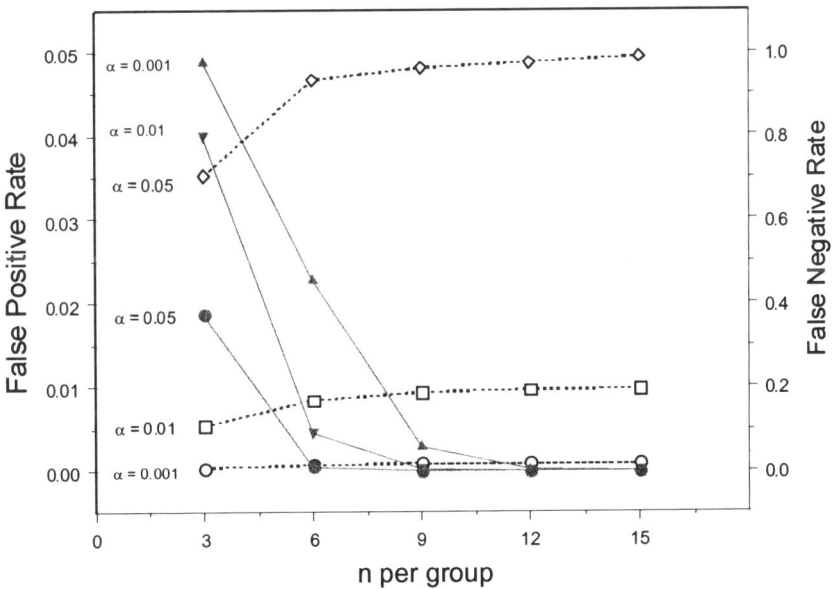

Figure 1. Plot of the simulated FPR (dotted lines) and FNR (solid lines) shown in Table 3.

each p-value with $\alpha^*=\alpha/m$ instead of α and make a decision for each individual test based on whether or not the observed p-value for that test is less than α^*. For example, suppose we are testing 5000 genes and we want to make sure that the probability of incorrectly claiming at least one gene is significantly differentially expressed (which is the *family-wise* type I error rate) is less than or equal to 0.05, then we need to set the α^* level as 0.05 / 5000 = 0.00001. Thus, there is only a 5% chance that a single false positive exists among the list of genes whose p values were $\leq$ .00001. This is very conservative since we may fail to detect many true positives with such a stringent adjusted α level. Equivalently, we can get the adjusted p-value by multiplying the observed p value by 5000 and then comparing the adjusted p value to 0.05. If the adjusted p-value exceeds 1, then it is set to 1.

Šidák Single Step Procedure

Compared to simple Bonferroni correction, this procedure is somewhat less conservative and the *family-wise* type I error rate is controlled at exactly α level if the α^* level for each individual gene is set at $1- (1- \alpha)^{1/m}$ for a total of m tests. Suppose $m = 5000$ and $\alpha= 0.05$, then $\alpha^* = 1 - (1-0.05)^{1/5000}=0.00001$. Note that this procedure may give different results than the Bonferroni procedure when the number of tests is not very big, but it does not make much difference with several thousands of tests. Note that in this example, both the Bonferroni and the Šidák procedures make the same stringent requirements for each individual gene to reject the null hypothesis. They are called *single step* adjustments because all p-values are compared to the same threshold. Equivalently, we can adjust the p-value by letting $p_{adj} = 1- (1-p)^{5000}$. If the adjusted p value exceeds 1, then it is set to 1.

Holm's Step Down Bonferroni Method

To control the experiment-wise type I error rate at α, this method is applied in multiple steps as follows: first sort the p-values of all tests in ascending order; then reject the null hypothesis for the gene with the lowest p-value if $p_{(1)} \leq \alpha / m$ where $p_{(1)}$ is the smallest p-value. If the first gene is not rejected, then all genes are claimed as non-significant; if rejected (claimed as positive), then the next gene is claimed as positive only if $p_{(2)} \leq \alpha /(m-1)$; Continuing in this fashion until we get $p_{(i)} > \alpha /(m - i +1)$. Then we stop and claim all the ordered genes before the ith gene as significant and the rest as non-significant. This method makes a difference for the first few genes in the list, but with thousands of tests, it does not make much difference when compared to the simple Bonferroni correction.

Duncan's and Modified Duncan's Procedures

Duncan (1955) proposed an adjustment of α level of the individual test that depends on the number of tests still under consideration. In particular, the test for the ith smallest p-value is to be rejected if the p-value is less than $\alpha_i = 1 - (1-\alpha)^i$ where α is the family wise nominal level. In a recent study (Wood et al, 2002), two modifications of Duncan's method for multiple comparisons were proposed that provide a more reasonable compromise between the false positive and false negative rates, as well as for the extremely large number of tests to be conducted in microarray experiments. Specifically, the p-values are sorted in ascending order and then sequentially tested. The ith hypothesis is rejected if the corresponding p-value $p_{(i)}$ is less than $\alpha_i = 1 - (1-\alpha)^{log(i)}$, or rejecting the ith largest if the p-value $p_{(i)}$ is less than $\alpha_i = 1 - (1-\alpha)^{sqrt(log(i))}$,where α is the family wise nominal α level. In the so called *Duncan's Log Test* the value of i in *Duncan's procedure* is replaced by *log(i)* which increases at a substantially slower rate than i and in *Duncan's Sqrt(Log) Test*, it is replaced by the square root of *log(i)*, which increases at an even slower rate. Although both of these modifications are essentially *ad hoc*, they are motivated by the asymptotic convergence rate of the maximum of independent normal random variables.

Šidák Step Down and Modified Šidák Step Down Procedures

Šidák step down method (Šidák SD) is a modification of the Šidák single step procedure. Rather than letting $\alpha^* = 1 - (1-\alpha)^{1/m}$ for all m tests, we choose a different α^*_i for each of the p-value $p_{(i)}$ in ascending order. First we sort the p-values in ascending order, then for $i = 1$ to m, let $\alpha^*(i) = 1 - (1-\alpha)^{1/(m-i+1)}$ and compare $p_{(i)}$ to α^*_i beginning with $p_{(1)}$. Continue as long as $p_{(i)} = \alpha^*_{(i)}$ and stop at the first k such that $p_{(k)} > \alpha^*_{(k)}$. Claim the first k genes as significant.

With several thousands of tests, Šidák step down procedure is as conservative as Bonferroni and Holm's procedure. Motivated by Wood's modification of Duncan's procedure, we proposed modification of Šidák step down procedure by taking logarithm and square root of logarithm of the order of the p-values. Namely, *Šidák step down Log procedure (Šidák SD Log)* differs from Šidák SD by letting $\alpha^*_{(i)} = 1 - (1-\alpha)^{1/(1+log(m-i+1))}$ and *Šidák step down Sqrt Log procedure(Šidák SD Sqrt Log)* differs from Šidák step down method only by letting $\alpha^*_{(i)} = 1 - (1-\alpha)^{1/sqrt(1+log(m-i+1))}$. It will be shown in Figure 2 that these modification result in better power while keeping type I error rate low.

Re-sampling Based Methods for P-value Adjustment

None of the above p value adjustment methods takes into account the dependence structure between the variables. In a microarray experiment, typically many genes are expected to be co-regulated and hence should have highly correlated expression levels. To address this problem, Dudoit et al. (2000) proposed a more general and less conservative step-down p-value adjustment procedure based on Westfall and Young (1993). It is a re-sampling based algorithm that controls the *family-wise* type I error. The algorithm (for 2 sample t tests) is as follows (adapted from Dudoit et al):

For the bth permutation
1. Permute the n columns of the data matrix X. The first (last) n_1 (n_2) columns now refer to the "fake" control (treatment) group.
2. For each gene, compute t-statistic: $t_1^{(b)},\ldots,t_m^{(b)}$
3. Next, compute

$$\mu_m^{(b)} \longleftarrow |t_j^{(b)}|$$

$$\mu_j^{(b)} \longleftarrow \max(\mu_{j+1}^{(b)},|t_j^{(b)}|)\ for\ 1 \le j \le m-1,$$

The above steps are repeated B times and the adjusted p-values are estimated by

$$\tilde{P}^*_{r_j} = \frac{\sum_{b=1}^{B} I(\mu_j^{(b)} \ge |t_{r_j}|)}{B}$$

with the monotonicity constraints enforced by setting

$$\tilde{P}^*_{r_1} \leftarrow \tilde{P}^*_{r_1}, \tilde{P}^*_{rj} \leftarrow \max(\tilde{P}^*_{r_j}, \tilde{P}^*_{rj-1})\ for\ 2 \le j \le m$$

In plain English, we sample the arrays randomly each time and assign the first n_1 arrays as the 'control' group and the last n_2 arrays as 'treatment' group; then we compute the test statistic $t_1, t_2, \ldots, t_m$; then we take their absolute values and sort them in descending order; repeat the above permutation B times and compute the proportion of times the observed t_j is less than or equal to the permuted t_j. This proportion is then used as an approximation of the p value.

One advantage of the permutation algorithm is that the correlations among genes are retained because arrays instead of individual genes are permuted. Another advantage is that it does not require assumption of underlying normal distributions. In the above example, the distribution is approximated by the empirical distribution of the permuted t. This is more robust to violations of the normality assumption than the usual t-test. It can be viewed as semi-parametric in this sense. Obviously, n needs to be reasonably large for this to work since the total possible number of permutations for a two sample experiment is $n! = n*(n-1)*...*2*1$. However, when we compute the test statistic, the order of chips within the same treatment group is not important. So there are actually $C_{n1+n2, n2}$ (read as: $(n_1 + n_2)$ choose n_2) different test statistics. Hence, if $n_1 = n_2 = 2$, then only $C_{4, 2} = 6$ different statistics are possible; if $n_1 = n_2 = 5$, then $C_{10, 5} = 252$ distinct statistics are possible; the possible number of distinct statistics become large if $n_1 = 9$ and $n_2 = 10$, since $C_{19, 10} = 92378$. To extend this to the cases with $g > 2$ groups, the possible number of distinct statistics is given by $(n_1 + n_2 + ... + n_g)! / (n_1!*n_2!* ... n_g!)$. Note that permutation test still requires reasonable sample size. By sampling from these possible permutations, we get the approximate distribution of the test statistic without assuming it follows any parametric distributions.

Similarly, we can also use *bootstrap* methods for the re-sampling algorithm. The only difference is that with permutation, no array will be sampled twice in each iteration (i.e. *sampling without replacement*); while with bootstrap, an array can be sampled more than once in one iteration while another array may not be sampled at all (i. e. *sampling with replacement*). Hence bootstrap should give more samples than permutation.

Both permutation and bootstrap methods are computationally intensive. It is more so with microarray data given that typically we have several thousand genes to test. Fortunately enormous advances in the computer industry during the past two decades have made it possible to implement these techniques even with a moderately powered PC. On the software front, experienced users can use powerful statistical software such as SAS, R, S-plus or SPSS; point-and-click users can use software specially developed to analyze microarray data, such as GeneSpring, GeneSight, Partek, etc., but may not be able to implement the most comprehensive or cutting edge techniques if they are not provided with these packages. However, for biologists interested in implementing some of the more advanced procedures, their implementation is only a statistician away.

Methods Based on Control of the False Discovery Rate (FDR)

All of the above procedures are based on control of false positive rate (i.e. the proportion of false positives out of all tests). Benjamini and Hochberg (1995) proposed an alternative in adjusting for multiplicity problems. The FDR is the proportion of false positives out of total positives (rather than the total number of tests). The argument in favor of controlling FDR rather than false positive rate is that controlling FDR takes into account how many discoveries are made. Consider two experiments with the same design but with different treatments. In experiment A, 100 out of 5000 genes have raw p value less than or equal to

0.05, while in experiment B, 500 out of 5000 genes have raw p value less than or equal to 0.05. If we control false positive rate, we may implement the same adjustment and end up with approximately the same number of significant genes; however, if we control FDR, then we may find more significant genes and we are willing to bear more false positive errors with the latter since more discoveries are made. This compromise usually gains some power. The implied assumption here is that the more genes are truly differentially expressed, the more genes will be found to be significant by statistical test. This assumption is usually reasonable.

The original Benjamini and Hochberg (BH) procedure is a simple stepwise procedure that c ontrols the F DR when t he test st atistics are st atistically independent. Recently FDR control when the test statistics are positively correlated has also been implemented. Benjamini and Liu (BL; 1999), as well as Benjamini and Yekutieli (BY; 2001) proposed modifications that control FDR even for the generally correlated test statistics.

The BH procedure works as follows (assuming FDR is to be controlled at 0.05 level): first sort the m p-values in ascending order and denote the ith smallest p-value by $p_{(i)}$ for each i between m and 1; next starting from the largest p-value $p_{(m)}$, compare $p_{(m)}$ with $0.05 * i/m$. Continue as long as $p_{(i)} > 0.05 * i/m$. Let k be the first time when $p_{(k)} \leq 0.05 * i/k$, stop and declare the test statistics corresponding to the smallest k p-values as significant. The BY procedure differs from the BH procedure only in that $p_{(i)}$ is compared to

$$0.05 * i/(m*a)$$

where

$$\alpha = (1 + \tfrac{1}{2} + \ldots + 1/m).$$

The BL procedure works as follows (assuming FDR to be controlled at 0.05 level): starting from the smallest p-value $p_{(1)}$, compare each $p_{(i)}$ with $h_{(i)} = 0.05 * m/(m+1-i)^2$. Reject the null hypothesis corresponding to $p_{(1)}$ if $p_{(1)} \leq h_{(1)}$ and continue to reject the null hypotheses as long as $p_{(i)} = h_{(i)}$. Let k be the first time when $p_{(k)} > h_{(k)}$, reject all of the null hypotheses corresponding to the (k -1) smallest p-values.

Comparisons of the Multiplicity Adjustment Procedures

From the following two examples, we compare the performance of all of the above procedures and make some recommendations based on their performance. The first example (Table 4) is a simulation study where we know exactly how many negatives and positives exist in the data. The second example uses data from a real experiment and we compare lists of genes found to be significant by different approaches (Figure 2).

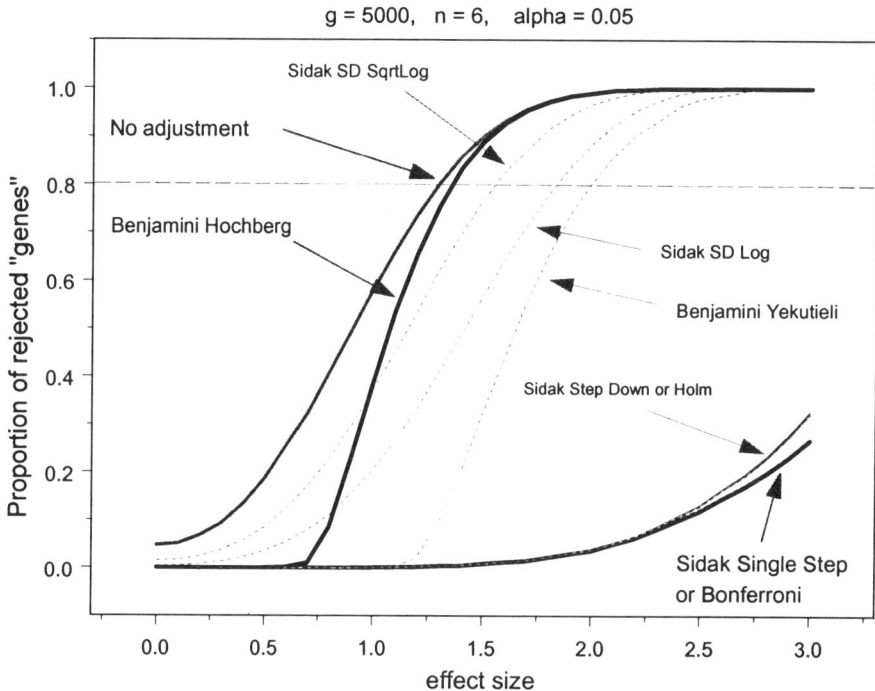

Figure 2. Simulated power curves (average proportion of rejection out of total number of tests) for parametric 2 sample t tests with different effect sizes for 5000 "genes" (n = 6 per group, nominal type I error rate = 0.05, based on 100 iterations).

Table 5 is based on a real microarray experiment with two groups of rats: one control group and one treatment group. The sample sizes are 10 and 9 respectively. Measures were made with Affymetrix RG-U34A rat 'gene chips'. There are 8799 probe sets representing genes and expressed sequence tags (ESTs- mRNA that is suspected of representing mRNA of a gene product, but that gene and/or product is not known) on each Gene Chip®. Two sample t tests were done both parametrically and non-parametrically, and different p-value adjustment procedures were then applied. The numbers of significant genes corresponding to each procedure are included in the table.

Comments and Suggestions on Multiplicity Adjustment Procedures

Based on the above examples and theoretical justifications, we make the following comments and recommendations:

Table 4. Comparison of different multiplicity adjustment procedures with two sample t-Test simulation

		Parametric			Permutation			Bootstrap		
		5	10	15	5	10	15	5	10	15
Raw	FPR	0.043	0.046	0.053	0.045	0.048	0.055	0.046	0.049	0.054
	FNR	0.565	0.380	0.301	0.553	0.371	0.291	0.551	0.373	0.295
Bonferroni	FPR	0.000	0.000	0.000	0.000	0.000	0.000	0.000	0.000	0.000
	FNR	0.985	0.766	0.686	0.980	0.756	0.675	0.967	0.749	0.677
Holm	FPR	0.000	0.000	0.000	0.000	0.000	0.000	0.000	0.000	0.000
	FNR	0.985	0.730	0.674	0.975	0.739	0.667	0.978	0.729	0.667
Šidák SS	FPR	0.000	0.000	0.000	0.000	0.000	0.000	0.000	0.000	0.000
	FNR	0.985	0.766	0.685	0.978	0.776	0.665	0.959	0.765	0.673
Šidák SD	FPR	0.000	0.000	0.000	0.000	0.000	0.000	0.000	0.000	0.000
	FNR	0.985	0.729	0.673	0.981	0.719	0.644	0.969	0.716	0.663
Šidák SD log	FPR	0.004	0.004	0.006	0.004	0.005	0.005	0.003	0.005	0.005
	FNR	0.748	0.545	0.462	0.735	0.555	0.480	0.758	0.563	0.452
Šidák SD square root log	FPR	0.013	0.014	0.016	0.014	0.013	0.016	0.013	0.015	0.017
	FNR	0.668	0.475	0.384	0.612	0.471	0.397	0.648	0.485	0.394
Benjamini Hochberg	FPR	0.000	0.000	0.000	0.000	0.000	0.000	0.000	0.000	0.000
	FNR	0.731	0.524	0.419	0.715	0.543	0.449	0.753	0.532	0.429
Benjamini Yekutieli	FPR	0.000	0.000	0.000	0.000	0.000	0.000	0.000	0.000	0.000
	FNR	0.840	0.640	0.557	0.806	0.618	0.571	0.824	0.661	0.567

Table 5. Comparison of different multiple testing adjustment procedures with data from real microarray experiment (nominal FPR or FDR controlled at 0.05, two sample t-Test).

Adjustment Procedure	Number claimed significant		
	Parametric	Permutated	Bootstrapped
None	1065	1096	1097
Bonferroni	9	24	24
Šidák Single Step	10	24	24
Šidák Step Down	10	24	24
Šidák SD log	206	250	251
Šidák SD Square root log	456	518	522
Holm	9	24	24
Benjamini Hochberg	38	66	64
Benjamini Yekutieli	10	25	24

1. As the false positive rate decreases, so does power. While false positives can be a serious problem in microarray experiments, we also need to worry about false negatives. No single method can solve these problems without compromise.

2. While Bonferroni, Šidák (single step and step down), and Holm adjustments control the false positive rate very well, they are too conservative and result in too many false negatives. On the other hand, many genes are claimed as positive falsely if we do not apply any adjustments. Šidák's log, and Duncan's square-root-log seem to present a good compromise between controlling false positives and false negatives. Benjamini Hochberg and Benjamini Liu procedures control FDR and perform fairly well. It is especially noteworthy that the Benjamini Hochberg procedure has a very attractive feature: when effect sizes are small (i.e. the differences between the two groups are small) it results in very few adjusted positive results, but when effect sizes increase, the power increases dramatically, and more results 'survive' the adjustment while the number of false negatives decreases drastically.

3. With normal distribution, bootstrap and permutation based adjustments perform as well as parametric t tests. When the distribution is not normal and sample size is reasonable, they fare better than parametric t tests. Many researchers have shown that microarray data seldom follow normal distributions, and relatively small sample sizes make the Central Limit Theorem (which states that the sample means will approximately follow normal distribution if the sample sizes are large) not readily applicable. With the additional advantage of preserving the correlation structure among the genes, re-sampling based algorithms should be preferred despite the fact that they are more computationally intensive.

4. Practically, confirmation (cross-platform validation) by other biological technology is desirable for at least some of the genes claimed as positive even if the p value is very significant. The reasons being: increased likelihood of cross hybridization in microarray experiments, the inability to sequence hybridized material to insure that it is indeed the material of interest, and that enormous variability that could be introduced in the experiment, at the various stages of data analysis (e.g., signal processing, normalization, etc.). In other words, the reliability of microarray technology in general is still not good enough for definite conclusions to be drawn from it without additional methods. Many journals require good statistical analysis as well as confirmation of some of the "significant" genes by other methods such as real-time polymerase chain reaction (RT-PCR). However, RT-PCR experiments are traditionally not replicated, at least not at the same level as is being recommended here for

microarray experiments. This can lead to some difficulties regarding interpretation of these validating procedures. Here, we recommend that the same replication standards recommended for microarray work be adhered to with regard to RT-PCR (and other) validating technologies. In this way, both technologies can be held to the same statistical standards. We think microarray technology is best used as an exploratory tool but not for confirmative purposes, at least in the near future, before it has been demonstrated to be more accurate and precise.

5. In the above examples, we used only two-sample designs to illustrate these ideas, but the methods based on p-value adjustment can be readily extended to other experimental designs such as 3 or more treatments, factorial design, time course design, etc. The re-sampling based algorithms may need more care, but should still be applicable in typical experimental design settings.

6. Other classical methods such as Scheffe, Tukey, SNK, etc. are not discussed above, but they should be considered when doing post hoc comparisons for individual genes that are found to be significant by overall F test in analysis of variance (ANOVA). For 1-Way ANOVA and t-test procedures, adjusting the p value is straight forward. However, many researchers faced with 2-Way (and greater) ANOVA results are unsure of how to proceed with p value adjustment. This is because >1 Way ANOVA tests yield multiple p values, some for main effects and one (or more) for interaction among main effects. The simplest way to adjust these is to adjust the often under-used 'overall' p-value for these ANOVA results. Then the various main effect and interaction p-values can be treated themselves as post hoc tests when interpreting the data.

7. For practical purposes, if the biological samples used are fairly homogenous within each treatment group, as is the case with cell lines, it is likely that the researcher will find more significant genes and hence we may want to implement stricter control of the false positive rate. On the other hand, if the samples show large variability, then we may want to guard against false negatives more strictly.

8. In any case, we should order the genes by significance levels and check the most significant genes first. Given that both false positives and false negatives are inevitable, rigorous bioinformatics procedures should follow statistical analysis to interpret the biological functions of the genes and make sense of the results. After all, scientific discoveries are obtained through many different routes.

ANALYSIS OF VARIANCE

As microarray technology evolves into a powerful tool for investigating gene expression on a genome wide scale, it has found applications in various fields. Although different disciplines have different problems to tackle, they share some common features with respect to experiment design. Historically, statisticians and other scientists working in agricultural or medical areas have developed many classical designs. While microarray experiments have many unique features, traditional statistical methodology and experimental design may serve as a good start in the design and analysis of such experiments. Since there are two major "denominations" of microarrays that are quite different from each other, we discuss them separately. Analysis of variance and experiment design issues with spotted cDNA array are discussed in the first subsection and those with Affymetrix oligonucleotide arrays are discussed in the next subsection. A summary is provided in the last subsection.

Spotted cDNA Color Arrays

The Churchill Group in the Jackson Laboratory have published several important papers on applying classical ANOVA approaches in design and analysis of experimental results using spotted cDNA arrays. The discussion in this session relies heavily on their research.

As described in previous chapters, there are four basic steps in spotted cDNA array experiments. First, DNA clones with known sequences are spotted and immobilized onto a glass slide. Next, pools of purified mRNA from tissue or cell populations under study are reverse-transcribed into cDNA and labeled with one of the two fluorescence dyes (e.g., Cy3 "Green" and Cy5 "Red"). Next, the two pools of mRNA are combined and applied to a microarray with many spots. Theoretically speaking, strands of cDNA will hybridize to the complementary sequences on the glass slide; any unhybridized cDNA will be "washed" off. Finally, the array will be scanned and the red and green signals from each spot will be read by a computer to give relative measurement of the expression levels of various genes in the original two samples. After image processing and data normalization, the ratio of red (R) to green (G) signal at each spot will be used for data analysis.

There are some unique features of spotted cDNA arrays. The first is that since generally two dyes are used for labeling the cDNA, the design for experiments with three or more treatments is not as intuitive as that with only two treatments. It is an incomplete block design since each block (array) can have only two treatment samples. The second is that many factors contribute to the measured signals. At least four factors can be identified even with a very simple treatment vs. control experiment: the treatment factor (T), the gene factor (G), the array factor (A), and the dye factor (D). Theoretically there could exist 16 possible experimental effects: the baseline effect (μ), the main effect of each of the four factors (T, G, A, D), the six two-factor interactions (TG, TA, TD, GA, GD, and AD); the four three-factor interactions (TGA, TGD, TAD, and GAD); and the four-factor interaction (TGAD). Since not all of these effects are identifiable, Kerr and Churchill (2000) thoroughly analyzed each of these effects and gave 3 plausible ANOVA models.

Sources of Variation in 2 color cDNA Array Experiments

Main effects: Array main effects measure the overall variation in fluorescent signal from array to array. These effects are mainly due to inconsistent conditions under which the arrays are probed, which make hybridization efficiencies of labeled cDNA vary from array to array. Gene main effects measure the gene-specific fluorescent signal. Some genes may emit higher or lower signals consistently compared to others. This may be due to biological reasons, e.g. for a cell to function appropriately, some gene's expression level may be consistently very high or low. Additionally, it may also be caused by differential labeling effects of different sequences. It is well known that hybridization efficiency (the binding power) is sequence specific. So even if the expression of two genes in the sample were exactly the same, their fluorescent signals could vary based on hybridization efficiency. Dye main effects measure differences in the two dye fluorescent labels. It has been reported that in some cases, one dye may be consistently 'darker' than the other. Treatment main effects of course measure the effects due to different treatments applied to the samples.

Interactions: For a particular tissue sample, Cy3- or Cy5-labeled cDNA is produced in separate runs of the reverse-transcription process. Differences in the runs can produce pools of cDNA of varying ranges of quality. This results in experimental treatment × dye (TD) interactions. Array × gene (GA) interactions occur because spots for a given gene on different arrays vary in the amount of cDNA available for hybridization. Thus GA effects are the 'spot' effects. Dye × gene (GD) interaction occurs if there are differences in the dyes that are gene-specific. For example, if the overall efficiency of incorporation of green dye were higher than that of red dye except for a few genes, this would cause GD interaction. While this is generally unlikely to occur, Kerr et al reported one extreme case in practice. They labeled one sample with red dye and the other sample with green dye in one array and then switched red and green dyes for the same 2 samples in the second array (this is so called 'dye-swap'). For most genes, the signals revealed consistent results independent of dye effect, but for one gene, higher green signals appeared in both arrays. This is most likely an experimental anomaly but would not be caught if only one array had been used. Treatment × gene (TG) interactions reflect differences in expression of particular treatment and gene combinations that are not parallel to the average effects of treatment main effect or gene main effect. Identifying genes whose expression changes in different treatments means identifying non-zero TG interactions. These are the effects that we are interested in. For practical purposes, the array × dye (AD), treatment × array (TA), and treatment × array × dye (TAD) interactions do not involve genes, and including or excluding them in the ANOVA will not change the estimates of TG effects, hence we can leave them for the residuals. The other high-order interactions: treatment × gene × array (TGA), treatment × gene × dye (TGD), gene × array × dye (GAD), and treatment × gene × array × dye (TGAD) effects are difficult to relate to the physical and chemical processes that make up this technology. They can be used to estimate the residuals under the fundamental justification that the results will not be reproducible if these high-order interactions between genes and other factors exist.

ANOVA Models in cDNA Array Experiments

Kerr et al (15) gave the following three models for experiments without replication:

$$y_{ijkg} = \mu + A_i + D_j + T_k + G_g + (TG)_{kg} + \varepsilon_{ijkg} \tag{1}$$

which includes only the main effects of each factor and the treatment × gene interactions; and

$$y_{ijkg} = \mu + A_i + D_j + T_k + G_g + (TG)_{kg} + (AG)_{ig} + \varepsilon_{ijkg} \tag{2}$$

which also accounts for spot-to-spot variation by including array × gene interactions; and

$$y_{ijkg} = \mu + A_i + D_j + T_k + G_g + (TG)_{kg} + (AG)_{ig} + (DG)_{jg} + \varepsilon_{ijkg} \tag{3}$$

which further considers dye × gene interactions. Later, they (2001) provided a similar model for an example with genes replicated by r spots within the same array:

$$y_{ijkg} = \mu + A_i + D_j + T_k + G_g + (TG)_{kg} + (AG)_{igr} + (DG)_{jg} + \varepsilon_{ijkgr} \tag{4}$$

An example taken from Kerr et al (2002) is illustrated as follows (Table 6). To study the in vitro effect of TCDD compound on the human hepatoma cell line HepG2, Martinez and Walker used 6 arrays with the 'dye-swap' design.

There are 1907 genes in the data set after removing 13 genes of artificial 'floor' values. They fitted the data with model (3) and provided the following ANOVA table (Table 7).

Proceeding with this model and examining the residuals separately for each array, they revealed systematic patterns of the residuals, which suggest that the assumption of constant variances is violated. To fix this problem, they attempted loess-transformation and shift-log transformation on the signals and found

Table 6. Example of spotted array 'dye-swap' experimental design.

	Cy3	Cy5
Array 1	TCDD	Control
Array 2	TCDD	Control
Array 3	Control	TCDD
Array 4	TCDD	Control
Array 5	Control	TCDD
Array 6	Control	TCDD

that the latter was a better fit to their data. More detailed explanation can be found in their paper (Kerr et al, 2002).

Table 7. Sources of variance estimated from 'dye swap' design .

Source	Sum of Squares	Degrees of Freedom	Mean Squares
Array	328.28	5	65.66
Dye	119.10	1	119.10
Array*Dye	128.42	5	25.68
Gene	35285.23	1906	18.52
Spot	1671.35	9530	0.18
Treatment*Gene	230.60	1906	0.12
Dye*Gene	316.84	1906	0.17
Residual	56.86	7624	0.0075
Adjusted Total	38136.69	22883	

In all of the above models, the residuals are assumed to be independently and identically normally distributed with mean 0 and constant variances. While this is rarely true for all genes, ANOVA methods are somewhat robust to violation of normality assumptions. However, the assumption of equal variances is more problematic. In order to tackle these problems, transformation of the raw data is suggested in many cases. Logarithm and loess transformation are the most recommended in published literatures. Although transformations do not solve the problem completely, they tend to make the ANOVA approach more robust. David Rocke (Durbin et al. 2002) proposed a two-component variance model which makes good sense.

In the aforementioned models, all effects are assumed to be fixed (i.e. the levels of each factor are considered as fixed). Wolfinger et al. (2001) also considered treating some of the effects as random effects. (A random effect is considered as a random sample from a population with certain probability distribution.) They proposed two interconnected ANOVA models, the "normalization" model and the "gene" model. The normalization model is:

$$y_{gik} = \mu + A_i + T_k + (AT)_{ik} + \varepsilon_{gik} \tag{5}$$

which does not include main effect for dyes since the dye effect was confounded with the treatment effect in their particular experiment. The residuals from the normalization model, r_{jik}'s, computed by subtracting the fitted values for the effects from the y_{jik} values are then fitted by the gene model:

$$r_{gik} = G_g + (GT)_{gk} + GA)_{gi} + \gamma_{gik} \tag{6}$$

The effects A_i, $(AT)_{ik}$, ε_{gik}, $(GA)_{gi}$ and γ_{gik} are all assumed to be normally distributed random variables with mean 0 and variance components σ_A^2, σ_{AT}^2, σ_ε^2, and σ_{gik}^2 respectively. These random effects are assumed to be independent both across their indices and with each other. The other effects are assumed to be fixed effects, thus both models are mixed models. A *mixed linear model* is a generalization of the standard linear model, the generalization being that the data are permitted to exhibit correlation and non-constant variability. The mixed linear model, therefore, provides us with the flexibility of modeling not only the data means (as in the standard linear model) but their variances and covariances as well. The major advantage of applying mixed models for microarray data is that we can now analyze the correlation among genes or samples. Details of mixed models and their advantages are beyond the scope of this book. Interested readers are referred to professional statisticians or statistical reference books such as McCulloch and Searle (2001). The main reason for the two-stage models is for computational efficiency.

Design of cDNA Spotted Array Experiments

In light of the above analysis of variance, several design schemes have been proposed and compared in the literature. The first is a *reference design*, which puts one reference sample on each array with the other sample being one of the many treatment samples. This design has been shown to be inefficient because half of the r esources are wasted on the reference sa mples (the potential advantage of the reference design is that, in the event that some 'universal' references were found for each tissue/ species, this would make spotted array data more amenable to inter-laboratory data sharing). The *loop design* proposed by Kerr et al. (2001) is better in at least three aspects: firstly, it collects twice as much data on the treatments of interest; secondly, the treatments are balanced with respect to dyes since each treatment is labeled once with red and green dyes and this balance means that the dye effect is no longer confounded with treatment effect; thirdly, more degrees of freedom are available for estimating error variation. When there are only two treatments in the experiment, the loop design becomes a Latin Square design (Figure 3).

Reference design:

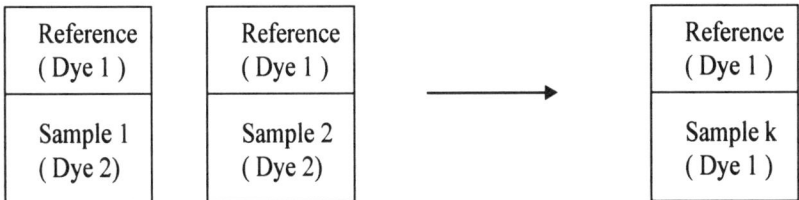

| Reference (Dye 1) | Reference (Dye 1) | → | Reference (Dye 1) |
| Sample 1 (Dye 2) | Sample 2 (Dye 2) | | Sample k (Dye 1) |

Loop design:

| Sample 1 | Sample 2 | → | Sample k |
| Sample 2 | Sample 3 | | Sample 1 |

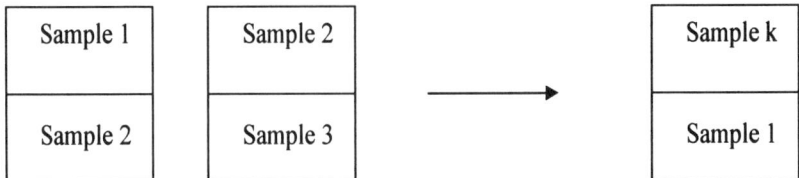

Figure 3. Different designs for cDNA array experiment.

The aforementioned ANOVA models fit well with experiments using a parallel treatment structure, but the models will be very complicated when there are two or more factors for the treatments. In some cases researchers are interested in such experiments. For example, it may be interesting to investigate the effects of different drugs and/or drug concentrations on gene expressions among animals with different genotypes or genders. Although it is theoretically possible to design such experiments with spotted cDNA microarray technology using loop design, it is practically cumbersome to identify and interpret all of the possible interactions. The oligonucleotide microarray, largely because of its single signal data reporting structure, serves better in such a scenario.

Affymetrix Oligonucleotide Arrays

Another major player in the microarray technology field, the high-density oligonucleotide microarray technology is different from the spotted cDNA array in many ways. Instead of using two dyes to label samples from different treatments, it labels all samples with the same dye, thus removing the dye effect. While there is only one probe for each gene in each sample in cDNA array, there are typically 11 to 20 unique probe pairs for the same "gene" (which is actually called probe set since some of them are ESTs). These probe pairs, while unique, may not necessarily be independent of each other. Since there is only one base pair difference between the perfect match and mismatch for each probe pair and the differential hybridization efficiency between the PM and MM is sequence specific, this introduces a new level of variance.

There are at least two different approaches to the analysis of Affymetrix microarray data depending whether or not one has faith in Affymetrix' commercial software in data normalization. A researcher would just take what he/she gets from the Affymetrix Microarray Analysis Suite (MAS Version 4 or 5) data analysis software and use the summary signal intensity for each probe set as the input data for further statistical analysis. The details of the Affymetrix algorithm can be found at the Affymetrix website or in the manual. For the sake of brevity, it suffices to say that the output signal for MAS5 is approximately a weighted average of the differences between PM and MM for the 11 to 20 probe pairs for each probe set. According to Affymetrix, this one number summary of each probe set is supposed to reflect the gene's relative expression level across arrays. An example of ANOVA model based on such signals is

$$y_{igl} = T_i + G_g + (TG)_{ig} + A_{l(i)} + e_{igl} \qquad (7)$$

where T_i is the ith treatment effect; G_g is the gth gene effect; $(TG)_{ig}$ is the treatment * gene interaction; all of which are considered as fixed effects. The other two terms: $A_{l(i)}$ is considered as normally distributed random effect with mean 0 and variance σ_A^2 and ε_{igl} is the random error assumed to be normally distributed with mean 0 and variance σ^2. This model can be readily extended to more complicated treatment structures, say a 2 x 2 factorial design. We just need to add another main effect of treatment and the interactions between the two treatment factors. The three way or four way interactions can be put into the error terms. An example of such model is

$$y_{ijkgl} = T_i + T_j^* + (TT^*)_{ij} + G_g + (TG)_{ik} + (T^*G)_{jk} + A_{l(ij)} + e_{ijkgl} \qquad (8)$$

where T and T* are two factors of treatments. This model can be fitted to the data generated by the following experiment in which there are 2 factors: one factor being age (Old vs. Young) and the other factor being medicine (drug A vs. placebo). There are 7 arrays for each combination of factors 1 and 2. After filtering out the probe sets that are not detectable across all 28 arrays, 6159 probe sets are left on each array. By analyzing all data in one model, we can identify the Age * Gene, Medicine * Gene, and Age * Gene * Medicine interactions. While this model may be desirable, it is extremely computationally intensive and not feasible for a desktop computer. One way to improve computing efficiency is to use two stage model (Wolfinger, 2001). First we can model the non-gene factors, then we can fit the residuals from first model to a model of the genes. For detailed procedure, refer to Wolfinger (2001). A more efficient approach is to analyze each probe set separately. Although this makes some of the above interactions unaccountable, the models are simpler and easier to interpret. An example model for one of the 6159 genes is as follows:

$$y_{ijkl} = T_i + T_j^* + (TT^*)_{ij} + A_{l(ij)} + e_{ijkl} \qquad (9)$$

and the ANOVA table (Table 8):

Table 8. ANOVA-based estimates of variance for a representative gene using Affymetrix GeneChips®

Source	Sum of Squares	Degrees of Freedom	Mean Squares
Age	235302	1	235302
Medicine	539099	1	539099
Age * Medicine	507388	1	507388
Residual	10318280	24	429928
Adjusted Total	11600070	27	

After fitting such a model to each of the 6159 genes, we can then order the p values of the overall model significance in ascending order and select genes with significant p values (adjusted by Benjamini Hochberg procedure as recommended in the multiple testing section for further analysis of the main effects and interactions of age and medicine.

While using the Affymetrix one number summary for each probe set per array makes things easier for downstream statistical analysis such as ANOVA and cluster analysis, some researchers argue that this one number summary does not account for the variability among the probe pairs appropriately and they recommend using the probe level data. Li and Wong (2001) proposed to fit the PM and MM data of each probe set with a multiplicative model and derive the PM - MM difference from such model. Such differences are then used as input for downstream analysis. Chu et al. (2002) suggested a systematic statistical linear mixed modeling approach using probe level data. In their examples, they identified cell line (L), treatment (T), and probe (P) effects as fixed, and array effects as random. The basic model for a single gene with multiple probe pairs is:

$$y_{ijkl} = L_i + T_j + (LT)_{ij} + P_k + (LP)_{ik} + (TP)_{jk} + A_{l(ij)} + e_{ijkl} \qquad (10)$$

The authors used $\log_2(PM_{ijkl})$ as the response variable and modified the basic model with or without including $\log_2(MM_{ijkl})$ as a covariate and concluded that it did not change the results very much for their example and hence MM might not be needed. However, whether or not this conclusion can be generalized is still not clear. At present, we still recommend using the one number summary for each probe set on each array by Affymetrix MAS 5.0 algorithm. Although it is not perfect, it provides a reasonable summary of the 11 or 20 measurements of the probe set with consideration of measurement error and outliers. It may help to look at the probe level data for some probe sets when there are major conflicts between the statistical analysis results and biological knowledge. In such cases, we want to check the details of the probe level data to make sure that the results are not messed up because of a few 'wild' probe pairs.

Design of High Density Oligonucleotide Microarray Experiments

The design of experiments with high-density oligonucleotide arrays is relatively simpler than that of spotted cDNA arrays. Since each sample is hybridized to one chip and all chips are measured the same way, there is no mixing of samples and hence no particular restrictions on the design except for the basic principles: randomization, replication and balance.

Summary

Analyzing the gene expression data with generalized linear model approach is a good start. While real data rarely satisfy the normality and homoscedasticity (equal variance) assumptions perfectly, the deviation are usually not so big as to prohibit the application of such models. By identifying the different sources of variation, we can get better estimates of the true effects of the factors of interest, and hence provide better input for downstream analysis such as hypothesis testing and cluster analysis. We can also identify outliers by analyzing the residuals. Appropriate normalization and transformation of the raw data can improve the model fitting. Major commercial statistical software such as SAS, S-Plus, SPSS, etc. have sophisticated procedures for implementing ANOVA with many options for modeling. For those who like free software and are willing to learn to write some code beyond point-and-click, the R package (www.r-project.org) is a perfect choice. It is almost as powerful and flexible as S-Plus and it is open source. An additional benefit of R is that people are sharing packages specifically written for microarray data analysis (e.g., www.bioconductor.org).

SUMMARY OF THE CHAPTER

In this chapter we attempt to convey three important messages to readers who are considering doing microarray experiments: First, microarrays are powerful but not omnipotent. Research hypotheses and careful experimental design are still essential for a successful microarray experiment. Second, statistical analysis (more generally, data analysis) is very complicated and challenging. Care must be taken since there are many pitfalls with those statistical methods that have been applied to microarray data analysis. Third, microarray data analysis, visualization and interpretation are currently a very active research area and many new methods or creative applications of old methods are being developed.

SOME USEFUL ONLINE SOURCES FOR MICROARRAY ANALYSIS:

R based expression analysis tools:
http://users.ox.ac.uk/~strimmer/rexpress.html
http://www.bioconductor.org
Jo DeRisi's software suite:
http://microarrays.org/software.html
Mike Eisen software:
http://rana.lbl.gov/EisenSoftware.htm
TIGR software:
http://www.tigr.org/softlab/

J express analysis suite:
http://www.molmine.com/index_j.html
EBI Expression profiler tool
http://ep.ebi.ac.uk/
Wentian Li's Microarray data and Analysis site
http://linkage.rockefeller.edu/wli/microarray/
Terry Speed's homepage
http://www.stat.berkeley.edu/users/terry/zarray/Html/index.html
SMA:
**http://www.stat.berkeley.edu/users/terry/zarray/Software/
smacode.html**
Plaid modeling:
http://www-stat.stanford.edu/~owen/plaid/
2HAPI microarray data analysis system:
http://www.sdsc.edu/mpr/2hapi/
SOTA
http://bioinfo.cnio.es/sotarray/
Kimono
http://whitefly.lbl.gov/~ihh/kimono/
Sanger center Expression browser
http://www.sanger.ac.uk/Users/mrp/java/ExpressionBrowser/
Rosetta's GEML conductor
http://www.rosettabio.com/products/conductor/default.htm
SwissProt:
http://www.expasy.ch/sprot
Pfam:
http://pfam.wustl.edu
GenMapp:
http://cancer.ucsd.edu/loslab
Gene Ontology:
http://www.geneontology.org
SAS:
http://www.sas.com
SPlus:
http://www.insightful.com
SPSS:
http://www.spss.com
GeneSpring:
http://www.silicongenetics.com
Partek:
http://www.partek.com

GeneSight:
http://www.biodiscovery.com
Spotfire:
http://www.spotfire.com
GOminer (not released yet up to Oct 2002)**:**
http://discover.nci.nih.gov
Onto Express:
http://www.openchannelfoundation.org/projects/Onto-Express

REFERENCES

1   Benjamini Y, Hochberg Y (1995) Controlling the false discovery rate: a practical and powerful approach to multiple testing. Journal of Royal Statistical Society Ser B 57, 289-300.

2   Benjamini Y, Liu W (1999). A distribution-free multiple test procedure that controls the false discovery rate. Tel Aviv. RP-SOR-99-3: Department of Statistics and O.R., Tel Aviv University.

3   Benjamini Y, Drai D, Elmer G, Kafkafi N, Golani I (2001) Controlling the false discovery rate in behavior genetics research. Behavioral Brain Research 125, 279-284.

4   Benjamini Y, Yekutieli D (2001) The control of the false discovery rate in multiple testing under dependency. The Annals of Statistics, in press.

5   Chu TM, Weir B, Wolfinger R (2002) A systematic statistical linear modeling approach to oligonucleotide array experiments. Mathematical Biosciences: 176, 35 - 51.

6   Dudoit S, Yang YH, Callow MJ, Speed TP (2000) Statistical methods for identifying differentially expressed genes in replicated cDNA microarray experiments. Department of Biochemistry, Stanford University, Technical report # 578.

7   Duncan DB (1955) Multiple range and multiple F tests. Biometrics 11, 1.

8   Durbin B.P. (2002) A variance-stablizing transformation for gene-expression microarray data. Bioinformatics 18, s105-s110.

9   Ewnes WJ, G R Grant (2001) Statistical Methods in Bioinformatics. New York: Springer-Verlag.

10   Hsu JC (1996) Multiple Comparisons.. London: Chapman and Hall.

11   Kerr, Martin and Churchill (2000) Analysis of variance for gene expression microarray data. Journal of Computational Biology, 7:819-83

12   Kerr and Churchill (2001) Statistical Design and the analysis of gene expression microarrays. Genetical Research, 77:123-128.

13   Kerr and Churchill (2001) Experimental design for gene expression arrays. Biostatistics, 2:183-201.

14   Kerr, Leiter, Picard and Churchill (2001) Analysis of a designed microarray experiment. Proceedings of the IEEE-Eurasip Nonlinear Signal and Image Processing Workshop.

15   Kerr, Afshari, Bennett, Bushel, Martinez, Walker and Churchill (2002) Statistical analysis of a gene expression microarray experiment with replication. Statistica Sinica, to appear.

16   Kohane IS, Kho, AT, Butte AJ (2002) Microarrays for an Integrative Genomics. Cambridge, MA: MIT Press.

17   Li C, Wong WH (2001) Model-based analysis of oligonucleotide arrays: Expression index computation and outlier detection. Proc. Natl. Acad. Sci. Vol. 98, 31-36.

18   Li C, Wong WH (2001) Model-based analysis of oligonucleotide arrays: model validation, design issues and standard error application, Genome Biology 2(8): research0032.1-0032.11

19   McCulloch CE, Searle SR (2001) Generalized, Linear, and Mixed Models. New York: Wiley.

20   Miller RA, Galecki A, Shmookler-Reis RJ (2001) Interpretation, Design, and Analysis of Gene Array Expression Experiments. Journal of Gerontology: Biological Sciences: 56A, N0. 2, B52-B5

21   Peng X, Wood C, Blalock E, Chen K, Landfield P, Stromberg A (2002) Sample pooling and fold change in microarray experiment: statistical perspectives. The 25th Midwest Biopharmaceutical Statistics Workshop.

22   Shaffer JP (1995) Multiple Hypothesis Testing. Annual Review of Psychology: 46, 561(24).

23   Westerfall PH, Young SS (1993) Resampling-based Multiple Testing. New York: Wiley.

24   Wolfinger RD, Gibson G, Wolfinger ED, Bennett L, Hamadeh H, Bushel P, Afshari C, Paules RS (2001) Assessing gene significance from cDNA microarray expression data via mixed models. Journal of Computational Biology: 8, 625-63

25   Wood CL, Stromberg AJ, Peng X (2002) Multiple Comparisons for Multiple Microarray Analysis. Unpublished.

Chapter 8

STRATEGIES FOR CLUSTERING, CLASSIFYING, INTEGRATING, STANDARDIZING AND VISUALIZING MICROARRAY GENE EXPRESSION DATA

Willy Valdivia Granda
North Dakota State University. Genomics and Bioinformatics Group.
Department of Plant Pathology.

INTRODUCTION

Over the last century, investigation of the anatomical and morphological characteristics of a small number of organisms has played an important role in the understanding of numerous biological processes. The rediscovery of Mendel's laws of heredity in the opening of the 20th century initiated a scientific quest to understand the mechanisms of how genetic information is transmitted and the biological consequences of genetic variation. With the emergence of molecular biology in the last thirty years, classical genetic research has shifted from understanding how visible traits are transmitted to the study of the genome structure at the molecular level. Innovations such as PCR and advances in robotics such as miniaturization and parallelization have lead to a rapid development of more accurate, sensitive and powerful devices used for the analysis of the molecular structure, function and interaction of gene products. With the development of ultrahigh throughput screening tools and the drive from 96-microwell plates to 384- and 1536-microwell plates, it is expected that the generation of whole sequences of different organisms will increase at a rate ~100 times higher than previously anticipated (HeadGordon and Wooley, 2001; Helfrich, 2002; Beson et al. 2002). As powerful, automated biological sampling and analytical tools are becoming available to more laboratories, an exponential and sometimes overwhelming accumulation of multi-format post-genomic datasets is produced. Consequently, modern biology is becoming a data driven multidisciplinary science in which biologists, mathematicians, statisticians, physicists and computer scientists are developing tools to identify 'in silico' the coding regions of genes, predict and model protein structural characteristics, define protein-protein interactions, construct biochemical networks and identify potential drug targets.

By measuring the expression of thousands of genes, DNA microarrays are increasing the level of understanding of complex biological systems. It has been argued that microarray technology is beginning to have the same impact on biological sciences that integrated circuits have already produced in physical sciences (Graves, 1999). Their small size, high density, and compatibility with fluorescence labeling allow researchers to use microarrays for the parallel analysis of different experimental samples. This technology offers enormous scientific potential for the exploration of genomic structure, function, and interaction related with temporal and spatial expression changes of thousands of genes. However, as more laboratories use microarrays, an exponential growth in the size and complexity of the data is beginning to be produced. This is the result of improvements in microarray technology, as well as the increasing multidisciplinary and inter-institutional collaboration among researchers. With the growth of this technology the extraction of knowledge from microarray data presents new opportunities and challenges. Previous chapters have discussed the fact that gene expression measurements using DNA microarrays are subject to different sources of variability. These include DNA damage during spotting, experimental errors during sample collection and mRNA isolation, RT-PCR variation, fluorescence labeling kits differences, microarray surface stability, and scanner calibration. Variability produces low quality data that obscures analysis and weakens inference of the biological process. Therefore, in order to provide a reliable dataset it is necessary to ensure the proper number of replications and a careful experimental design. Although the optimum number of replicates necessary to obtain an accurate measure for one gene may not be the same as number needed for another gene, the replication of microarray experiments increases the precision in the estimation of gene expression level and provides information about the uncertainty on these estimates. On one hand, replications can ensure a reliable result; on the other hand microarray experiments are expensive in terms of reagents, labor, time and equipment. A description of these issues is discussed in chapters five, six, and seven.

Before deciding which microarray data analysis technique should be used, it is necessary to determine how the data should be normalized. Normalization strategies are aimed at reducing data variability and dimensionality. Chapter five discussed different normalization methods, their advantages, limitations and the fact that normalization methods can help in the removal of inaccurate measurements. After this process, the next step is to use different mathematical strategies to find gene expression patterns and to assign gene function and co-regulation. Before and during the microarray data analysis process most biologists confront a series of critical questions: How to determine which genes are induced or repressed? Which similarity measure is appropriate for a given data analysis algorithm? How can the previously accumulated biological information for the gene or genes under study be used in the microarray data classification process? Which is the best technique to analyze large microarray datasets? What are the computational issues related with performance of different algorithms? How will these analyses reveal functions of unknown genes? All these questions can be summarized by asking: Which statistical or computational techniques should be chosen to establish gene expression patterns that can be related with the study in question or to place different genes into organized pathways or phylogenomic maps? Answering the above questions is difficult due to the inherent complexity of biological systems and the large number of data points generated by microarray technology. Therefore, in order to appreciate different techniques applicable for the analysis of microarray data, biologists need some exposure to statistical analysis and computational principles that will help in the election of the proper algorithm, commercial or public software. However, the commonality between statistical analysis and microarray data mining has caused some confusion among biologists and there is still some debate on the subject among statisticians and computer scientists. The size of microarray data means that it is no longer feasible for statisticians to interact directly with the data. Therefore, computers and database queries are essential in microarray data manipulation and analysis. Nevertheless, data mining techniques depend on the ability to consider statistical issues. It is beyond the scope of this chapter to clarify these arguments.

This chapter is aimed at introducing the reader to the basic concepts underlying the statistical and data mining methods used for the analysis of microarray data. While I attempt to provide an introductory review and a basic guide on microarray data analysis strategies, complex mathematical equations and their computational implementations are minimized. When it is necessary, the reader is pointed using the symbol "*" to references describing the statistical and computational details. This chapter does not include early comparative gene expression analysis using microarrays where gene expression was established in terms of fold change. An inherent problem with this criterion is that genes with low absolute expression levels have a greater inherent error in their measurements and are more likely than higher expressing genes to meet any fold change cut-off (Mutch et al. 2002). Different concepts related to the microarray data analysis process including microarray gene expression matrix, outliers, missing values, distance functions, unsupervised and supervised methods, advantages, limitations and considerations to estimate their reliability are presented. When it is necessary, biological examples are provided with the aim to highlight the relevance of some microarray data analysis methods. This chapter also introduces information about software (public and private) that can help the reader to choose suitable tools for the analysis of their particular microarray gene expression data. Some aspects about the implementation of microarray data repositories, the development of standards including the Minimum Information About Microarray Experiments (MIAME) and its computational implementation through MAGE-OM, MAGE-ML and MAGE-stk are highlighted. Finally, current trends and future challenges that microarray technology will present are also summarized.

MICROARRAY GENE EXPRESSION MATRIX

Microarray gene expression data consist of the measurement of the expression of many genes (gene profiles) with a limited number of conditions or time points ordered in a table. This data structure contains features uncommon in other datasets including thousands of genes (variables) and tens or hundreds of microarray samples (observations). These datasets are known in the computational and statistical community as high dimensionality or multivariate data. The table is defined as a microarray gene expression matrix (GEM) generated from different microarray collections in which rows represent genes and columns represent different microarrays. Thus GEM $= g \times n$, where g is the number of genes whose expression is measured in each n array. A microarray gene expression data matrix can be transposed if the goal of the analysis is to find relevant microarrays based on the expression levels of different genes. In this case, the variables are the microarrays and the different genes are the observations. In addition to the gene expression values, outliers, empty and missing values are part of every gene expression matrix. The implications of such values and approaches for estimating them and reducing their negative effect are discussed below.

Gene Expression Outliers

In a microarray gene expression matrix, we will encounter certain values that are regarded as outliers. Outliers are important since the performance of some analysis tools can be affected negatively by their presence (analysis techniques that are not affected by outliers are often termed 'robust'). Gene expression outliers are measurements that appear to be inconsistent compared to other members of the sa me matrix. These values are outside of some standard distribution and cannot be ascribed to chance or natural variability. The origin of these values can be due to experimental artifacts and can account up to 15% of a typical microarray experiment (Lee et al. 2000; Nadon and Shoemaker et al. 2002). To determine which gene expression levels are outliers; it is necessary to measure, or make valid assumptions of, the variability of the system (Mutch et al. 2002). Different data analysis techniques discussed in this chapter can be used to determine outlier status.

Empty and Missing Values

Other components of a microarray gene expression matrix are empty and missing values. Empty values have no corresponding real-world values, while missing values were not captured for some reason. Establishing if a particular value is empty rather than missing is difficult because it may require information that is not available within the dataset. Missing data are a common problem in essentially every microarray study, and occur for a variety of reasons including: omissions in the data entry process, experimental equipment malfunction, insufficient scanning resolution, dust or scratches on the microarray glass surface. According to Rubin (1976) missing data can be characterized as: 1) Missing completely at random, where cases with complete data are indistinguishable from cases with incomplete data. 2) Missing at random, where cases with incomplete data differ from cases with complete data, but the pattern of data missing is traceable or predictable from other variables. 3) Informatively missing, where the probability that a missing gene expression value depends on both observed and unobserved values.

The relevance of missing values in the data analysis process is due to their negative effect in the performance of many analysis methods and because this loss of information leads to the distortion of the final analysis and a false assumption about the biological process. The more values that are missing, the greater is the potential distortion, and the difficulty in the estimating of these values increases. One of the most used microarray datasets is derived from the analysis of 60 cancer cell lines (also know as the NCI 60 microarray dataset (Ross et al. 2000)) and the lymphoma samples (Alizadeh et al. 2000). According to Dudoit et al. (2000), the mean percentage of unreliable or missing data of NCI-60 dataset was around 3.3% while for the lymphoma it was around 6.6%.

Handling Missing Values

The previous section introduced briefly the origins of missing data and their negative effect in the interpretation of microarray data. To minimize their effect, different statistical methods and their computational implementations can be used prior to the analysis process. When analyzing a gene expression matrix, most computational approaches either will reduce the collected data by discarding missing records or will estimate their value. From this point arises the question: How can missing gene expression values be estimated using the available data? Numerous statistical solutions have been proposed. However, here we only present a brief overview of some of these methods in order to provide the reader with a more concrete understanding of this issue. This list is not exhaustive, but it covers some of the more widely recognized approaches.

List-Wise or Case-Wise Data Deletion

If a record has missing data for any variable used in a particular analysis, the computer program will omit that entire record from the analysis. This means eliminating the complete row. This practice therefore leads to excessive loss of data points and the resulting analysis may represent a very skewed view of the biological process. Therefore, the user should try to minimize the use of this approach.

Pair-Wise Data Deletion

This is an alternative method that can be used to estimate gene expression values when there are a small number of missing data and a considerable number of microarray experiments. This method estimates the proximity between any two cases from only those variables that have valid entries. All available observations are used to compute the means and variances, while all available pairs of values are used to compute covariance (Raymond and Robert, 1987). Thus, correlations are computed using only those observations that have non-missing values on both variables. When data are missing for either (or both) variable(s) for a subject, the case is excluded from the computation. The problem with this method is the potential inconsistency of the covariance matrix in a multivariate microarray dataset.

Mean Substitution

The mean substitution assumes that a missing value for a particular gene can be estimated by the mean (expected value) from the non-missing observations. Row averaging assumes that the expression of a gene in one of the column is similar to the expression of the same gene in another column. Several limitations have been highlighted for these methods (Anderson et al. 1983*, Little and Rubin, 1987). A main argument against the use of this method is that it produces biased results of both variances and co-variances. This situation is relevant in microarray experiments where different samples o r treatments m ake unpractical t he imputation of missing values.

Regression Methods

Regression methods rely on the information contained in the non-missing values of variables to provide estimates of the missing values for the variable of interest. A regression equation is based on complete case data for a given variable, treating it as the outcome and using all other relevant variables as predictors. Then, for cases where a gene expression value is missing, the available data is replaced into the regression equation and used as predictors to substitute the missing value by the value predicted by the equation. Using regression-estimated scores for missing data does not attenuate the relationships among variables, because a regression-estimated score will not equal the mean, unless: 1) The R-squared for the regression equation is exactly 0; or 2) The predictor variable scores for non-missing data for cases with missing data all exactly equal the respective means of these predictor variables.

Other Methods

Imputation of missing values using k-nearest neighbors has been reported by Dudoit et al. (2000). In this application, the neighbors are the genes and the distance between neighbors is based on their correlation. For each missing data point, the algorithm computes its correlation with all other p-1 genes and for each missing entry, identifies the k-nearest genes having a complete entry, and imputes the entry by averaging the corresponding entries for the k-neighbors. Using a variety of missing conditions (a gene expression matrix with 1 to 20% of missing values) the accuracy of four different missing expression values estimators has been reported by Troyanskaya et al. (2001). This evaluation compared single value decomposition (SVDimpute), weighted k-nearest neighbor (KNNimpute), row average and filling missing values with zeros. It was concluded that k-nearest neighbor provided a more robust imputation (6-26% average deviation from the real values) than SVDimpute. Both methods outperformed row averaging and filling missing values with zeros. The authors highlighted that KNNimpute should not be use in a gene expression matrix with less than four columns and that SVDimpute is more effective in time series experiments with low noise level.

DISTANCE FUNCTIONS

As discussed previously, a gene expression matrix is composed of different elements including outliers, missing values, and gene expression measurements. Many microarray data analysis methods use distance functions to separate gene expression values into different categories, assign them functional roles, and/or map them in specific biological processes. The selection of the distance function represents the first step towards the analysis of microarray data since the performance of some computational tools, the quality of the results and the graphical representations that they produce, will depend on its selection. In general, distance functions can be divided in two classes: (a) similarity based on positive correlations, which may identify co-regulation (b) similarity with positive and negative correlations, which may help to i dentify control p rocess that antagonistically regulate the expression of other genes (D'haseleer et al. 2000). Before presenting different distance functions, the reader should be aware of some basic criteria for their selection.

Criteria of a Distance Function

Lets us assume that the level of expression of two genes (i) and (j) are available. How do w e establish t he d istance between them (d), or t heir degree of relationship? The proper selection of a distance function not only allows the computational implementation to group or classify the genes as members of one category or another, but also can give some interpretation of how genes interact in the co-regulation process. The distance between two genes can be:

a) **Positive definite:** If the points i and j are different, the distance between them must be positive. If the points are the same, then the distance must be zero.

i.   if $(i \neq j)$, $d(i, j) > 0$
ii.   if $(i = j)$, $d(i, j) = 0$

b)   **Symmetric:** The distance from i to j is the same as the distance from j to i. That is, for any two points i and j,

$$d(i, j) = d(j,i)$$

c)   **Satisfies the triangle inequality:** The distance between two points can never be more than the sum of their distances from some third point. That is, for any three points i, j and z,

$$d(i,j) \leq d(i,z) + d(z,j)$$

In this chapter, only well-known distance functions used for gene expression pattern estimation are presented.

Euclidean Distance

Euclidean distance (also know as L^p distance) is appropriate for variables that are uncorrelated and have equal variances. These distance functions can often be used to reflect dissimilarity between the values of two gene expression levels calculated using the Pythagorean theorem.

Mahalanobis Squared Distance

This is the most popular parametric measure of discrepancy between two multidimensional samples. Mahalanobis Squared Distance (MSD) gives a one-dimensional measure of how far a gene expression value is from a functional class where scale independence is needed. This distance measure is useful to detect outliers and to deal with covariance among microarray experiments (Hardin and Rocke, 2000; Chilingaryan et al. 2002). In this context, an outlier can be thought of as a point with a large MSD. A measure of the distance (D) between two groups of individuals is given by

$$D^2 = (\bar{x}_1 - \bar{x}_2)'S^{-1}(\bar{x}_1 - \bar{x}_2)$$

where $\bar{x}_1$ and $\bar{x}_2$ are the mean vectors of the two groups and S is a weighted average of the variance-covariance matrices of the two groups, S_1 and S_2

$$S = \frac{n_1 S_1 + n_2 S_2}{n_1 + n_2}$$

where n_1 and n_2 are the samples sizes in the two groups. The Mahalanobis distance effectively performs a transformation of the Euclidean distance.

Minkowski Distance

The general distance form is the weighted Minkowski distance. Considering a point, X, in n-dimensional space as a vector $<x_1, x_2, x_3, ..., x_n>$, the weighted Minkowski distance,

$$d_p(X,Y) = \left\{ \sum_{i=1}^{n} w_i |x_i - y_i|^p \right\}^{\frac{1}{p}}$$

where, p is a positive integer, x_i and y_i are the ith components of X and Y, respectively, and w_i (≥ 0) is the weight associated to the ith dimension or i^{th} feature.

Associating weights allows some of the variables to dominate the others in similarity matching. This is useful when it is known that some functional categories are 'more important' than others in a particular microarray experiment. Otherwise, the Minkowski distance is used with $w_i = 1$ for all i.

$$d_p(X,Y) = \left\{ \sum_{i=1}^{n} |x_i - y_i|^p \right\}^{\frac{1}{p}}$$

Manhattan Distance

This method is also known as 'city block' distance. In this case, after the reordering of genes and experiments, gene expression values are partitioned into homogenous rectangular blocks.

$$d_{ij} = \sum_{k=1}^{q} |x_{ik} - x_{jk}|$$

where q is the number of variables and $x_{ik}, x_{jk}, k=1,...,q$ are the observations on the gene expression of i and j.

Pearson Correlation Coefficient

The Pearson correlation measures the strength of the linear relationship between two variables. Therefore, the correlation coefficient can be used to associate two or more gene expression values and to find gene expression trends and identify potentially co-regulated genes. The strength of the relationships between gene expression values is expressed by the *correlation coefficient*. Pearson correlation coefficients give *r* values between -1 and 1, with 1 meaning that the two series are identical in pattern and relative magnitude, 0 value meaning they are linearly uncorrelated, and -1 value meaning they are perfect opposites. The more profiles that have the same trend; the closer the *r*-value is to 1 (Ross et al. 2000; Scherf et al. 2000; Bittner et al. 2000; Herrero et al. 2001). Let *x* and *y* be *n*-component vectors for which we want to calculate the degree of association. For pairs of quantities (x_i, y_i), $i=1,\ldots,n$ the correlation coefficient *r* is given by the formula:

$$r = \frac{\sum_{i=1}^{n}(x_i - \bar{x})(y_i - \bar{y})}{\sqrt{\sum_{i=1}^{n}(x_i - \bar{x})^2 \sum_{i=1}^{n}(y_i - \bar{y})^2}}$$

where $\bar{x}$ and $\bar{y}$ are the means of the vectors of *x* and *y* respectively.

Considerations in the Selection of Distance Functions

It is necessary to clarify that there is no one 'right' distance function for grouping gene expression values. Euclidean distances are additive in the sense that variables contribute independently to the measure of distance, and this may not always be appropriate (Hand et al. 2001). If some gene expression values have a wider range, they will tend to dominate the metric. A small change in the expression level of some genes may result in a significant change in the structure of the matrix. An option to reduce this effect is to standardize the variables by dividing the measurements by the standard deviation. A limitation of the Mahalanobis distance is that it becomes more computationally intensive as the number of microarrays in the gene expression matrix increases. While it is argued that Manhattan distances are not powerful enough to capture relevant features found in complex microarray data experiments, a limitation of using the correlation coefficient is that their reliability depends on absolute expression level of the compared genes. A positive correlation between two highly expressed genes is much more significant than the same correlation between two poorly expressed genes (Getz et al. 2000). If gene expression measurements are not normalized, correlations measurements can be missed. Using a rank correlation (e.g., Spearman's correlation) may be more useful in establishing statistically significant associations.

UNSUPERVISED ANALYSIS AND CLUSTERING OF MICROARRAY DATA

Unsupervised methods have been the focus of considerable research for several decades in the analysis of satellite and aerial images, telecommunication and market data. In the field of microarray analysis, these methods are defined as unsupervised because, prior to the analysis, existing knowledge about the functional roles of different genes is not considered. Instead, this information is introduced after the analysis process. Therefore, unsupervised microarray data analysis can be described as an exploratory process where the system establishes the existence of related gene categories without knowledge or care for an imposed structure. The system is given a microarray dataset and it is left to find patterns, regularities, or groups. In this process it assumes that each gene fits into a category that can be associated with function or co-regulation. As a result, it is expected that unsupervised analysis will find a potential new explanation or representation revealing gene expression associations not previously evident. Currently, unsupervised analysis constitutes the main technique for the interpretation of microarray data. This is due in some way to their simplicity, graphical representation and faster execution.

Clustering represents unsupervised methods that can be used to decide if the elements of a gene expression matrix fall into distinct groups. The natural basis for organizing gene expression data this way is the assumption that similar expression levels might indicate related biological function or co-regulation. Under these circumstances, the clustering process can be helpful to identify the function of unknown genes (Eisen et al. 1998). During the clustering process, expression values are grouped based on a distance function and microarrays are grouped based on similar global expression profiles. In other words, it is possible to use gene expression groups to establish the degree of relationship among different microarrays experiments. The members of each gene expression cluster are similar to other members within that cluster, but they are dissimilar from members of another cluster. The first step in the clustering process is to select an appropriate description of similarity or dissimilarity by choosing a distance function. This choice may be as important as the selection of the clustering algorithm (D'haeseleer et al. 2000; Jain et al. 1999). Since there are many possible similarity measurements that can be used, there is no one clustering technique that is universally applicable in uncovering the variety of structures present in microarray datasets. Moreover, the availability of a vast collection of clustering algorithms in the literature can easily confound a user's attempt to select an algorithm suitable for the problem at hand (Jian et al. 1999). In general, it can be said that clustering methods start with as many clusters as there are individual gene expression values, and then sequentially join or separate objects based on the distance function. There are two main methods to cluster data: Hierarchical and Partitive (Figure 1).

Hierarchical Clustering

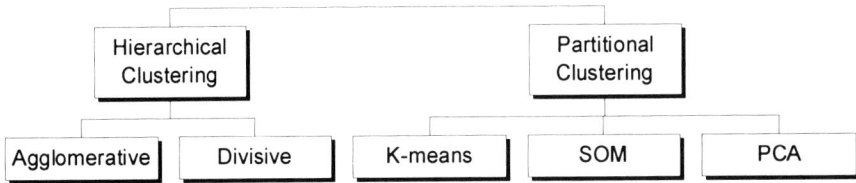

Figure 1. Main unsupervised methods for microarray data analysis.

Hierarchical clustering, proposed by Sneath and Sokai (1973*), is one of the most widely used techniques for the analysis of microarray gene expression data and has been used to identify genes with similar profiles and thus with similar functions (Eisen et al. 1998; Spellman et al. 1998; Seo and Shneiderman, 2002). In the clustering process, each gene expression value is assigned to coordinates reflecting the distance(s) to other genes by using a pairwise similarity measure. Hierarchical clustering algorithms can be further divided into two subgroups: 1) based on a criterion of dissimilarity (divisive) or 2) based on a criterion of similarity (agglomerative). A divisive approach begins with all gene expression values in a single cluster and performs splitting until a criterion is met. This process is also known as 'top-down.' The agglomerative approach begins with each gene expression value in distinct (singleton) sets, and successively merges the clusters until a criterion is reached. This method is also called 'bottom-up' where the high hierarchical levels are resolved before going to the details of the lowest levels.

The result of the hierarchical clustering process is a representation as a tree shaped graphic called a dendrogram (Eisen et al. 1998; Brazma and Vilo, 2000; Herrero et al. 2001). The dendrogram (the Greek word dendron means 'tree', gramma means 'drawing') provides a highly interpretable visual summary of the clustering process. The dendrogram is represented as a color code graph in which each gene expression value is called leaf. A red square represents an increase in gene expression level, and a green square represents a decrease. Some applications represent gray colors as missing data and in other cases gene expression values as blue and yellow squares. The intensity of the color is a measure of the relative difference between other values and clusters. A cluster will always have two directly connected components, either two other clusters, two vectors (leaf nodes), or one of each. The length of the horizontal lines that connect two clusters (nodes) is a measure of their relative closeness. The hierarchy shown in the dendrogram and the linear presentation of the colors helps the user to identify shared expression patterns just by looking at the color codes of closely clustered elements (Figure 2). Since outliers may finish as singletons or as small clusters removed from the valid clusters, hierarchical clustering can be used for their detection and elimination.

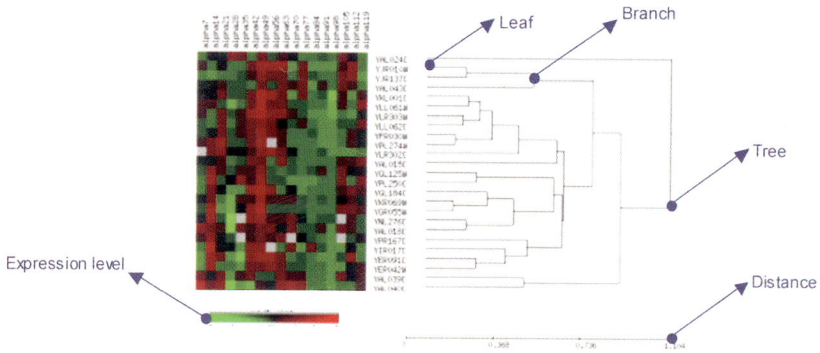

Figure 2. Elements of a dendrogram.

When applying clustering techniques, it is important to determine if the microarray experiments have independent information or if they are highly correlated. The terms 'bi-clustering' and 'two way clustering' have been used to describe a clustering technique capable of grouping both genes and microarray subsets simultaneously (Figure 3). Alon et al. (1999) have applied two-way clustering to study 6500 genes of 40 tumors and 22 normal colon samples. The idea of the two-way clustering approach is that a small subset of genes participates in a cellular process, which takes place only in a subset of the samples (Getz et al. 2000). Under the same context and using the super-paramagnetic hierarchical clustering algorithm, Getz et al. (2000) analyzed two datasets, one from an experiment on colon cancer (40 colon tumor samples and 22 normal colon samples (Alon et al., 1999)) and another from leukemia (72 samples from acute leukemia patients (Golub et al. 1999)). From these, 47 cases were diagnosed as acute lymphoblastic leukemia and the other 25 as acute myeloid leukemia. The algorithm searched for pairs of a relatively small subset of genes and samples yielding stable and significant partitions. Other terms associated with the same idea include 'direct clustering' and 'box clustering'

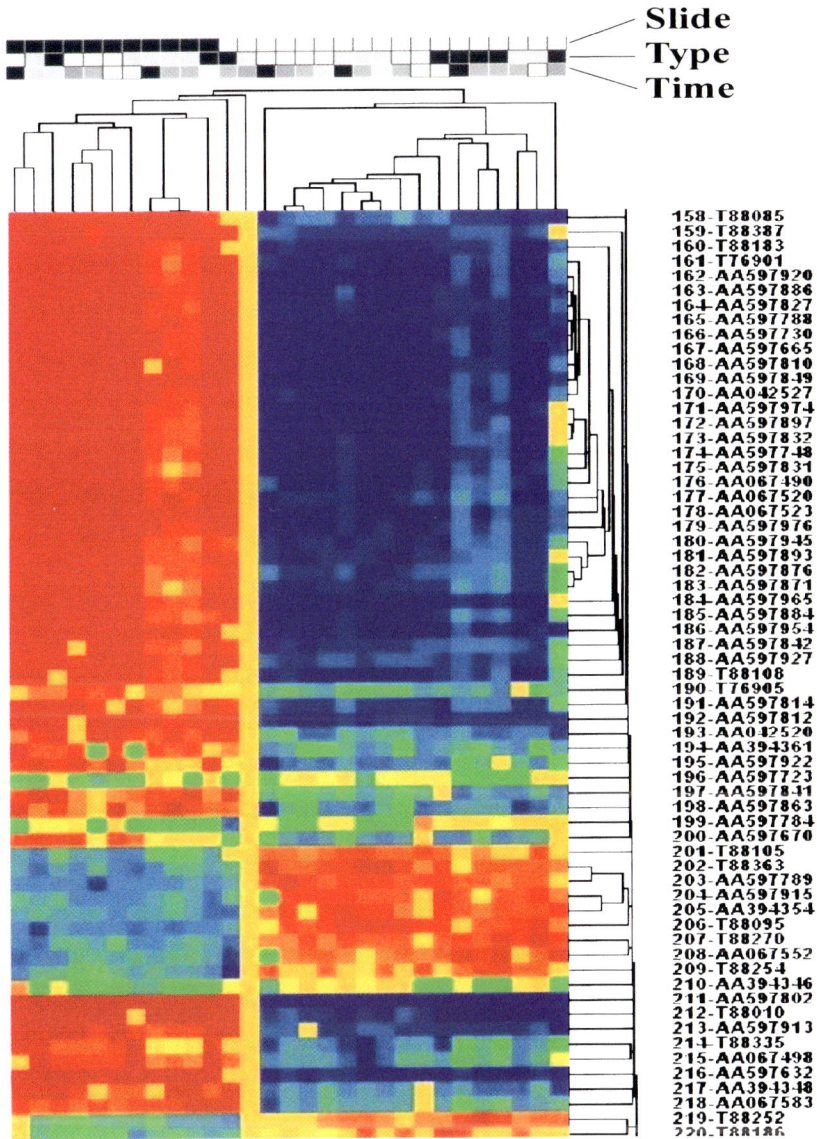

Figure 3. Clustering of the circadian rhythms of two *Arabidopsis thaliana* mutant ecotypes (Col and Lan) and the wild type Columbia using the expression profiles of 220 genes. At the top is a summarized description of the three factors involved in classification for each column: **Slide**- slide 1 in black and slide 2 in white; **Type**- wild type is black, Lan mutant is gray, and Col mutant is white; **Time**- as determined by a 24 hour (white) to 44 hour (black) range with intermediate values in grayscale. Differential expression is shown from >6 fold increase (blue), to > 6 fold decrease (red) and no change (gold). CLUSFAVOR with a correlation distance function and single link pair-wise distance were used. Notice that the grouping is more dependent on slide number (Slide) than ecotypes (Type) and time series (Time).

(Cheng and Church, 2000). These approaches were also used to separate cancer tumors and to find new tumor subclasses (Alizadeh et al. 2000). Using hierarchical clustering with a Pearson correlation distance function, Ross et al. (2000) identified systematic variation in gene expression patterns for 60 human cell lines and 8000 different human transcripts. The most notable property of the clustered data is that the cell lines with common presumptive tissues grouped together.

Pair-wise Distance Metrics

Single Link

This technique is also referred as the 'minimum method' or 'nearest neighbors.' Using the single distance function, gene expression groups are formed from the minimum distance between two gene expression values. When applying this type of distance measure, 'chaining effects' may be observed. This refers to the tendency to incorporate intermediate gene expression groups into an existing group rather than form a new one. Because the tendency of single link to maximize connectivity of gene expression values, in most cases dendrograms often are not amenable for a substantive interpretation unless gene expression values are well separated (Heyer et al. 1999; Zhang and Zhao, 2000).

Complete Link

This technique is also referred as 'furthest neighbors' or 'diameter.' Complete link produces compact groups that do not exceed some threshold (Heyer et al. 1999). The maximum distance between gene expression values determines the difference between two groups. Complete link algorithms require well-separated gene expression values. However, this is rarely the case. Consequently, this method usually performs well in cases when the gene expression values form naturally distinct 'clumps.'

Average Link

Average link often is used as an alternative to the two previous options (Heyer et al. 1999; Scherf et al. 2000). This method relies on the average value of the pair within a cluster, rather than the maximum or minimum similarity as with the single link or the complete link methods. The basic assumption regarding this method is that all the gene expression values in a cluster contribute to the inter-cluster similarity.

Unweighted pair-group average (UPGMA)

Sokal and Michener (1958) introduced the unweighted pair-group average method. It is possible to calculate the distance between two clusters as the average distance between all pairs of gene expression values in the two different clusters. This method is very efficient when the objects form naturally distinct 'clumps,' and it performs equally well with elongated, 'chain' type clusters.

Weighted pair-group average (WPGMA)

Weighted pair-group average is identical to the unweighted pair-group average method, except that in the computations, the size of the respective clusters (i.e., the number of expression values contained in them) are used as a weight to make proper allowance for their relative importance. Thus, this method (rather than the UPGMA) should be used when the cluster sizes are suspected to be greatly uneven.

Unweighted pair-group centroid (UPGMC)

Often the centroid or average of the gene expression levels belonging to a particular cluster can be considered as a representative point for a particular cluster. Unweighted pair-group method uses the centroids or averages in the multidimensional microarray space to estimate the distance between two clusters.

Weighted Pair-Group Median Centroid (WPGMC)

Weighted pair-group method using the centroid average is identical to the previous one, except that weighting is introduced into the computations. This takes in consideration differences in cluster sizes (i.e., the number of gene expression values contained in them). Thus, when there are (or one suspects there to be) considerable differences in cluster sizes, this method is preferable to the previous one.

Ward's Method

This method is distinct from all other methods because it uses an analysis of variance to evaluate the distances between expression value centroids. In short, this method attempts to minimize the sum of squares of any two (hypothetical) clusters that can be formed at each step. Ward's method is regarded as very efficient, however, it tends to create clusters of small size and it is sensitive to outliers.

Partitional Clustering

Partitional clustering makes implicit assumptions on the form and number of clusters since the user usually chooses the cluster number *a priori*. Given a gene expression matrix, partitional algorithms will divide gene expression values (g) into k groups in an iterative process until each partition represents a cluster and $k \le g$. The process must satisfy two basic requirements 1) each cluster must contain at least one gene expression value, and 2) each expression value must belong to one cluster.

K-means Clustering

The k-means clustering algorithm is one of the most popular partitive unsupervised methods and was introduced by MacQueen (1967*) and applied to microarray data (Tavazoie et al. 1999; Soukas et al. 2000; Aronow et al. 2001). With some prior knowledge about the number of possible clusters in a microarray dataset, k-means clustering can be used as an alternative to hierarchical methods. It is argued that k-means gives good results in microarray datasets where the clusters are expected to be compact (similar gene expression values) and therefore with similar biological functions. The k-means clustering algorithm simply partitions gene expression values into different clusters without trying to specify the relationship between individual genes. No dendrogram is produced and prior to running the algorithm, the number of clusters is defined by the user.

In the context of microarray data analysis, the first step of the k-means algorithm involves the representation of each gene expression value as a point. Then, the algorithm identifies k-points (seed points) and computes them as the centroids. At each iteration the k-points remain fixed for an entire pass through the microarray dataset (Celis et al. 2000). In the next iteration, the remaining points are assigned to one of nearest k-points in a way that minimizes the sum of the distance between all points and all seed points (Figure 4). The computation process is terminated when there are no further changes in the assignment of gene points to the centroids.

$$E = \sum_{k=1}^{C} \sum_{x \in Qk} \left\| x - ck \right\|^2$$

where C is the number of clusters and c_k is the center of the cluster k.

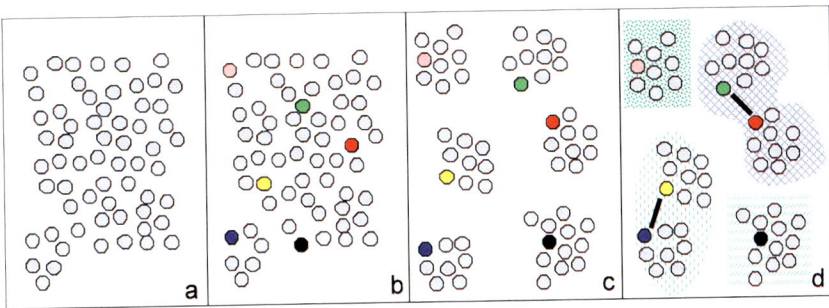

Figure 4. The k-mean clustering process: (a) representation of gene expression as points; (b) identification of the seed points; (c) cluster of points to the seed points; (d) formation of new clusters.

Fuzzy clustering

In a process known as 'fuzzyfication' precise numbers such as gene expression levels are transformed into qualitative descriptors. While very similar to the process of cluster formation, a main difference between fuzzy k-clustering and standard k-clustering is that fuzzy clustering considers each gene point as member of every cluster with certain probability (discrete membership) (Woolf and Wang, 2000, Gasch and Eisen, 2002). This allows fuzzy k-means algorithm to identify overlapping groups of genes and to recognize the role of a gene in different pathways. The analysis of 83 microarray experiments performed in yeast, suggest that fuzzy k-clustering is superior to hierarchical and k-means clustering in forming gene expression clusters while allowing genes to maintain memberships in multiple clusters.

Self Organizing Maps

Self Organizing Maps (SOM) or Kohonen networks were introduced by Kohonen (1981, 1997*). This approach is one of the best-known unsupervised methods and it has been found to be superior in both robustness and accuracy to other clustering techniques when analyzing microarray data (Heyer et al. 1999; Tamayo et al. 1999, Törönen et al. 1999; Herrero et al. 2001). Although SOM are related to the k-means clustering, they contrast with the rigid structures of k-means clustering by imposing partial memberships in the elements of a cluster while at the same time providing a scalable solution for the analysis and visualization of large and noisy microarray datasets (Herrero et al. 2001).

Before initiating the analysis the user defines a geometric configuration for the partitions, typically a two dimensional rectangular or hexagonal grid of 'neurons.' Then the SOM algorithm finds prototype vectors representing the input microarray datasets, and at the same time provides continuous mapping to the neurons (Törönen et al. 1999). This process is also know as 'constrained topological mapping' since the high-dimensional observations can be mapped in the two coordinate system (Hastie et al. 2001). If a grid of centroids is imposed in the multi-dimensional space, then the centroids are allowed to drift towards collections of points. Completed centroids reflect clusters of genes demonstrating similar time course or behavior. In this process, SOMs may reveal conditions under which particular genes are regulated. Because regulatory genes appear to predict functional similarity, comparison with other genes in the same neuron may give valuable hints about possible functional roles of unknown genes (Törönen et al. 1999).

A variation of this method has been proposed as SOTA. Self Organizing Tree Algorithm (SOTA) is a binary tree topology approach based on SOM and growing cell structures (Herrero et al. 2001). It grows as a binary tree at different gene expression levels and at the same time reduces the dimensionality of a gene expression matrix (Figure 5). When the expression of less than 600 genes was analyzed the hierarchical clustering using UPGMA distances was faster than SOTA. However, the running time of SOTA when analyzing 5000 genes was three orders of magnitude faster than the hierarchical clustering UPGMA distances (Herrero et al. 2001).

Principal Component Analysis

Principal component analysis (PCA) (also called single value decomposition or Karhunen-Loève expansion) is an exploratory multivariate technique for determining the key properties in a multidimensional dataset. In this sense PCA

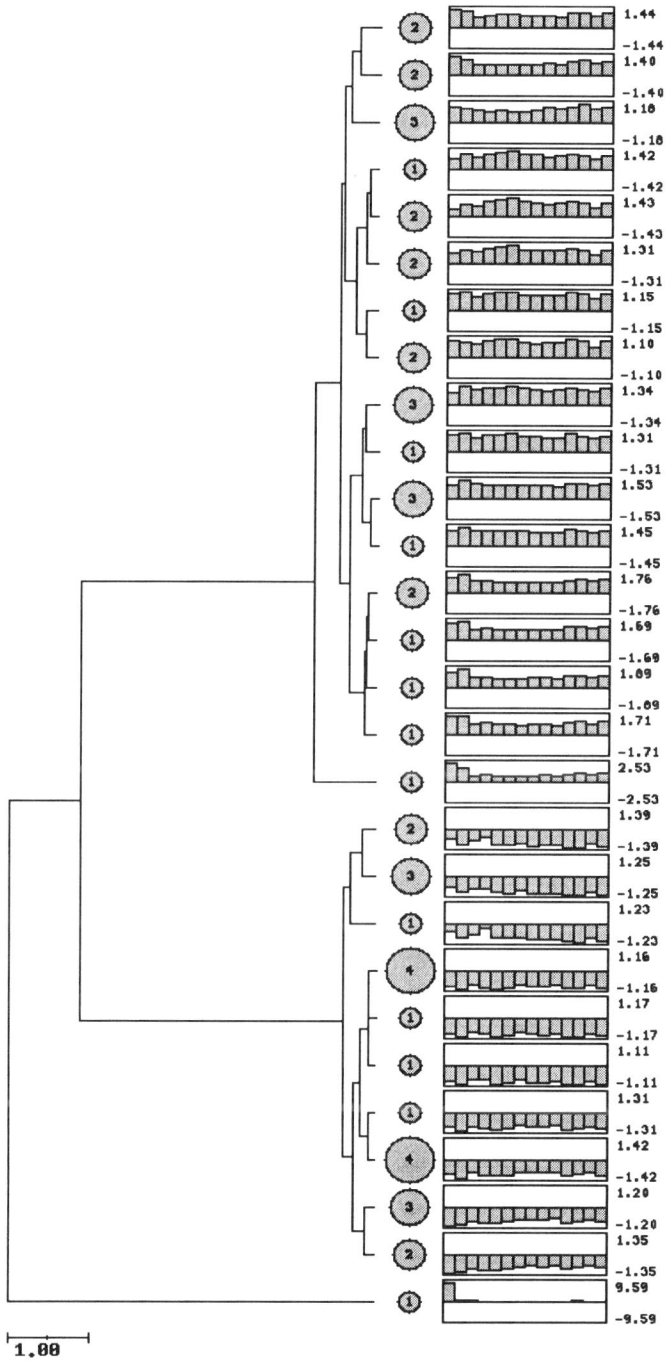

Figure 5. A dendrogram generated with SOTA-TREE using a Euclidean distance function and a single link pairwise distance matrix.

can be used effectively to compress microarray data and to find relevant features about the differences of diverse experiments. Since PCA is aimed at reducing the dimensionality of microarray data by trimming away data that is statistically irrelevant to the explanation of a particular biological process, this method has an additional benefit related with its ability to decrease information to the levels that can be visually inspected (Hilsenbeck et al. 1999; Raychaudhuri et al. 2000; Alter et al. 2000; Rifkin et al. 2000, Chapman et al. 2001; Quackenbush, 2001; Armstrong et al. 2002). This is because the graphical representations are restricted to one, two or three dimensions.

When using PCA to analyze microarray gene expression data, it is important to determine w hether the objective is to c ompare different m icroarray experiments or to compare different genes. This consideration has an important effect in the dimension of the data matrix and the computational cost. Using genes as variables means a far greater computational cost. The principal component analysis of a microarray gene expression matrix begins with the reduction of its dimensionality. This process is completed by finding new variables (r) accounting for much of the variance in the original matrix. For this purpose, the PCA transforms the microarrays into n-factors that subsequently are arranged by significance. When genes are variables, the analysis creates a set of 'principal gene components' or 'eigengenes' that reveal what genes explain the observed experimental responses. When experiments are the variables, the analysis creates a set of 'principal experiment components' or 'eigenarrays' that best explain the expression level of certain genes (Raychaudhuri et al. 2000; Gilbert et al. 2000; Alter et al. 2000; Mendez et al. 2002, Peterson, 2003). The k-means and SOM algorithms consider that each gene has a discrete membership to a particular cluster. Using PCA, each gene is a member of every component (group), and the degree of membership is reflected in its eigen vector.

Correspondence Analysis

Correspondence analysis (CA) is similar to PCA and can be used to reveal the principal axes of high dimensional spaces produced by microarray data. However, unlike PCA, correspondence analysis is able to account for genes in hybridization-dimensional space and gene-dimensional space at the same time. This p roduces associations both within a nd between t hese two variables (Fellenberg et al. 2001). The process starts from a table of frequencies of two categorical variables and attempts to quantify columns (or rows) of the table in such a way that the between c olumns (between rows) sum of squares i s maximized.

Gene Shaving

The gene shaving method is a PCA block clustering technique proposed by Hastie et al. (2001a*). The algorithm uses an iterative procedure to identify subsets of highly correlated genes that vary greatly between samples. Gene shaving differs from hierarchical clustering and *k*-means methods in the sense that for the gene shaving process genes may belong to more than one cluster. One of the characteristics of this approach involves the ability of the algorithm to be configured as a supervised microarray data analysis technique.

METHODS FOR VALIDATING UNSUPERVISED ANALYSIS

Unsupervised clustering methods have been successfully applied to microarray gene expression profiling. However, by their nature, they are exploratory rather than confirmatory analyses of microarray data. Since the effects of variation and uncertainties in measured gene expression levels across samples and experiments have been ignored in the literature, it can be argued that many unsupervised algorithms provide limited information about the accuracy of the results. This aspect is complicated because in many cases the data is derived from microarray experiments with poor quality and/or unknown variability. Since the main limitations of unsupervised methods are related to their sensitivity to outliers and missing data, it is necessary to apply supplementary validation methods attempting to provide some confidence that the formed clusters are real and not due to random chance. In this chapter some of the most important validation techniques are highlighted.

Bootstrapping

Bootstrapping is a 'resampling' based technique widely used to study the variability in microarray experiments (Zhang and Zhao, 2000; Scherf et al. 2000; Kerr and Churchill, 2001; Ambroise and McLachlan, 2002; Brody et al. 2002). This technique can provide some estimate of statistical accuracy in situations where these estimates are difficult or impossible to derive analytically (Efron and Tibshirani, 1993). The basic idea involves sampling the gene expression matrix to produce a random set of samples. Each of these samples is known as a bootstrap and each provides an estimate of the parameter of interest. Repeating the sampling a large number of times provides information on the variability of the clustering process (Hastie et al. 2001*). When accuracy is high, the bootstrap estimates will be more like the original estimates and the bootstrap clustering will be more like the original clustering (Kerr and Churchill, 2001). However, some limitations of this approach include its dependence on experimental replication and the computational intensity required to perform the validation process.

Jackknifing

Jackknife is a cluster validation procedure that can be used to reduce bias in the estimation of functional categories using DNA microarray data. The principle behind this method is to omit each sample member in turn from the data thereby generating n samples each of size $n-1$. The parameter of interest can be estimated from each sub-sample giving a series of estimates. The jackknife approach has been used to estimate the validity of hierarchical and k-mean algorithms using the single, average, complete and correlation distance functions (Yeung et al. 2001). This validation method suggests that single link has poor performance and that k-means with average link (see Pairwise distance metrics) achieves results similar to k-means with random initialization.

Other Validation Methods

Using both simulated and gene expression data, Dudoit and Frydlyan (2002*) compared a new, prediction-based resampling method (Clest). Clest estimates cluster number based on the reproducibility of cluster assignments with eight different methods (gap, gapPC, hart, sl, ch, kl and sil). According to the report, Clest was more accurate and robust than the other approaches. Additional methods for the validation of unsupervised methods include using Silhouette, Dunn's based index, and Davis-Boulding index (Bolshakova and Azuaje, 2002; Azuaje and Bolshakova, 2002).

Considerations when Using Unsupervised Techniques

Confusion has emerged when comparing and contrasting different clustering algorithms because of the different contexts in which unsupervised methods are used, and because research efforts have been concentrated on scalability and applicability to different data formats. At this point in time, there is a limited number of studies validating or comparing the results of different unsupervised algorithms (Yeung et al. 2001, Duogherty et al. 2002; Lanagrebe et al. 2002). The reader should be aware that comparing unsupervised algorithms entails several considerations: 1) the mathematical principles involved 2) different distance functions 3) algorithm processing time 4) quality of the graphical representation of the results. Since gene expression patterns can be derived from the analysis of multiple experiments, unsupervised methods are more successful if there is already a wealth of knowledge about the biological process in question. When this knowledge is limited, unsupervised methods may not be the most useful approach. It is difficult to determine the accuracy of the unsupervised analysis process unless methods for estimating the confidence of the results are applied and there is a biological basis for the interpretation of the results.

When clustering microarray data, it is implied that genes with similar patterns or expression levels belong to the same functional group or are co-regulated. However, one of the greatest challenges in analyzing microarray data involves the capture of complex relationships among genes that, while having similar expression levels, participate in different metabolic, regulatory or signaling pathways. This situation is complicated by the fact that there is no unsupervised method universally applicable in uncovering the variety of structures present in different microarray datasets. Critics of hierarchical clustering argue that it is an unstructured approach since it imposes a hierarchical structure whether or not this truly exists in a particular microarray dataset. As result, the expression representing a cluster may no longer represent genes with the same biological function (Tamayo et al. 1999, Herrero et al. 2001; Quackenbush, 2001).

The quality of a clustering result depends on both the similarity function and its computational implementation. Therefore, the interpretation of dendrograms should be treated with caution because in analyzing the same data, different distance functions will lead to changes in the density, distribution, and shapes of the dendrograms (Figure 6). In addition, while dendrograms are best suited to the description of true hierarchical descent (e.g., taxonomical differences across species), they are not designed to reflect the multiple ways in which genes interact.

While k-means clustering can be easily scaled to analyze large datasets, a major difficulty with this algorithm is its sensitivity to the selection of the initial number of clusters. Both k-means and SOM algorithms require a priori specification of the number of clusters. Thus, each gene is allocated to only one of the pre-defined clusters. In many cases, gene expressions values that do not fit a particular group are assigned to a cluster that is not necessarily biologically relevant. This results in a reduction in the effective scope of the analysis. The k-means clustering is very sensitive to outliers and noise because a small number of such data can substantially influence both seed points and the interpretation of results. If the computational implementation of k-means selects seed points

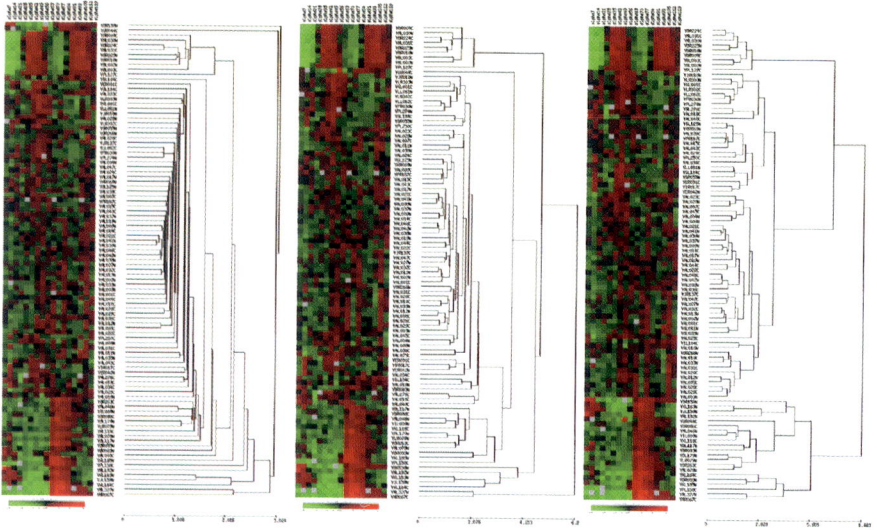

Figure 6. A microarray gene expression matrix clustered with EPCLUST using hierarchical clustering with (a) single, (b) average and (c) complete link pairwise distance matrix.

at random, the cluster formation is arbitrary and the obtained memberships might be difficult to replicate.

Similar to k-means and SOM, a main consideration in the use of PCA involves determining the number of components needed to represent gene expression data in groups that allow for a meaningful interpretation. The quality of the results using PCA on both real and synthetic data suggests that PCA is not necessarily a better performer than other clustering methods and that there is no obvious relationship between the cluster quality and the number of principal components used (Yeung and Ruzzo, 2001). This method also requires a gene expression matrix without missing values (Troyanskaya et al. 2001).

In general, clustering massive, high-dimensionality genomic datasets demands both computational power and intuitive understanding of the problem in hand (Table 1). Most clustering algorithms do not work optimally in high dimensional spaces as normally found in large microarray datasets. This problem has been referred as the 'dimensionality curse' (Aggarwal, 2002). When the data size increases, the computational load also becomes a problem, especially if some of the distance functions and the validation methods are already computationally expensive. Consequently, there is a trade-off between the speed of a clustering algorithm and its quality. A good clustering method produces high quality clusters with high intra-class similarity and low inter-class similarity. However, an algorithm with such characteristics may be unable to handle large microarray datasets, making it suitable only for the analysis of a gene expression matrix with limited number of experiments.

SUPERVISED MICROARRAY DATA ANALYSIS

A more complex statistical approach for the analysis of microarray data consists in the incorporation of human direction within a particular computational implementation. This approach, known as supervised classification or machine learning, requires training sets previously assembled by an expert and uses them as models to classify the elements of a gene expression matrix. Supervised analysis methods are often referred as discriminatory or multivariate classification techniques (Vapnik, 1995*; Baldi and Brunak, 1998*, Hastie et al. 2001*; Baldi and Brunak, 2001*). A supervised analysis of microarray data requires the construction of a *classifier* capable to assign a *class label* to instances described as *attributes* (Barash and F riedman, 2002). Because of the mathematical foundations and the incorporation of domain knowledge, supervised learning methods are more robust than unsupervised approaches when presented with complex data variables (features) such as those found in large, noisy and incomplete microarray datasets.

Unlike traditional statistical data analysis, which is usually aimed at the estimation of population parameters, the emphasis in machine learning is to develop a classifier with a good generalization performance. This is achieved by assessing

Table 1. Comparison of different unsupervised microarray data analysis methods.

Characteristic	Hierarchical Clustering	K-means	SOM	PCA
Dendrogram	Yes	No	No ‡	No
Cluster formation	After analysis	Fixed beforehand	Fixed ‡ beforehand	Fixed beforehand
Speed	Slow†	Fast	Fast	Slow
Outliers	Sensitive	Very Sensitive	Robust	Sensitive
Unequal cluster size effect	No†	Sensitive	Very sensitive	Very Sensitive
Affected by missing values	No	Very Sensitive	Robust	Very Sensitive

† Depending on the selected pairwise distance.
‡ See Figure 4.

its performance using a training dataset were the class label attribute is known. After the evaluation of misclassification (rate of false positives and false negatives belonging to a particular class label) the classifier can be reapplied to microarray gene expression matrices where the class label attribute (functional category) is unknown. This can lead to establish the function of unknown genes (Hvidsten et al. 2001). In addition, supervised methods can be used to establish the relevant transcriptional profiles that can allow the differentiation of different microarrays.

Supervised microarray data analysis process typically involves four phases:
- The construction of a classifier or model
- A learning phase
- A testing phase
- An application phase

The inputs and stages required for the construction of a classifier are: 1) a set of genes (training set), 2) the functional classes to which these genes belong (dependent variable), and 3) a set of variables describing different characteristics of the genes (independent variable). In the learning phase, training data is analyzed by a classification algorithm. Each gene expression value of a training dataset is a training sample. A class label attribute is identified, whose values are used to label the training dataset. In the testing phase, the test data is used to assess the accuracy of classifier. Each gene expression value is randomly selected from the test dataset with the additional condition that no training gene expression value should be in both the learning set and the test set. In the application phase, the classifier predicts the class label of the unknown gene expression values that cannot be ascribed to the class label attribute. There are numerous classification methods and hybrids between unsupervised and supervised approaches. In this section we present some of the most relevant methodologies.

Linear Discriminant Analysis

A linear discriminant analysis estimates whether or not a set of variables are capable of distinguishing, or discriminating, between two or more groups of individuals. Under these circumstances, a dataset is said to be "linearly separable" if a linear discriminant function can differentiate it without error. Linear or Fisher's discriminant analysis is a multivariate technique concerned with separating distinct sets of objects (or observations) and allocating new objects (observations) to previously defined groups. The motivation behind Fisher's discriminant analysis is to obtain a reasonable representation of the gene expression matrix with few linear combinations of the observations (Bø and Jonassen, 2002). Consequently, this approach can be used for the classification of either gene expression values or microarrays.

The basic assumption of this approach is to seek a linear transformation, Z of the gene expression measurements

$$Z = a_1x_1 + ... + a_qx_q$$

such that the ratio of the between-group variance of Z to its within group is maximized. The solution coefficient $\mathbf{a}' = [a_1,...,a_q]$ is

$$a = S^{-1}(\bar{x}_1 - \bar{x}_2)$$

where S is the pooled within groups *variable-covariance matrix* of the two groups, and $\bar{x}_1$ and $\bar{x}_2$ are the group mean vectors. However, limited data and a large number of genes may result in unstable matrices of between- and within-groups sum of squares, and may provide poor estimates of the corresponding population (Dudoit et al., 2000).

Decision Trees

A decision tree is flowchart-like structure representing alternatives of a decision making process (Figure 7). These methods are attractive in the analysis of microarray data because the classifier can represent symbolic rules that can be understood more easily by humans (Dubitzky et al. 2001). Decision trees are most commonly used for classification, but can also be used for predicting missing values. Since the paths traced from root to leaf hold the functional class prediction for a particular gene expression profile, the function of unknown genes can be classified by testing them against the tree.

The construction of decision trees consists of two phases: tree building and pruning. In the building phase, the training set is recursively partitioned until all the records have the same class (also known as pure tree). The decision to declare a leaf or continue to split is based with in attribute with the highest information gain. Once an attribute

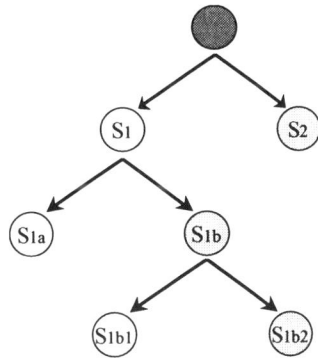

Figure 7. Each node denotes a test on an attribute, each branch represents an outcome (S=solution) of the test and the leaf nodes represent classes or class distributions.

has been used, it is not considered in descendent nodes. The decision tree algorithm stops when all samples for a given node belong to the same class or when there are no remaining attributes. The pruning phase is used to improve the classification accuracy of the tree. A number of different strategies including MDL pruning, cost-complexity pruning and pessimistic pruning can be use (Garofalis et al. 2001*). Computational implementations of this analysis method include C5.0, CART, SLIQ, PUBLIC, and BOAT (Garofalis et al. 2001).

NEAREST NEIGHBORS

The k-nearest neighbors were proposed by Fix and Hodges (1951) as non-parametric estimates of maximum likelihood of a discriminant analysis. Although the nearest-neighbor technique can be used as an unsupervised method, it is commonly used in a supervised fashion. This method is preferred because it is easy to implement computationally, has high classification accuracy, and is not affected greatly by missing and noisy microarray datasets (Golub et al. 1999; Dudoit et al. 2002; Hand et al, 2001; Li et al. 2001). The k-nearest neighbors classifier is *memory based* (it keeps the training set in memory) and does not require a model to fit. The classification can be decided according to a simple majority decision among the most similar or 'nearest' training set examples. These estimates are obtained by first reducing the dimension of the feature space using a distance function (e.g. Euclidean distance). The confidence level for each class is calculated by k_i/k, where k_i is the number of training samples of the neighborhood (Qi, 2002). According to Qi (2002), the choice of k should be an odd number that is less than the square root of the number of samples in the microarray dataset. The mathematical process of this technique can be summarized as follows:

$$Y = \frac{1}{k} \sum_{x_i \in N_k(x)} y_i$$

where $N_k(x)$ is the neighborhood of x defined by the k closest points in the training sample. This classification method computes the similarity between a test object (gene expression level) and all the objects of a training set. It then selects the k most similar training set objects, and determines the class of the test object based on the classes of these k-nearest neighbors by majority vote (Dudoit et al., 2000; Kuramochi and Karypis, 2001; Hand et al. 2001).

The k-nearest neighbor classification method has been successfully applied in microarray data. Golub et al. (1999) used this technique to construct class predictors capable to categorize leukemia samples. Evaluation of k-nearest neighbors against discrimination methods (Fisher linear discriminate and classification trees) suggests that k-nearest neighbors performs better (reduced error) in the classification of cancer tumors (Dudoit et al. 2000). The combination of k-nearest neighbors with genetic algorithms to classify gene expression data derived from colon cancer (Alon et al. 1999) and lymphoma microarray experiments (Alizadeh et al 2000) have been reported by Li et al. (2001). An interesting aspect of this research was that genetic algorithms improved the classification performance of *k*-nearest neighbors and allowed the estimation of the relative predictive importance of the genes used in the classification process. While the method *k*-nearest neighbors alone (Golub et al 1999) classified correctly 29 of 34 samples, *k*-nearest neighbors with genetic algorithms correctly classified 33 of 34 samples. More recently the Behavior-Knowledge Space method combined with the *k*-nearest neighbor approach resulted in a 25% improvement in accuracy for the classification of microarray data of cancer tumor types than both methods alone (Qi, 2002).

SUPPORT VECTOR MACHINES

The support vector machine (SVM) classification method was developed by Vapnik (1995) and belongs to a larger class of algorithms known as kernel methods. The mathematical advantages of SVM include flexibility in choosing a distance function, ability to detect outliers, and capacity to deal with high dimensional datasets. As applied to gene expression analysis, SVMs begin with a set of genes that have a common function (same functional class label attribute) and with genes that are known not to be members of the specified functional class. This training set is assembled from the domain knowledge, and the SVM algorithm will use it and learn to discriminate gene expression values and assign them as members and non-members of particular functional category (Burges 1998; Mukherjee 1999; Moler et al. 2000, Brown et al. 2000; Guyon et al. 2000; Furey et al. 2001; Model et al. 2001; Su et al. 2001; Ambroise and McLachlan, 2002; Pavlidis et al. 2002).

To understand the classification process of a SVM let us assume that it is necessary to classify genes of an expression matrix into two different functional classes (which are labeled with + or - if they belong or not to such categories). A training set is assembled with gene expression levels $\{x_1, ... x_k\}$ with known functional class labels $\{y_1, ... y_k\}$. These patterns are used to build a decision function (or discriminant function) $D(x)$, that is scaled to the size of an input dataset. New gene expression levels are classified according to the sign of the decision function:

$D(x) > 0$ then $x \in$ class (+)
$D(x) < 0$ then $x \in$ class (-)
$D(x) = 0$ is decision boundary

SVMs are inherently binary (Figure 8), however, it is possible to combine different binary classifiers and separate multiple classes of genes. One solution to the

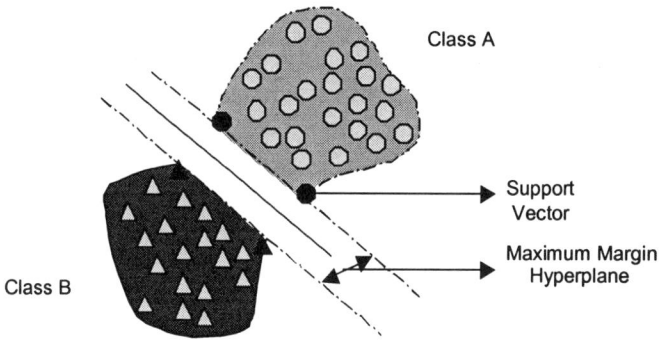

Figure 8. Elements of a support vector machine.

inseparability problem is to map the data into a higher dimensionality space and define a separating hyper-plane. SVM maps the input vectors x into a high-dimensional feature space Z through non-linear mapping chosen *a priori*. In this space, an optimal separating hyperplane (support vector) is constructed (Vapnik, 1995). A *kernel function* defines the feature in which the training set examples will be classified (Brown et al. 2000). A linear kernel is sufficient to classify microarray data. If more detailed classification of microarray data is desired, a gaussian kernel can be used (Muejerke et al. 2002).

The best decision hyperplane is the one with the maximum margin (Kuramochi and Karypis, 2001; Pavlidis et al. 2002). A SVM can locate a separating hyperplane in the feature space and classify points in the space without ever representing the space explicitly (Brown et al. 1999). The quality of a decision hyperplane is determined by the distance (referred as margin) between two hyperplanes that are parallel to the decision hyperplane and touch the closest data points of each class (Figure 8). The location of the hyperplane in the feature space depends on weights in the training data.

For microarray data analysis, SVM has proven to be superior to other classification methods including Parzen windows, Fisher's linear discriminating, decision trees, and k-nearest neighbors in both accuracy and robustness (Brown et al. 2000; Ramaswamy et al. 2001; Pavlidis et al. 2002). An additional advantage of the SVM classifier is that it can be reapplied to the original training data to identify outliers, as well as to find false positives and false negatives that were previously unrecognized (Brown et al. 2000). Also, SVM can be retrained on the reduced feature set space until the feature set is empty. Such successive feature removal is called backward elimination (Model et al. 2001; Ambroise and McLachlan, 2002).

Artificial Neural Networks

Artificial neural networks have attracted considerable attention in recent years (Reed and Marks, 1998*; Hand et al. 2001). These algorithms are based on principles observed in functioning brain neurons. The response of each neuron is modeled by a non-linear function of its inputs and an internal state. Under these assumptions, many simple neurons linked together can be used to compute complex problems. By varying the connections and the connecting weights, it is possible to achieve a proper differentiation of gene categories. In its simplest form, the network topology can be represented as:

$$y = f\left(\sum_k w_k x_k\right)$$

where x_k are the output signals of other nodes or external system outputs, w_k are the weights of connecting links, and $f(o)$ is a simple nonlinear function. The unit computes a weighted linear combination of its inputs and passes this through the nonlinear f to produce a scalar output, which is used to make a logical decision. The strength of this method is that it does not require genes to be exclusively associated with a single sample type. This approach has been successfully to classify and predict cancers types (Khan et al. 2001) and to distinguish among subtypes of neoplastic colorectal lesions (Selaru et al. 2002).

Bayesian Classifiers

Bayesian methods can incorporate external information into the analysis process so that the posterior distribution is conditional on the total information available. Suppose that we have different gene functional categories $X = \{X_1,...,X_n\}$. In addition, we have a microarray dataset $D = \{x_1,...,x_n\}$, which is a random sample from an unknown probability distribution for X. We can assume that each microarray data point x in D consists of an observation of all the variables in X. A Bayesian classifier will use the Bayes' theorem to predict class membership as a conditional probability that a given microarray dataset falls into a particular functional class. In its simplest form, the Bayes' theorem can be written in terms of the conditional probability as:

$$\Pr(Bj \mid A) = \frac{\Pr(A \mid Bj)\Pr(Bj)}{\sum_{j=1}^{k} \Pr(A \mid Bj)\Pr(Bj)}$$

where $Pr(A|B_j)$ denotes the probability of event A conditional on event B_j and

$B_1, B_2,..., B_k$ are mutually exclusive and exhaustive. The theorem gives the probability P of B_j when A is known to have occurred. The quantity $Pr(B_j)$ is termed the *prior probability* and $Pr(B|A)$ the *posterior probability*. When both the prior and posterior probabilities have the same functional form, the prior is a *conjugate prior* (Baldi and Long et al. 2001). Baldi and Long (2001) reported that this approach overcomes the limitations in the analysis of gene expression matrices derived from experiments with low replications.

Naïve Bayesian Classifiers

Naïve Bayesian classifiers can be used to compute a point estimate of the parameters, or to make probability statements as credibility intervals, conditional on the available information (Ramoni and Sebastiani, 1998). These methods are characterized by the assumption that by computing its posterior probability a hypothesis or model can be assessed. The assumptions made by this model give its 'naive' name. This method performs well in classification problems with large number of attributes. An additional advantage is that the acquisition of new data does not require re-processing data considered so far (Ramoni and Sebastiani, 1998; Thomas et al. 2001). When coupled with the conditional independence of the data given the parameters, the inference of the procedure can be processed as a whole (*in batch*), or one at a time (sequentially).

The naïve Bayesian classification predictor variables $X_1,...,X_k$ are assumed to be independent of each other when conditioned on the class variable C. The model is constructed by the joint probability as follows:

$$P(x,c) = P(C=c)\prod_{i=1}^{k} P(X_i = x_i \mid C=c)$$

Before incorporating any data, this formula assumes that all classes are equally probable (i.e. the probability that a chemical will belong to a certain class is the same for each class) and that within each class, the gene expression values are equally probable (Thomas et al. 2001). Under these assumptions the Bayesian probability theory can be use to calculate the predictive distribution of a class c given x and the gene expression data D by:

$$P(c \mid x, D) = \frac{P(c,x \mid D)}{P(x \mid D)}$$

where the numerator is calculated as:

$$P(c,x \mid D) = \frac{t_c + 1}{N_t + NC}\prod_{i=1}^{k}\frac{f_{cxi}+1}{F_c + V_{xi}}$$

where t_c is the number of treatments in class c, N is the total number of treatments, NC is the total number of classes, f_{cxi} is the number of cases in class c having a value equal to x_i, F_c is the number of treatments in class c, and V_{xi} is the number of possible values of x_i. The denominator is the same for all c and can be calculated as:

$$P(x = (x \mid D) = \sum P(c', x \mid D)$$

The results of this conditional predictive distribution can then be used to classify the data vector (Keller et al. 2000; Thomas et al. 2001).

Bayesian Belief Networks

A Bayesian network is a graph-based model of joint multivariate probability distributions that can be useful in capturing the conditional independence between microarray experiments. Also, they have the ability to disclose probabilistic relationships among genes (Hwang et al. 2002). The Bayesian network approach can be used to infer if the expression level of a particular gene depends on the experimental condition and if this dependence direct, or indirect (Friedman et al. 2000). To answer these questions, this method uses domain information to build a conditional probability table and to represent dependencies between data and parameters associated with the sampling model (Ramoni and Sebastiani, 1998). Another characteristic of this technique is robustness when handling missing and noisy microarray data (Friedman et al. 2000).

For a given microarray dataset D, it is important to derive the posterior probability. The probability space is defined as the set of possible assignments to the set of attributes $A_1,...,A_n$ of a relation **R**. Bayesian networks can compactly represent a joint distribution over $A_1,...,A_n$ by utilizing a structure that captures conditional independences among attributes:

$$P(X) = \prod_{i=1}^{n} P(X_i \mid Pa_i),$$

where Pa_i is the set of parents of X_i in the network structure, $P(X_i|Pa_i)$ is the local probability distribution related to node X_i (Hwang et al. 2002).

A Bayesian network consists of two components. The first component is a directed acyclic graph whose nodes correspond to attributes $A_1,...,A_n$. The edges in the graph denote a direct dependence on the attribute A_i. The graphical structure encodes a set of conditional independence assumptions: each node A_i is conditionally independent of its non-descendants. The second component of the Bayesian network describes the statistical relationship between each node and its parents. It consists of a conditional probability distribution (CPD; $P(A_i|$ Parents (A_i)) for each attribute, which specifies the distribution over the values of A_i given a possible assignment of values to its parents.

METHODS TO IMPROVE CLASSIFIER PERFORMANCE

As presented in this chapter, supervised methods appear robust and flexible for the analysis of microarray data since domain knowledge is incorporated in the classification of the functional categories of genes. As with unsupervised techniques, the classification error of supervised methods depends on the particular characteristics of a gene expression matrix, the training set, and the computational implementation. To reduce misclassification, it is possible to apply statistical and computational techniques capable to provide a better understanding of the biological process. Re-sampling methods such as bagging and boosting have been applied successfully to improve the prediction accuracy of supervised learning algorithms (Dudoit et al. 2000; Skurichina and Dubin, 2002).

Bagging

Bootstrap aggregating (bagging) was proposed by Breiman (1994*) and it is based on bootstrapping concepts. The algorithm uses bootstrap samples to build the classifiers. Each bootstrap sample of m instances is formed by sampling m from the training set with replacement. The usefulness of bagging depends upon the small sample size properties of the classifier and seems to be useful in linear discriminators such as Fisher's linear discriminator or decision trees (Skurichina and Duin, 1998; Dudoit et al. 2000). In other words, bagging is not useful for classifiers whose generalization error increases with an increase in the training sample size (this referred as classifiers with a decreasing learning curve; Skurichina and Duin, 1998). When applying bagging to classifiers like nearest neighbors and Fisher's linear discriminator, it is reasonable to use no more than 20-100 bootstrap replications (Skurichina and Duin, 1998).

Boosting

Boosting was proposed by Freund and Schapire (1997) as a technique combining weak classifiers in order to get a classification rule with better performance. In the literature the term 'weak classifier' refers to badly performing classifiers, unstable classifiers, classifiers with a low complexity, or classifiers assuming certain models that are not necessarily true (Skurichina and Duin, 2002). In the process of boosting, both training and classifiers are obtained randomly and independently (contrarily to bagging). Boosting uses a weighted voting of the training set where the weights are increased for those cases that are most often misclassified (Dudoit et al., 2000; Skurichina and Duin, 2002). While boosting performs well when large training datasets are provided, as the number of data points used in the training set decreases, the performance of the boosting is also negatively affected (Skurichina and Duin, 2002).

Considerations When Using Supervised Techniques

Before using machine learning techniques, it is critical to identify the exact problem to be solved and select a proper training dataset. This selection is as important as the choice of the classification technique. In some cases, assembling the training set requires considerable effort. Once available, this particular data will be applicable and useful for a particular microarray gene expression matrix. In some cases, a classifier built using small training datasets has difficulties adjusting the parameters and might be biased with a high probability of misclassification. In this regard, the poor performance of Fisher linear discriminator is due to the covariance matrices within small training sets and large gene expression matrices. A desirable characteristic of a classifier is good generalization performance, which on one hand is complex enough to capture the essential features within a gene expression matrix, and on the other hand avoids over-fitting the data (Model et al. 2001). However, the learning community is aware that the more variables one models, the more difficult the classification and more computationally intensive the task becomes. This is because the space of the models to be searched increases exponentially with the number of parameters of the model, and therefore with the number of variables (D'haeseleer 2001).

Methods like nearest neighbors do not build the model for classification and the training is stored in memory. This is why k-nearest neighbors are considered 'lazy classifiers' and sometimes perform well under large memory requirements. Since it must maintain all training data in memory, nearest neighbors do not identify the genes most useful for class distinctions. While support vector machines have been reported to perform better that other classifiers including decision trees and k-nearest neighbors, a limitation of SVM in regard to these methods is that it requires a large training dataset (Dudoit et al. 2002). For some datasets, SVM may not achieve a proper separation since the kernel function may be improperly defined, or there may be a problem in the training set. It is also difficult to choose the kernel function, parameters and penalties (Quackenbush, 2001). The higher dimensionality space incurs both computational and learning-theoretic costs.

The complexity of computing the conditional probability values can become prohibitive for most of the microarray datasets with a large attribute space. Bayesian models become infeasible, and heuristic methods have to be used to limit the search process. In some cases, exact Bayesian analysis requires computing the posterior distribution; this approach becomes infeasible as the proportion of missing data increases. Bayesian Belief Networks require build time and domain knowledge whereas the Naïve approach looses accuracy if the assumption is not valid.

Bagging and boosting are designed to be applied to decision tree algorithms and could be beneficial for k-nearest neighbors. However, the reader should be aware that bagging and boosting techniques are dependent on, and sensitive to, changes in the training datasets. When the training set is large bagging is useless (Skurichina and Duin, 2002*). Both bagging and boosting are unable to take into account the heterogeneity of the distribution of microarray data. Although bagging and boosting techniques improve classification performance, their use has been not widely made in microarray datasets. Using other datasets it has been demonstrated that boosting gives a better performance than bagging (Skurichina and Duin, 2002).

GENETIC AND BIOCHEMICAL NETWORKS

To counteract environmental perturbation, biological entities self-organize as complex systems in which their gene interactions form properties capable of change at different time scales. Functionally diverse sets of genes interact selectively to produce coherent complex patterns. The degree of gene interconnection within a system defines it as an entity and determines its dynamics and interactions. The components of the biological system can have a number of homogeneous (*e.g.,* single gene copies) and heterogeneous (*e.g.,* gene duplications, mutations) components. Each component resides in a certain state determined by the system's current state as well as the current states of its neighbors. However, these interaction mechanisms are apparently based on preferential attachment (Ravasz et al. 2002, Koonin et al. 2002, Kitano 2002, Bose et al. 2002). The complex communication seems extremely robust. For example, the failure of a network element rarely produces a breakdown in the complete network.

Various approaches for the computational modeling of genetic and biochemical networks have been compiled by Bower and Bolouri (2001*) and Kitano (2002a*). These mathematical techniques use gene expression data and represent biological processes as Boolean, graphical, continuous and free scale networks models (Liang et al. 1998; Akutsu et al. 2000; Butte and Kohane, 2000; Butte et al. 2000; D'haeseleer et al. 2000; Silvescu and Hanovar, 2001). Mathematical analysis of gene expression data with the aim to model gene expression behavior and predict part of the cellular process provides an extraordinary opportunity for the interpretation of the biological systems. However, most of these approaches are still in their infancy and to use them, it is necessary to take into consideration that the transition from transcriptome to proteome, and from proteome to complex systems, demands large amounts of data and computational power. It also can be argued that some of the issues restricting the construction of biochemical networks are: 1) limited definition of the structure of the system (e.g. biological function of a gene), 2) difficulty specifying a mathematical model to compute a system's dynamics without a definition of the system structure, 3) inherent noise of biological systems make it likely that biochemical network components at the transcriptional level are different at the proteomic level.

ADDITIONAL METHODS FOR MICROARRAY DATA ANALYSIS

Phylogenomic Microarray Data Analysis

This approach argues that a better inference of the biological process using DNA microarrays can be achieved by incorporating evolutionary relationships among organisms. The main biological considerations represented in the mathematical model are: 1) genetic constitution of an organism (K) has genes belonging to two different groups (node and fluctuating genes), 2) most nodes/genes are conserved across species, 3) node genes can be represented as Boolean, 4) fluctuating genes are represented as very highly expressed, highly expressed, very highly repressed, or highly repressed. Then each gene expression data point is assigned to a corresponding signaling, metabolic, regulatory, or unknown functional class (Valdivia-Granda et al. 2002). Then Boolean contributions of *node genes (genes X)* can form *input switches* modulating the level of expression of *fluctuating genes (genes Y)*. Then $K = \{X, Y\} = n$. The maximum number of *input switches (S)* required for the computing of a pathway each represents individual assumptions. This function is based on the number of occurrences of the *node genes* across E. Rules (R) are derived from all possible variations of the X, and its consequences in Y. Then S is constructed from the independent assumptions underlying Y, in which consequences 1) are needed (AND), 2) consequences are not (NOT) needed, or 3) similar consequences are considered (OR) in E. Since the genomic complexity is large, we use the information of K_l to determine S in K_n to derive rules for the phylogenomic classification.

Emerging Approaches

In the last three years an increasing number of papers are reporting new methods or variations of classical statistical and computational approaches that can be used for the analysis of microarray data (Hang et al. 2002; Pan et al. 2002; Holter et al. 2002; Steffen et al. 2002). Statistical modeling has been proposed for searching large microarray datasets for genes that have a transcriptional response to a stimulus (Zhao et al. 2001). The single-pulse model makes accommodations for the systematic heterogeneity in expression levels. There are emerging trends to represent gene expression data as a biological pathway associated with particular tissues.

Techniques involving pattern recognition, language processing, and robot control combined with different optimization methods are likely to find their applications in the analysis of microarray data. There are already some potential applications that are not covered in chapter for lack of space. However, some links of potential interest for the reader are provided (Table 2).

Table 2. List of some research areas of potential interest in their application for microarray data analysis.

Applications	Website
Kernel methods	www.kernel-machines.org
Genetic Algorithms Archives	www.aic.nrl.navy.mil/galist
Artificial Life Resources	www.scs.carleton.ca/~csgs/resources/gaal.html
The R Project	www.r-project.org
Boosting research	www.boosting.org
Data mining group	www.dmg.org

MICROARRAY DATA VISUALIZATION

Visualization tools make possible human visual pattern recognition and can allow the identification of important regularities in the data without any further processing. The use of visualization tools not only can save hundreds of hours in the interpretation of microarray experiments but also can lead to more ambitious microarray data analysis projects. Using different graphical user interfaces (GUIs) it is possible to represent different views of the results after the application of a microarray data technique. These interfaces not only can provide visual components of the results, but also can facilitate the biological interpretation and representation of gene interactions and dynamics.

A number of display options are useful for gene expression data analysis. These include scatter plots, graph plots, histograms, bar charts, pathway diagrams and genomic positional maps. Generally, scatter plots are used to compare the distribution of mRNA levels in one sample versus another (i.e. when comparing control and treated samples). They are often used to assess data quality, since normally most genes are distributed around the central line. A graph plot, on the other hand, provides information about the overall behavior of each gene across many different samples. Scatter plots are a highly efficient way of detecting interesting trends among specific genes and identifying outliers. The pathway diagram allows researchers to look at groups of genes and their expression levels (indicated by color) within a specific biological context, frequently a biochemical network (Figure 9). Using genomic expression maps, it is possible to correlate behavior of specific groups of genes with their corresponding genomic regions. Such a tool is particularly useful when an organism's complete sequence information is available

MICROARRAY DATA STANDARDIZATION AND INTEGRATION

The Internet has become one of the most important forms of communication. With the push of a button, millions of transactions, documents, images, sound and other forms of data are shared across the globe. Microarray data is also part of this information sharing age. While some microarray data is exchanged among researchers in the public sector, the wealth of many pharmaceutical companies is represented by their private microarray databases that also need to be transferred from researcher to researcher within the same institution. As advances in the collection of data continue, many research laboratories are becoming overwhelmed with different formats of microarray gene expression

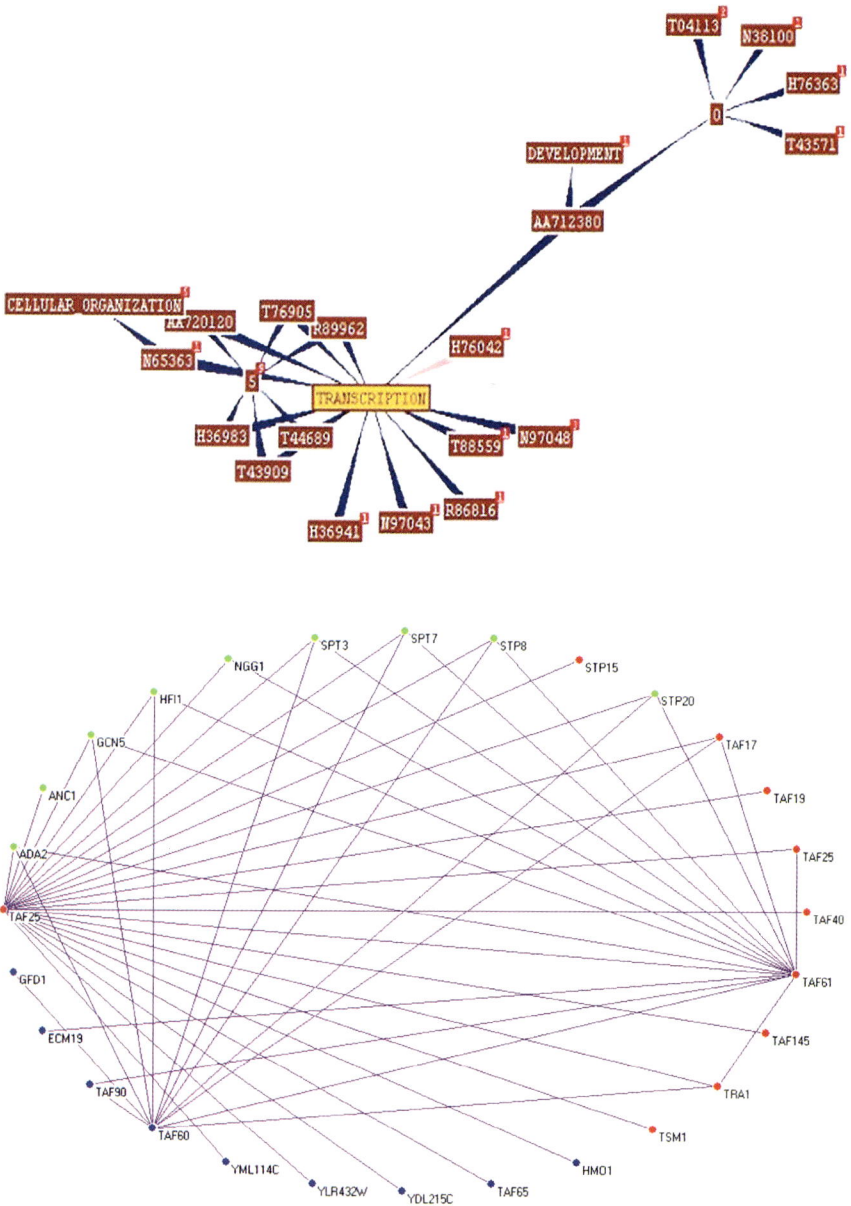

Figure 9. Representation of gene expression data as networks. **Top**- BISON-ARRAY using functional categories. **Bottom**- Network representation of gene expression using PAJEK.

information. This situation has generated an urgent need for new techniques and tools that can assist in transforming this data into useful knowledge.

Until recently there was a not a clear explanation of the meaning of the content of microarray data, and several factors contributed to the barrier in the widespread access of microarray data (Brazma et al. 2001). To deal properly with this situation, it became evident that five main issues needed to be considered, including: 1) agreement on the minimum information that a microarray experiment should contain, 2) development of standard nomenclature for an effective and seamless application of computational and statistical analysis, 3) development of universal and upgradeable languages (ontologies) allowing the standard description microarray experimental details, 4) development of computational tools for searching documents within a database, 5) use of standard reference probes for the comparison of experiments performed by different laboratories (Brazma et al. 2000). Although the best way to approach and solve these issues is still in active development, considerable progress has been made in the standardization of both experimental results and database models.

Minimum Information About a Microarray Experiment (MIAME)

It is widely agreed that there is a need for microarray data repositories (Brazma et al. 2001; Spellman et al. 2002). However, since the data generated from microarrays is dependent on the experimental process and technology used, it is relevant to define standard ways to describe biological samples and experimental conditions. This system can allow querying and perhaps comparing microarray data across laboratories, permitting the work of other researchers to be re-analyzed, validated and extended. Additionally, querying and retrieving microarray data stored in different repositories can generate new hypotheses while at the same time avoiding the repetition of some microarray experiments. This not only can save experimental resources but also more importantly, time. To facilitate the adoption of standards for the description and data representation of microarray experiments, the Microarray Gene Expression Database (MGED) society was founded in (1999). The work of this public and private effort resulted in two main initiatives that recently were unified to be come the Minimum Information about a Microarray Experiment (MIAME) standard. MIAME concentrates on recorded information about each microarray experiment and structures this data in a way that enables useful querying and analysis (Brazma et al. 2001). The MIAME standard is aimed at serving as a basic guideline for the development of microarray databases and data management software by describing the process of biological and experimental relevance within a microarray experiment.

In this regard, the MIAME format has six major sections trying to cover all relevant aspects of the transcriptional profiling using DNA arrays including: 1) experimental design 2) array design 3) samples 4) hybridization 5) measurements and 6) normalization.

Experiment Description

This section describes the experiment as a whole. An experiment is understood as a set of one or more hybridizations related to the same laboratory, experiment, or publication. Each experiment should have author and contact information including URL and a description of the experiment in a single sentence. The minimum information of this section includes a description of the following parts.

- Type of Experiment
- Experimental variables
- Samples used, extract preparation and labeling protocols
- Hybridization procedures and parameters
- Measurement data and specifications of data processing

Array Design

This section describes details of the physical design of the microarrays used in a particular experiment. The annotation in this section should include the details on the construction and an explanation of the common features of the array and design elements (e.g., each spot). Three levels of array design elements should be described: *feature* - the location on the array, *reporter* - the nucleotide sequence present in a particular location on the array, and *composite sequence* - a set of reporters used collectively to measure an expression of a particular gene, exon, or splice-variant. This section also should include at least the following information:

- A simple description of each microarray ID
- The array type definition
- Description of the physical content of each spotted element

Sample Annotation

This information can be useful in the interpretation of microarray data since it defines the biological nature of a primary sample (i.e., organism taxonomy, strain,

tissue, cell line) as well details of the treatment and its derived sample. The derived sample section is sub-divided into three parts:
- Biological treatment applied
- Technical details about the extraction of nucleic acids
- Technical details about cDNA labeling.

Hybridization Annotation

This section specifies the extraction, labeling and hybridization protocols. This includes choice of hybridization, solution, wash procedure, hybridization time, volume, temperature and description of the hybridizations experiments.

Measurement Annotation

This section describes the process converting a raw image to final data. Although MIAME does not require a particular image format, the data should be provided as raw scanner image accompanied by scanning information and relevant information about the laboratory protocols. This section should also provide information to understand how the image analysis was carried out (software used), all relevant parameters and definitions of the quantifications used.

Normalization Controls Annotation

This section includes information regarding the normalization protocols and control elements. This specification should include
- Normalization strategy
- Normalization and quality control algorithms
- Identification, localization and type of the control elements (spiking, housekeeping genes, etc)
- Hybridization extract preparation indicating how control samples are included in sample targets prior to hybridization

A more advanced specification also includes samples compared with a common reference sample to facilitate inferences about the relative expression changes between samples.

MICROARRAY GENE EXPRESSION MARKUP LANGUAGE (MAGE-ML)

An important component of the microarray process is the ability to manage the data so it can be distributed to different researchers interested in a particular experiment. To achieve this it is necessary to define how the data should be stored, indexed, queried, and retrieved. Computer scientists have developed programming languages and database management systems (DBMS) that provide the basic building blocks for unifying, exchanging and analyzing microarray data from unrelated repositories. The data stored in a database is structured according to a schema (database definition) specified in a data definition language (DDL) and manipulated with the operations specified by the data manipulation language (DML). Schema constructs provide the basis for the specification of implicit information in a microarray database. Their purpose is to define and describe a class of documents by constraining meaning, usage and relationships of their constituent parts. Schemas are more valuable if they represent a clear view of the component databases and they should not be affected by implementation considerations, such as limits on the number of classes, tables, or attributes (Kazic, 2 000). Constructing g lobal schemas f or heterogeneous d atabase management systems usually requires detecting semantic conflicts between schemas, which range from naming conflicts and inconsistencies to detecting identical entities of interest that are represented differently. Both DDL and DML are languages based on the data model that defines the semantics. Semantic data integration can be achieved using ontologies. Ontologies are formal descriptions of the concepts and entities for a domain of interest and are open to many types of relations between concepts. By having a set of well-defined terms it is possible develop well-defied relationships. Therefore to analyze the wealth of highly heterogeneous biological information, it is necessary to define common terms and meanings that will integrate data models, schemas, query languages and terminologies that will result in improved transactional capabilities. Automated methods of interconnection and a systematic knowledge management with proper representation, integration and exchange are being developed (Stoecker et al. 2002; Spellman et al. 2002).

Unlike sequence data, microarray data is more complex because the steps involved in the fabrication and experimental process contribute considerably to the analysis, interpretation and visualization of the microarray gene expression data. The description of these steps allows a better understar•ling of gene expression profiles since the sources of errors can be detected. Microarray Gene Expression Object Management (MAGE-OM) is a data-centric unified modeling language (UML) that contains 132 classes grouped into 17 packages, containing in total 123 attributes and 223 associations between classes reflecting the core requirements of MIAME (Spellman et al. 2002; Stoeckert et al. 2002). MAGE-ML is computationally implemented as a suite of software libraries under MAGE-STK. The function of this suite is to provide tools in different computer languages including MAGE-Java, MAGE-C++, and MAGE-Perl to annotate MAGE data (Spellman et al. 2002).

MAGE-OM allows users to describe protocols (as well experimental materials) using free-text descriptions. MAGE-ML is an XML representation of MAGE-OM based on the tag sets describing the contents of the microarray data rather than the format or presentation of the data itself. The eXtensible Markup Language (XML) is derived from the Standard Generalized Markup Language (SGML), the international standard for defining descriptions of the structure and content of different types of electronic documents. XML provides a framework for tagging structured data that can be used for specific tag sets, such as gene expression data, along with other attributes for each tag and how the tags can be nested. An XML element is either well-formed obeying the syntax of XML or XML valid conforming the logical structure defined by document type description (DTD) (Achard et al. 2001). The DTD is the classification system that defines the different types of information pieces in any document. DTD is the 'rule set' for any number of XML documents. Any Web page that indicates the DTD to which it conforms will instantly allow the user of an XML-enabled search engine to restrict query to that DTD-defined space. XML technology offers instrumentation for the index creation of genomic databases. The relevance of the XML framework is particularly useful for re-ordering and sharing information.

A major advantage of the MAGE-ML in an XML format is that, while it supports information from a variety of gene expression measurements, it does not impose any particular data analysis method. MAGE-ML also has advantages in the sense that many laboratories can verify the same microarray experiments with other more sensitive and accurate computational methodologies.

What to Share?

When microarray data is shared, researchers have the opportunity to analyze this information in different ways. Multiple and different evaluations of the same data not only will re-assure previous conclusions, but can give the opportunity to provide a more wise sense of what is happening at the biological level. Nevertheless, sharing data begs the question- what type of information should be shared? In this regard, at least four relevant levels of microarray data sharing can be recognize: 1) scanned images (raw data), 2) quantitative outputs from the image analysis, 3) derived measurements after normalization, 4) list of most relevant genes presented in a publication table (Brazma et al. 2001). Sharing microarray raw images is the most flexible option, allowing the testing of different procedures for data extraction, normalization, and analysis. However, this route poses practical limitations. Due to the size of the image files, there is no consensus as to whether image data storage would be cost-effective and whether this should be the task of public repositories or the primary authors (Brazma et al. 2001; Geschwind, 2001). Sharing the extracted (but not normalized) spot intensity and background data solves some of the practical limitations that are related with raw images. This level of sharing is well suited for many microarray data repositories. However, it requires the implementation of appropriate database management systems (DBMS) to allow the query and retrieval of this information. Another form of microarray data sharing before analysis is simply sharing the expression ratios of each spot. However, in this form, much information about the experiment, the normalization process and image analysis and calibration is lost. The last form of microarray data sharing consists of providing the list of genes that significantly differ between experimental samples. The publisher or the researcher normally accomplishes this as part of a publication or in many cases as supplementary information that is maintained.

MICROARRAY DATA REPOSITORIES

There are around 25 different microarray repositories and databases (Baxevanis, 2003). This section describes the main features of some microarray data repositories (see Table 3).

NCBI Gene Expression Omnibus (GEO)

NCBI Gene Expression Omnibus (GEO), is a public gene expression data repository created in response to the need to store high-throughput gene expression data. GEO has three main goals: 1) develop and maintain a durable repository for gene expression data, 2) balance requirements for standards and flexibility, 3) provide this service without limiting the development of new initiatives. The GEO has three accessioned data entities: platform, samples and series (Edgar et al. 2002). A platform is a list of probes that define which molecule may be detected. A sample describes the derivation of the set of molecules that are being probed, and utilizes platforms to generate molecular abundance data. A series organizes samples into meaningful datasets bound by a common attribute. Series accession numbers have a 'GSE' prefix (Edgar et al. 2002)

Table 3. Characteristics of some microarray databases.

Database Name	MIAME Supportive	Type of Implementation
AMAD	No	Flat files
ArrayDB	No	Sybase
ArrayExpress	Yes	MySQL, Oracle
BASE	Yes	MySQL and PHP
Dragon	No	MySQL
GeneX	Yes	PostgreSQL
GEO	Yes	Flat files
MAXBD	Yes	MySQL, Oracle
MEPD	No	IBM DB2
NOMAD	Yes	MySQL
SMD	Yes	PostgreSQL
StressDB	Yes	Oracle
yMGV	No	PostgreSQL and PHP4

KEGG/EXPRESSION Database

In the DBGET/LinkDB system, the EXPRESSION database the original information is stored and managed in a relational database and the flat file version for the microarray data is automatically generated from it as flat files. Each entry starts with a line for the ENTRY and ends with a line of triple slashes (// /). The In addition to the numerical and text information of each experiment, a graphical view of the array data is provided. Clusters of co-expressed genes are also clustered in the pathways or in the chromosomal locations by mapping to the KEGG/PATHWAY and KEGG/GENOME databases. In addition to the numerical and text information of each experiment, a graphical view of the array data is provided. This particular view is created by taking the ratios of the two channels, namely Cy5- and Cy3-labeled spots, and is useful for identifying clusters of co-expressed genes that are also clustered in the pathways or in the chromosomal locations by mapping to the KEGG/PATHWAY and KEGG/ GENOME databases, respectively (Keneshita et al. 2002)

ArrayExpress Database

ArrayExpress is a microarray database implemented by the European Molecular Biology Laboratory. This repository is implemented from the MAGE-ML and MIAME standards. There are two forms to submit data to this repository: 1) directly by using MAGE-ML files or 2) using a web tool that allows the user to annotate their experiments (known as MIAMExpress). While MIAMExpress consists of a MySQL database, ArrayExpress is implemented in Oracle (Brazma et al. 2003).

Stanford Microarray Database

This database is one of the most important, offering microarray data from raw images to publication tables. The system runs in a Solaris operating system with a DBMS using Oracle 8. This database provides the opportunity to load new data produced by GenePix and ScanAlyze (Gollub et al. 2003).

CHALLENGES IN MICROARRAY GENE EXPRESSION DATA ANALYSIS

As discussed in previous chapters, gene expression measurement using DNA microarrays without a careful experimental design and lacking the necessary number of replications will produce datasets littered with errors and missing values. In most published studies, the effects of experimental errors and biological noise are not appreciated properly since the measurements are treated as precise estimations of gene expression levels. The variance and noise across different microarrays can be the result of different factors including contamination of clones, mislabeling and incorrectly annotated probes and errors during the spotting process (Yue et al. 2001; Halgren et al. 2001, Kuo et al. 2002, Kothapalli et al. 2002; Spruill et al. 2002). For example, Lee et al. (2000) found that under relatively controlled conditions false positives dominate cDNA microarray experiments and that in any single replication as much as 5% of the mRNA in a sample tissue either fails to be represented as a probe or fails to hybridize. Kuo et al. (2002) compared the data from cDNA arrays and oligonucleotide microarrays measuring the same samples and found very little correlation. The above factors contribute to a variance that mis- or over-represen a particular experimental process and data analysis technique, leading to erroneous interpretation of the data.

Since the information needed to make a protein is contained in mRNA, a researcher using DNA microarray technology assumes that by quantifying the abundance of mRNA, the biological process can be inferred. However, the reader should be aware that the genome structure of each organism is dominated by many sources of variation (natural noise) influencing the measurement of gene expression levels. It is known that multiple proteins can arise from a single gene or the mRNA can be subject to post-translational modification. Therefore, the clusters formed by different microarray gene expression analysis techniques might not represent the true interactions at the protein level. Another source of biological noise results from DNA sequences producing disabled transcripts that obviously hybridize to the probe. For example, pseudogenes result in the loss of gene function at the transcription or translation level (or both) as the accumulation of stop codons result in a sequence no longer capable of producing a functional protein (Proudfoot 1980; Harrison et al. 2002). The initial representation of the sequence data from human chromosome 22 indicates that 19% of the coding regions are pseudogenes and of these, 82% were processed (Wagner 1998). Recent analysis suggests that the human genome could contain approximately 20,000 pseudogenes, with more of half of them transcribed (Zhang et al. 2002). If these DNA regions are producing transcriptional products, what is the level of their expression? Are they activated or regulated as regular genes? Homology analysis shows that some of these pseudogenes are very similar to functional genes. How much of their products hybridize to the microarray probes? Certainly the contribution of disabled pseudogenes will be higher in cDNA microarrays than in oligonucleotide arrays. The difficulty in interpreting the physiological significance of microarray data is exacerbated by the fact that there is an incomplete understanding of post-translational modifications of transcript gene products. The reader should be aware that the level of gene expression captured in a microarray is just a snapshot of a particular experimental time point. A collection of genes may exhibit similar expression patterns over various experimental time points. However, at the proteomic level, since protein folding and interaction operates at the microsecond scale, the levels of translation at each experimental time point may not necessarily coincide with the patterns observed at the transcriptional level.

Gene expression experiments are increasingly conducted between multidisciplinary and inter-institutional groups. However, gene expression measurements using different microarray technologies or from different laboratories may not be quantitatively comparable unless two basic considerations are implemented. First, sample preparations and experimental protocols should be rigorously controlled. Second, with the goal of creating reference points allowing comparison of microarray data across laboratories, multiple control probes should be used. As standardization allows the query of large datasets (terabyte-scale), laboratories will need to implement databases and develop new terminologies (ontologies) to describe more complex experiments. If microarrays become a regular clinical diagnostics tool coupled with the development of nanotechnology capable of constantly sampling gene expression levels, it can be expected that future microarray data will arrive in the form of data streams requiring real-time analysis. For some laboratories with restricted internet bandwidth, transfer or access of large microarray datasets may be very difficult to achieve in a reasonable time frame. In addition, many of the current microarray data analysis methods lack a configuration to optimally perform in a high performance computing setting (e.g. distributed computing). Consequently, managing, analyzing and visualizing this information will continue to present challenges in areas of data mining, data retrieval, supercomputing and bandwidth.

CONCLUSIONS

This chapter presented the reader with a diversity of strategies for clustering, classifying, integrating, standardizing and visualizing microarray gene expression data. A combination of appropriate microarray and sample replications, with rigorous experimental design is necessary for a clear understanding of the biological process. The elements of a gene expression matrix can be grouped using descriptive models (unsupervised) or predictive (supervised) methods. Unsupervised methods through agglomerative or partitive techniques can be useful to find patterns representing gene co-regulation or can be used to assign the functional role of an unknown gene. Within these exploratory methods, there is no 'best technique' for the analysis of microarray data. This situation is due to the large number of combinations between distance matrices and clustering algorithms.

The large number of data points measured in a microarray experiment, the lack of detailed biological information about many genes, and the likely interaction among gene products suggests that unsupervised clustering is appropriate for the exploratory assessment of multidimensional microarray data. Robust and reproducible clustering results may remain elusive for many datasets and are dependent of the distance functions. The application of validation techniques such as bootstrapping and jackknifing can provide a confidence measure of the obtained results. However, successful techniques generally are the result of the combination of both mathematical methods and biological expertise in order to build a reasonable model that can be trained to classify gene expression data.

An alternative to unsupervised methods can be the construction of classifiers assembled from training datasets and biological expertise. These techniques seem more robust and accurate than unsupervised approaches when presented with large gene expression matrices that are complex not only for their size, but also are noisy and contain incomplete values. These techniques also can be useful in assigning unknown genes to a functional biological category or to isolate genes that identify a particular sample.

As the volume of microarray data grows, there is a rising need to provide data summarization of gene expression patterns. With the large volume of genomic information available today, there is an increasing need to integrate microarray data analysis tools with other genomic analysis tools including sequence, protein families and ontology definitions. This means defining the relationships among pathways, motifs, metabolites, mRNA and other cellular components. Also, in order to make these microarray data analyses possible, it is necessary to describe and integrate experimental results from different laboratories employing different DNA chip fabrication methods and experimental designs. This will require the employment of initiatives such as MIAME and MAGE-ML.

DNA microarray technology as well as modeling techniques to represent complex biological systems are still in their infancy and will continue to improve as high performance computational methods mature. At the current time, no single best technique for microarray data analysis has emerged. It is clear that the proper understanding of complex biological systems will require the fundamental integration of training in computer science, statistics, classical genetics and molecular biology.

REFERENCES

1   Aach J, Rindone W, Church GM (2000) Systematic management and analysis of yeast gene expression data. Genome Res (10)431-345.

2   Achard F, Vaysseix G, Barilot E (2001) XML, Bioinformatics and data integration. Bioinformatics (17)2:115-125.

3   Aggarwal CC (2002) Towards effective and interpretable data mining by visual interaction. SIGKDD explorations (3)2:11-34.

4   Akutsu T, Miyano S, Kuhara, S (2000) Inferring qualitative relations in genetic networks and metabolic pathway. Bioinformatics 16:727-734.

5   Alizadeh AA, Eisen MB, Davis RE, Ma C, Lossos IS, Rosenwald A, Boldrick JC, Sabet H, Tran T, Yu X, Powell JI, Yang L, Marti GE, Moore T, Hudson J, Lu L, Lewis DB, Tibshirani R, Sherlock G, Chan WC, Greiner TC, Weisenburger DD, Armitage JO, Warnke R and Staudt LM et al. (2000). Distinct types of diffuse large B-cell lymphoma identified by gene expression profiling. Nature 403:503-511.

6   Alon U, Barkai N, Notterman DA, Gish K, Ybarra S, Mack D, Levine AJ (1999) Broad
     patters of gene expression revealed by clustering analysis. Proc. Natl. Acad. Sci. USA
     (96)12:6745-6750.

7   Alter O, Brown P, Botstein D (2000) Singular value decomposition for genome-wide expres-
     sion data processing and modelling. Proc. Natl. Acad. Sci. USA (97)18:10101-10106.

8   Ambroise C, McLachlam G (2002) Selection bias in gene extraction on the basis of microarray
     gene-expression data. Proc Nat Acad Sci USA (99)10:6562-6566.

9   Anderson AB, Basilevsky A, Hum DPJ (1983) Missing Data: A review of the literature.
     (Rossi PH, Wright JD, Anderson AB Eds). Handbook in Survey Research (pp. 415-494).
     Academic Press.

10   Aronow BJ, Richardson B, Handwerger S (2001) Microarray analysis of trophoblast differ-
     entiation: gene expression reprogramming in key gene function categories. Physiol Genomics
     6:105-116.

11   Azuaje F, Bolshakova N (2002) Clustering genomic expression data: Design and evaluation
     principles. In: Understanding and Using Microarray Techniques. A practical Guide. (Bubitzky
     BD, Granzow M Eds) London: Spring Verlag.

12   Baldi P, Brunak S (2001) Bioinformatics: the Machine Learning Approach. Cambridge: MIT
     Press.

13   Baldi P, Long A (2001) A Bayesian framework for the analysis of microarray expression data:
     regularized t-test and statistical inferences of gene changes. Bioinformatics (17) 6:509-519.

14   Baldi P, Natfield W (2002) DNA microarrays and gene expression. From experiments to data
     analysis and modelling. Cambridge: Oxford UP.

15   Barash Y, Friedman N (2002) Context-specific Bayesian clustering for gene expression data.
     J Comput Biol 9(2):169-191.

16   Barillot E, Achard F (2000) XML: a lingua franca for science. TIBTECH 18:331-333.

17   Benson DA, Karsch-Mizrachi I, Lipman D, Ostell J, Rapp BA Wheeler D (2002) GenBank.
     Nucleic Acids Res (30):17-20.

18   Bittner M, Meltzer P, Chen Y, Jiang Y, Seftor E, Hendrix M, Radmacher M, Simon R,
     Yakhini Z, Ben-Dor A, Dougherty E, Wang E, Marincola F, Gooden C, Lueders J, Glatfelter
     A, Pollock P, Gillanders E, Leja D, Dietrich K, Berens M, Alberts D, Sondak V, Hayward N,
     Trent J (2000) Molecular classification of cutaneous malignant melanoma by gene expres-
     sion profiling. Nature 406:536-440.

19   Bø TH, Jonassen I (2002) New feature selection procedure for classification of expression profiles. Genome Biology 3(4);research0017.1-0017.11.

20   Bolshakova N, Azuaje F (2003) Cluster validation for genome expression data. Technical Report TCD-CS-2002-33 Computer Science Department. Trinity College Dublin http://www.cs.tcd.ie/publications/tech-reports/reports.02/TCD-CS-2002-33.pdf

21   Bower JM, Bolouri H (2001) Compuational modelling of biochemical networks. Massachusetts: MIT Press.

22   Brazma A, Hingamp P, Quackenbush J, Sherlock G, Spellman P, Stoeckert C, Aach J, Ansorge W, Ball C A, Causton HC, Gaasterland T, Glenisson P, Holstege FCP, Kim I, Markowitz V, Matese JC, Parkinson H, Robinson A, Sarkans U, Schulze-Kremer S, Stewart J, Taylor R, Vilo J, Vingron M (2001) Minimun information about a microarray experiment (MIAME)-toward standards for microarray data. Nature Gen 29:365-371.

23   Brazma A, Parkinson H, Sarkans U, Shojatalab M, Vilo J, Abeygunawardena N, Holloway E, Kapushesky M, Kemmeren P, Lara GG, Oezcimen A, Rocca-Serra P, Sansone SA (2003) ArrayExpress-a public repository for microarray gene expression data at the EBI. Nucleic Acids Res 31(1):68-71.

24   Brazma A, Robinson A, Cameron G, Ashburner M (2002) One-shop for microarray data. Nature 403:699-700.

25   Brazma A, Vilo J (2002) Gene Expression Data Analysis. FEBS Lett (480)1:17-24.

26   Breiman L (1998) Bagging Predictors. Technical Report No. 421. Department of Statistics University of California Berkeley.

27   Brody J.P., Williams B.A., Wold B.J., Quake S.R. (2002) Significance and statistical errors in the analysis of DNA microarray data. Proc. Nat. Acad. Sci. USA (99):20:12975-12978.

28   Brown MPS, Grundy WN, Lin D, Cristianini N, Sugnet CW, Furey TS, Ares M, Haussler D (2000) Knowledge-based analysis of microarray gene expression data by using support vector machines. Proc Natl Acad Sci USA 97:262-267.

29   Burges, C. (1998) A Tutorial on Support Vector Machines for Pattern Recognition. Data Mining and Knowledge Discovery (2)2:1-43.

30   Butte AJ, Tamayo P, Slonin D, Golub T, Kohane I (2000) Discovering functional relationships between RNA expression and chemotherapeutic susceptibility using relevance networks. Proc. Nat. Acad. Sci. USA 97(22):12182-12186.

31   Celis JE, Kruhoffer M, Gromova I, Frederiksen C, Ostergaard M, Thykjaer T, Gromov P, Yu J, Palsdottir H, Magnusson N, Orntoft TF (2000) Gene expression profiling: monitoring transcription and translation products using DNA microarrays and proteomics. FEBS Lett 480(1):2-16.

32   Cheng, Y, Church GM (2000) Biclustering of expression data. Proc Int Conf Intell Syst Mol Biol 8:93-103.

33   Chilingaryan A, Gevorgyan N, Vardanyan D, Jones D, Szabo A (2002) Paper title. Mathematical Biosciences (176):59-72.

34   D'haeseleer P (2001) Beyond Co-Expression: Gene Network Interference. www.cs.unm.edu/~patrick/networks/diss.pdf

35   D'haeseleer P, Liang S, Somogyi R (2000) Genetic network interference: from co-expression clustering to reverse engineering. Bioinformatics (16)8:707-726.

36   Dudoit S, Fridlyand J (2002) A prediction-based resampling methods for estimating the number of clusters in a dataset. Genome Biology (3)7:research0036.1-0036.21.

37   Dudoit S, Fridlyand J, Speed TP (2000) Comparison of discrimination methods of tumors using gene expression data. Department of Statistics Technical Report 576. University of Berkeley.

38   Dubitzky W, Granzow W, Berrar D (2001) Data Mining and Machine Learning Methods for Microarray Analysis. In: Methods of Microarray Data Analysis (Lin SM, Johnson KF eds) (pp 5-22). Massachusetts: Kluwer Academic Publishers.

39   Duogherty E, Barrera J, Brun M, Kim S, Cesar RM, Chen Y, Bittner M, Trent M (2002) Inference from clustering with application to gene-expression microarrays. J Comp Biol (9)1:105-126.

40   Edgar R, Domrachev RM Lash AE (2002) Gene Expression Omnibus: NCBI gene expression and hybridization array data repository. Nucleic Acids Res (30)1:207-210.

41   Efron B, Tibshirani RJ (1993) An introduction to the bootstrap. New York: Chapman & Hall.

42   Eisen MB, Spellman PT, Brown PO, Botstein D (1998) Cluster analysis and display of genome-wide expression patterns. Proc Natl Acad Sci USA 95(25):14863-14868.

43   Fellenberg K, Hauser NC, Brors B, Neutzner A, Hoheisel JD, Vingron M (2001) Correspondence analysis applied to microarray data. Proc Natl Acad Sci USA 98: 10781-10786.

44   Fix E, Hodges J (1951) Discriminatory analysis non parametric discrimination: consistency properties. Technical Report Randolph Filed Texas. USAF School of Aviation Medicine.

45   Freund Y, Schapire RE (1997) A decision-theoretic generalization of online learning and an application to boosting. J Comp Syst Sci 55(1):119-139.

46   Friedman N, Linial M, Nachman I, Pe'er D (2000) Using Bayesian Networks to Analyze
      Expression Data. J Comput Biol 7(3-4):601-20

47   Furey TS, Cristianini N, Duffy N, Bednarski DW, Schummer M, Haussler D (2000) Sup-
      port vector machine classification and validation of cancer tissue samples using microarray
      expression data. Bioinformatics (16)10:906-914.

48   Garofalakis M, Hyun D, Rastogi R, Shim (2000) Efficient Algorithms for Constructing
      Decision Trees with Constraints. Proc. Sixth ACM SIGKDD. Paper 296.

49   Geschwind DH (2001) Sharing gene expression data: an array of options. Nature Rev
      Neuroscience. (2):435-438.

50   Getz G, Levine E, Domany E (2000) Coupled two-way clustering analysis of gene microarray
      data. Proc. Nat. Acad. Sci. USA (97)22:12079-12084.

51   Gilbert DR, Schroeder M, van Helden J (2000) Interactive visualization and exploration of
      relationships between biological objects. TIBTECH (18):487-494.

52   Golub TR, Slonim DK, Tamayo P, Huard C, Gaasenbeek M, Mesirov JP, Coller H, Loh ML,
      Downing JR, Caligluri MA, Bloomfield CD, Lander ES (1999) Molecular classification of
      cancer: class discovery and class prediction by gene expression monitoring. Science 286:531-
      537.

53   Gollub J, Ball CA, Binkley G, Demeter J, Finkelstein DB, Hebert JM, Hernandez-Boussard
      T, Jin H, Kaloper M, Matese JC, Schroeder M, Brown PO, Botstein D, Sherlock G (2003)
      The Stanford Microarray Database: data access and quality assessment tools. Nucleic Acids
      Res (1):94-96

54   Graves DJ (1999) Powerful tools for genetic analysis come to age. TIBTECH (17) 127-134.

55   Guyon I, Weston J, Barnhill S, Vapnik V (2000) Gene selection for cancer discrimination
      using support vector machines. Machine Learning 46(1/3):389.

56   Halgren RG, Fielden MR, Fong CJ, Zacharewski TR (2001) Assessment of clone identity
      and sequence fidelity for 1189 IMAGE cDNA clones. Nucleic Acids Res. 29(2):582-8.

57   Han J, Kamber M (2001) Data mining. Concepts and applications. San Francisco: Morgan
      Kaufmann Press.

58   Hand DJ, (1999) Statistics and Data Mining: Intersecting Disciplines. Proc. Fifth ACM
      SIGKDD (1)1:16-19.

59   Hand DJ, Mannila H, Smyth P (2001) Principles of data mining. Cambridge: MIT Press.

60   Harding J, Rocke DM (2002) Robust Model-Based Clustering of Genes in Microarray Data: Are there Gene Clusters? www.camda.duke.edu/CAMDA00/Abstracts/Presentations/ Poster_13.pdf

61   Harrison P, Kumar A, Lan N, Echols N, Snyder M, Gerstein M (2002) A small reservoir of disabled ORFs in the yeast genome and its implications for the dynamics of proteome evolution. J Mol Biol 316: 409-19.

62   Hastie T, Tibshirani R, Eisen MB, Alizadeh A, Levy R, Staudt L, Chan WC, Botstein D, Brown PO (2001a) Gene shaving as a method for identifying distinct sets of genes with similar expression patterns. Genome Biology 1(2):research0003.1-0003.21.

63   Hastie T, Tibshirani R, Friedman J (2001) The elements of statistical learning. Data Mining Inference and prediction. Berlin: Springer-Verlag.

64   Hawng KB, Cho DY, Park S, Kim SD, Zhang BT (2002) Applying machine learning techniques to analysis of gene expression data: Cancer diagnostics. Methods of Microarray Data Analysis. (Lin SM, Johnson, KF eds.) (pp 167-182). Massachusetts: Kluwer Academic Publishers.

65   HeadGordon T, Wooley J (2001) Computational challenges in structural and functional genomics. IBM System Journal. (40)2: 265-296.

66   Helfrich JP (2002) Raw Data to Knowledge Warehouse in Proteomic-Based Drug Discovery: A Scientific Data Management Issue. Biotechniques Supp. on Comp Proteom 48-53.

67   Herrero J, Valencia A, Dopazo J (2001) A hierarchical unsupervised growing neural network for clustering gene expression patterns. Bioinformatics (17)2:126-136.

68   Heyer LJ, Kruglyak S, Yooseph S (1999) Exploring expression data: identification and analysis of coexpressed genes. Genome Res 9:1106-1115.

69   Hilsenbeck SG, Friedrichs WE, Schiff R, O'Connell P, Hansen RK, Osborne CK, Fuqua SAW (1999) Statistical analysis of array expression data as applied to the problem of tamoxifen resistance. J Nat Cancer Inst 91: 453-459.

70   Holter NS, Maritan A, Cieplak M, Federoff NV, Banavar JR (2002) Dynamic modelling of expression data. Proc Nat Acad Sci USA (98)4:193-1698.

71   Hvidsten TR, Komorowski J, Sandvik AK, Legreid AL (2001) Predicting gene function from gene expressions and ontologies. In: Pacific Symposium on Biocomputing pp. 299-310 (Altman RB Dunker AK Hunter L Lauderdale K and Klein TE eds) Mauna Lani Hawaii World Scientific Publishing Co.

72   Jain AK, Murty MN, Flynn PJ (1999) Data clustering: a review. ACM Computing Surveys 31(3):264-323.

73   Jamil HM, Modica GA, Teran MA (2001) Towards a Visual Query Interphase for Phyloge-
      netic Databases. CIKM'01:57-64.

74   Kanehisa M, Goto S, Kawashima S, Nakaya A. (2002) The KEGG databases at GenomeNet.
      Nucleic Acids Res. 30(1):42-6.

75   Kauffman, SA (1998) Investigations. New York: Oxford UP.

76   Kazic T (2000) Semiotes: a semantics for sharing. Bioinformatics 16(12):1129-1144.

77   Keller DA, Schummer M, Hood L, Ruzzo WL (2000) Bayesian Classification of DNA
      Array Expression Data. Technical Report UW-CSE-2000-08-01.

78   Kerr MK, Churchill GA (2001) Bootstrapping cluster analysis: Asessing the reliability of
      conclusions from microarray experiments. Proc. Nat. Acad. Sci. USA (98)16:8961-8965.

79   Khan J, Wei JS, Ringner M, Saal LH, Ladanyi M, Westermann F, Berthold F, Schwab M,
      Antonescu CR, Peterson C, Meltzer PS (2001) Classification and diagnostic prediction of
      cancers using gene expression profiling and artificial neural networks. Nat Med 7(6):658-
      659.

80   Kitano H (2002) Computational system biology. Nature 420:206-210.

81   Kitano H (2002a) Foundations of system biology. Massachusetts: MIT Press.

82   Kohonen T (1981) Automatic formation of topological maps of patterns in a self-organizing
      system. In Proc. Second Scandinavian Conf. on Image Analysis 214-220.

83   Kohonen T (1997) Self-organizing maps. Berlin: Springer-Verlag.

84   Kothapalli R. Yoder SJ, Mane S, Loughram Jr TP (2002) Microarray results: How accurate
      they are? BMC Bioinformatics (3):22

85   Kuo WP, Jenseen T, Butte AT, Ohno-Machado L, Kohane IS (2002) Analysis of matched
      mRNA measurements from two different microarray technologies. Bioinformatics (18):405-
      412.

86   Kuramochi M, Karypis G (2001) Gene Classification using expression profiles: A feasibility
      study. Department of Computer Science/Army HPC Research Center. Technical Report 01-
      029.

87   Landgrebe J, Wurst W, Welzl G (2002) Permutation-validated principal components analy-
      sis of microarray data. Genome Biol 3(4):research0019.

88   Lee MT, Kuo FFC, Whitemore GA, Sklar J (2000) Importance of replication in microarray gene expression studies: Statistical methods and evidence of repetitive cDNA hybridisations. Proc. Nat. Acad. Sci. USA (97)18:9834-9839.

89   Li L, Weinberg CR, Darden TA, Pedersen LA (2001) Gene selection for sample classification based on gene expression data: study of sensitivity to choice of parameters of the GA/KNN algorithms. Bioinformatics 12(12):1131-1142.

90   Liang S, Fuhrman S, Somogyi R (1998) REVEAL. A genereal reverse engineering Algorithm for the Interference of Genetic Network Architecture. Pac. Symp. Biocomputing 18-29.

91   Little RA, Rubin DR (1987) Statistical analysis with missing data. New York: John Wiley & Sons.

92   Lockhart DJ, Winzeler EA (2001) Genomics gene expression and DNA arrays. Nature (405):827-836.

93   MacQueen J 1967. Some methods for classification and analysis of multivariate observations. Proceedings of the Fifth Berkeley Symposium on mathematical statistics and probability 1:281-297.

94   Mendez MA, Hodar C, Vulpe C, Gonzalez M, Cambiazo V (2002) Discriminant analysis to evaluate clustering of gene expression data. FEBS Letts 522(1-3):24-28.

95   Model F, Konig T, Piepenbrock C, Adorjan P (2002) Statistical process control for large scale microarray experiments. Bioinformatics 155-163.

96   Moler EJ, Chow ML, Mian JS (2000) Analysis of molecular profile data using generative and discriminative methods. Physiol. Genomics 4:109-126.

97   Mukherjee S (2002) Classifying Microarray Data Using Support Vector Machines. Berrar DP, Dubitzky W, Granzow M (Eds). A Practical Approach to Microarray Data Analysis. Boston: Kluwer Academic Publishers.

98   Mukherjee S, Tamayo P, Mesirov JP, Slonim D, Verri A, Poggio T (199) Support Vector Machine Classification of Microarray Data. CBCL Paper182/AI Memo-1676, Massachusetts Institute of Technology. Cambridge.

99   Mutch DM, Berger A, Mansourian R, Rytz A, Roberts MA (2002) The limit of the fold change: A practical approach for selecting differentially expressed genes from microarray data. BMC Bioinformatics 3:17

100   Nadon R, Shoemaker J (2002) Statistical issues with microarrays: processing and analysis. Trends in Genetics 18(5):265-271.

101 Pan K, Lih C, Cohen SN (2002) Analysis of NDA microarrays using algorithms that employ rule-based expert knowledge. Proc Nat Acad Sci USA 99(4):21118-2123.

102 Pavlidis P, Weston J, Cai J, Grundy WN (2001) Gene functional classification from heterogeneous data. RECOMB 2001: Proc Fifth Ann Int Conf Comp Biol 249-255.

103 Peterson LE (2003) Partitioning large-sample microarray-based gene expression profiles using principal components analysis. Comput Methods Programs Biomed 70(2):107-119

104 Proudfoot N (1980) Pseudogenes. Nature 286(5776):840-841.

105 Qi. H (2002) Feature Selection and kNN fusion in molecular classification of multiple tumor types. Proc. Intern. Conf. on Mathematics and Engineering Techniques in Medicine and Biological Sciences (METMBS'02) http://aicip.ece.utk.edu/publication/02metmbs.pdf

106 Quackenbush J. Computational analysis of microarray data. (2001) Nat Rev Genet 2(6):418-427.

107 Ramoni M, Sebastiani P (1998) Bayesian methods for intelligent data analysis. Kmi Technical reportKMi-TR-67. The Open University.

108 Ramoni M, Sebastiani P, Kohane I.S. (2002) From the cover: Cluster Analysis of Gene Expression Dynamics. Proc Nat Acad Sci USA 99(14):9121-9126.

109 Ravasz E, Somera L, Mongru DA, Oltvai N, Barabasi AL (2002) Hierarchical organization of modularity in metabolic networks. Science 297:1551-1555.

110 Raychaudhuri S, Stuart JM, Altman RB (2000) Principal components analysis to summarize microarray experiments: application to sporulation time series. Pac Symp Biocomput 5:452-463. (Altman RB Dunker AK Hunter L Lauderdale K and Klein TE eds) Mauna Lani Hawaii World Scientific Publishing Co.

111 Ramaswamy S, Tamayo P, Rifkin R, Mukherjee S, Yeang CH, Angelo M, Ladd C, Reich M, Latulippe E, Mesirov JP, Poggio T, Gerald W, Loda M, Lander ES, Golub TR (2001) Multiclass cancer diagnosis using tumor gene expression signatures. Proc Natl Acad Sci USA 98(26):15149-15154

112 Raymond, MR, Roberts DM (1987) A comparison of methods for treating incomplete data in selection research. Educational and Psychological Measurement 47:13-26.

113 Reed RD, Marks II RJ (1998) Neural smithing. Supervised learning in feedforward artificial neural networks. Cambridge: MIT Press.

114 Rifkin SA, Atteson K, Kim J (2000) constrain structure analysis of gene expression. Funt Integr Genomics 1:174-185.

115 Ross DT, Scherf U, Eisen MB, Perou CM, Rees C, Spellman P, Iyer V, Jeffrey SS, Van de
     Rijn M, Waltham M, Pergamenschikov A, Lee JC, Lashkari D, Shalon D, Myers TG Weinstein
     JN, Botstein D and Brown PO (2000) Systematic Variation in Gene Expression patters in
     human cancer cell lines. Nature (24):224-235.

116 Rubin DB 1976. Inference and missing values. Biometrika. 63:581-592.

117 Scherf U, Ross DT, Waltham M, Smith LH, Lee JK, Tanabe L, Ko hn KW, Reinho ld WC,
     Myers TG, Andrews DT, ScudieroDA, Eisen MB, Sausville EA, Pommier Y, Botstein D,
     Brown PO, Weinstein JN (2000) A gene expression database for the molecular pharmacology
     of cancer. Nature (24):236-244.

118 Selaru FM, Xu Y, Yin J, Zou T, Liu TC, Mori Y, Abraham JM, Sato F, Wang S, Twigg C,
     Olaru A, Shustova V, Leytin A, Hytiroglou P, Shibata D, Harpaz N, Meltzer SJ (2002)
     Artificial neural networks distinguish among subtypes of neoplastic colorectal lesions. Gas-
     troenterology 122(3):606-613.

119 Seo J, Shneiderman B (2002) Interactively exploring hierarchical clustering Results. IEEE
     Computer (35)7:80-86

120 Sherlock G, Hernandez-Boussard T, Kasarskis A, Binkley G, Matese JC, Dwight SS, Kaloper
     M, Weng S, Jin H, Ball CA, Eisen MB, Spellman PT, Brown PO, Botstein D, Cherry JM
     (2001) The Stanford Microarray Database. Nucleic Acids Res (1):152-155.

121 Silvescu, A., and Honavar, V. (2001). Temporal Boolean Network Models of Genetic Net-
     works and their inference from gene expression time series. Complex Syst (13)1:54-75.

122 Skurichina M, Duin RPW (1998) Bagging for linear classifiers. Pattern Recognition 31(7):909-
     930.

123 Skurichina M, Duin RPW (2002) Bagging, boosting and the random sample method for linear
     classifiers. Pattern Analysis & Appli (5):121-135.

124 Sneath PHA. Sokal RR (1973) Numerical Taxonomy. San Francisco: Freeman & Co., Pub-
     lishers.

125 Sokal RR, Michener CD, (1958) A statistical method for evaluating systematic relation-
     ships. Sci. Bull. University of Kansas 38:1409-1438.

126 Soukas A, Cohen P, Socci ND, Friedman JM (2000) Leptin-specific patterns of gene expres-
     sion in white adipose tissue. Genes & Development 14:963-980.

127 Spellman PT, Miller M, Stewart J, Troup C, Sarkans U, Chervitz S, Bernhart D, Sherlock G,
     Ball C, Lepage M, Swiatek M, Marks WL, Goncalves J, Markel S, Iordan D, Shojatalab M,
     Pizarro A, White J, Hubley R, Deutsch E, Senger M, Aronow BJ, Robinson A, Bassett D,
     Stoeckert CJ Jr, Brazma A (2002) Design and implementation of microarray gene expression
     markup language (MAGE-ML). Genome Biology 3(9):research0046.1-0046.9.

128 Spellman PT, Sherlock G, Zhang MQ, Iyer VR, Anders K, Eisen MB, Brown PO, Botstein D, Futcher B (1998) Comprehensive identification of cell cycle-regulated genes of the yeast Saccharomyces cerevisiae by microarray hybridization. Mol. Biol. Cell 9:3273-3297.

129 Spruill SE, Lu J, Hardy S, Weir B (2002) Assessing sources of variability in gene expression data. Biotechniques 33:916-923.

130 Stoeckert CJ, Causton HC, Ball CA (2002) Microarray databases: standards and ontologies. Nat Genet. Suppl 2:469-73.

131 Strohman R (2002) Maneuvering in the complex path from genotype to phenotype. Science 296:701-702.

132 Su AI, Welsh JB, Sapinoso LM, Kern SG, Dimitrov P, Lapp H, Schultz PG, Powell SM, Moskaluk CA, Frierson HFJr, Hampton GM (2001) Molecular Classification of Human Carcinomas by Use of Gene Expression Signatures. Cancer Res 61:7388-7393

133 Tamayo P, Slonim D, Mesirov J, Zhu Q, Kitareewan S, Dmitrovsky E, Lander ES, Golub TR (1999) Interpreting patterns of gene expression with self-organizing maps: methods and application to hematopoietic differentiation. Proc Natl Acad Sci. USA 96(6):2907-2912.

134 Tavazoie S, Hughes JD, Campbell MJ, Cho RJ, Church GM (1999) Systematic determination of genetic network architecture. Nat Genet 22:281-285.

135 Thomas R (1991) Regulatory networks seen as asynchronous automata: A biological Description. J Theor Biol (153):1-23.

136 Thomas RS, Rank DR, Penn SG, Zastrow GM, Hayes KR, Pande K, Glover E, Silander T, Craven MW, Reddy JK, Jovanovich SB, Bradfield CA. (2001) Identification of toxicologically predictive gene sets using cDNA microarrays. Mol. Pharmacol 60:1189-1194.

137 Törönen P, Kolehmainen M, Wong G, Castrén E (1999) Analysis of gene expression data using self-organizing maps. FEBS Lett 451(2):142-146.

138 Troyanskaya O, Cantor M, Sherlock G, Brown P, Hastie T, Tibshirani R, Botstein D, Altman RB (2001) Missing value estimation methods for DNA microarrays. Bioinformatics 17(6):520-525.

139 Tusher GV, Tibshirani R, Chu G (2001) Significance analysis applied to ionizing radiation response. Proc. Nat. Acad. Sci. USA (98)9:5116-5121.

140 Valdivia-Granda WA, Deckard E, Perrizo W (2002) Peano Count Trees (P-Trees) and Rule Association Mining for Gene Expression Profiling of DNA Microarray Data. Proc. Inter Conf in Bioinformatics. Bangkok, Thailand OstraAna08.

141 Vapnik V (1995) The Nature of Statistical Learning Theory. Berlin: Springer-Verlag.

142 Wagner A (1998) The fate of duplicated genes: loss or new function? BioEssays 20 785-788.

143 Wolf PJ, Wang Y (2002) A fuzzy logic approach to analysing gene expression data. Physiol Genomics 3:9-15.

144 Yeung KY, Haynor DR, Ruzzo W (2001a) Validating clustering for gene expression data. Bioinformatics (17)4:309-318.

145 Yeung KY, Ruzzo W (2001) Principal component analysis for clustering for gene expression data. Bioinformatics (17)9:763-774.

146 Yue H, Eastman PS, Wang B, Minor J, Doctolero MH, Nuttal R, Stack R, Becker JW, Montgomery JR, Vainer M, Johnston R. (2001) An evaluation of the performance of cDNA microarrays for detecting changes in global mRNA expression. Nucleic Acids Res (29) 8:e41.

147 Zhang K, Zhao H (2000) Assessing reliability of gene clusters from gene expression data. Funct Integr Genomics 1(3):156-173.

148 Zhang, Z, Harrison P, Gerstein M (2002) Identification and analysis of over 2000 ribosomal protein pseudogenes in the human genome. Genome Res 12(10):1466-1482.

149 Zhao L.P, Prentice R, Breeden L (2001) Statistical modeling of large microarray datasets to identify stimulus-response profiles. Proc. Nat. Acad. Sci. USA (98)10:5631-5636.

Index

A

absence call 201, 212
Affymetrix 179, 182, 191, 193, 195, 201, 202, 203, 205, 207, 219, 231, 259, 263, 269
algorithm 202, 204, 205, 207, 208, 210, 211, 212, 216, 256, 269, 271
amplification 126, 135, 144, 146
annotation 230, 231
ANOVA 190, 195, 211, 220, 221, 235, 262, 265, 266, 267, 269, 270, 272
array design 211, 320
average difference 203, 213
average link 292, 300

B

bagging 312, 314
Bayesian Belief Network 311, 314
Bayesian classifier 309
bias 159, 163, 164, 166
bioconductor 207, 208, 236
biological function 245, 248, 249, 262
biological samples 189, 243, 246, 262
Bonferroni 251, 252, 254, 255, 261
boosting 312, 314
bootstrap 257, 261
bootstrapping 299

C

calculating fold change 183
classifying 328
CLEANING 1, 3, 4, 5, 6, 7, 8, 9, 34
  acid 3, 4, 5, 6
  alkaline 6
  physical 3, 6
    laser 6, 8
    ozone 6, 8
    plasma 6, 7
    pyrolysis 6, 7, 8
    ultrasonic 6
    UV 6, 8

O
oligonucleotide (oligo) 43, 101, 102, 124, 126, 133, 135
  length 52
  melting temperature 52, 57
  set design 53
outliers 191, 279, 280, 284, 285, 290, 308, 317

P
pair-group 293
pair-wise 282, 292
partitional 294
PCR 44, 46, 66, 67, 68, 71, 79, 80
PCR products 98, 100, 101, 102
Pearson 208, 286, 292
phylogenomic 279, 315, 316
pitfalls 201, 249, 272
plate 95
pooling 191, 192, 199
post-hoc pattern matching 217
power estimation 247
principal component analysis 296, 298
probe level 202, 203, 211
probe pair 203, 204, 205, 207, 208, 211, 212
probe set 201, 202, 203, 204, 207, 209, 211

Q
quantile 208

R
R 207
re-sampling 256, 257, 261, 262
replicates 187, 189, 190, 193, 199

S
sample annotation 320
sample preparation 94, 98, 101, 102, 105, 108, 110, 116, 119, 120
sample size 246, 247, 252, 257
scan 74
scanner 125, 127, 139, 142, 143
scanning
  confocal 76
  detection 75

scanning (cont'd)
field size 75
non-confocal 76
resolution 74
sensitivity 75
throughput 75
self organizing maps (SOM) 296
set-up 124, 128, 131, 137
Šidák 251, 254
signal intensity 201, 203, 212, 219
single link 292, 293, 300
slide 1, 4, 9, 11, 13, 14, 29
slides 123, 126, 129, 133, 134, 140
sodalime 1
sources of variation 264, 271
spotted 124, 125, 126, 134, 142, 179, 181, 184, 194, 201, 202, 263, 266, 268, 269
spotting buffer 98, 100
standardization 317, 319, 327
standardizing 328
Stanford Microarray Database 326
statistical analysis 243, 244, 248, 271, 272
sub-pooling 189, 192, 193, 194
supervised 279, 299, 304, 306, 312, 313, 328
support vector machine 307, 333
surface 63
SURFACE ANALYSIS 2, 32
  analysis 2, 32, 34
    atomic force microscopy (AFM) 3, 32, 34
    contact angle 2, 3, 32
    ellipsometry 35
    vibrational spectroscopy 35
    X-ray Photoelectron Spectroscopy (XPS) 32, 34
surface chemistry 94, 97, 98, 99, 100, 101, 104, 105, 106, 110, 116, 119, 120
surface hydrophobicity 3, 32

T
target preparation 45, 66
treatment structure 243, 246, 269, 270
troubleshooting 109, 118